10th Anniversary of *Nanomaterials*—Recent Advances in Environmental Nanoscience and Nanotechnology

10th Anniversary of *Nanomaterials*—Recent Advances in Environmental Nanoscience and Nanotechnology

Editor

Ioannis V. Yentekakis

MDPI • Basel • Beijing • Wuhan • Barcelona • Belgrade • Manchester • Tokyo • Cluj • Tianjin

Editor
Ioannis V. Yentekakis
Physical Chemistry and
Chemical Processes
Laboratory, School of
Environmental Engineering,
Technical University of Crete
(TUC)
73100 Chania, Crete, Greece

Editorial Office
MDPI
St. Alban-Anlage 66
4052 Basel, Switzerland

This is a reprint of articles from the Special Issue published online in the open access journal *Nanomaterials* (ISSN 2079-4991) (available at: https://www.mdpi.com/journal/nanomaterials/special_issues/Environmental_Nanoscience).

For citation purposes, cite each article independently as indicated on the article page online and as indicated below:

LastName, A.A.; LastName, B.B.; LastName, C.C. Article Title. *Journal Name* **Year**, *Volume Number*, Page Range.

ISBN 978-3-0365-4161-7 (Hbk)
ISBN 978-3-0365-4162-4 (PDF)

Contents

About the Editor

Ioannis V. Yentekakis is a Full Professor of Physical Chemistry and Catalysis in the School of Chemical & Environmental Engineering, Technical University of Crete, Greece. He received his Chemical Engineering Diploma (1983) and PhD (1988) from the University of Patras (UP). His academic career and research is associated with Princeton University, USA; ICE-HT/FORTH (Post Doc); Cambridge University, UK (visiting professor collaborating with Prof. R.M. Lambert); Dept. of Chemical Engineering, University of Patras (Lecturer: 1996 and Assistant Professor: 2000); and finally, Technical University of Crete (Associate Professor: 2001, and Full Professor: 2006–current). His current research interests concern the development of novel nanostructured materials and heterogeneous catalytic processes for green chemistry, environmental protection, natural gas and biogas valorization, hydrocarbon processing, and renewable/sustainable energy generation. So far, he has co-authored 128 publications in high-impact peer-reviewed journals, more than 150 conference proceedings, 3 patents, 10 scientific and technical books, and 5 invited chapters in international books. His published work has received ¿5200 citations, h-factor=43 (Source: Google scholar). He is the Section Editor-in-Chief of Nanomaterials MDPI, the Specialty Chief-Editor of Frontiers in Environmental Chemistry: Catalytic Remediation, and an Editorial Board Member in eight additional journals. Prof. Yendekakis's research has been funded by various private and public bodies, including the European Union and the Hellenic Ministry of Development-GSRT, with 36 research projects (in 21 of which he was the coordinator).

Preface to "10th Anniversary of *Nanomaterials*—Recent Advances in Environmental Nanoscience and Nanotechnology"

The rapid development of nanoscience and nanotechnology, including advanced methods of fabrication and characterization of nanostructured composites and hybrid materials, has been the driving force of strong progress in all areas of classical science and engineering (e.g., chemistry, physical chemistry, chemical engineering, catalysis, biology, medicine, optoelectronics, etc.), including current topical issues related to environmental protection, renewable and low carbon footprint energy production, and sustainable development.

Rationally designed nano-structured catalysts particularly at the nano-to-atom scale, the so-called nano-catalysts, offer great advantages that provide innovative, cost-effective, and durable materials with exceptional well-tailored surface chemistry properties and performance optimization in environmental and energy applications. Among others, these applications include emission-control catalysis, waste treatment, photocatalysis, bio-refinery, CO_2 utilization, and fuel cell applications as well as hydrocarbon processing for H_2, value-added chemicals, and liquid fuel production.

Celebrating its 10th anniversary, the journal Nanomaterials organized a series of Special Issues with state-of-the-art topics that cover and highlight the aims and scope of the journal as well as the future scientific and technical perspectives of this field. The present SI entitled "Advances in Environmental Nanoscience and Nanotechnology" was one such SI that finally succeeded in compiling some important and high-quality contributions that cover the recent research progress in various sub-directions of the title and can become guides for further research in the field.

Ioannis V. Yentekakis
Editor

Editorial

The 10th Anniversary of Nanomaterials—Recent Advances in Environmental Nanoscience and Nanotechnology

Ioannis V. Yentekakis [1,2]

[1] Laboratory of Physical Chemistry & Chemical Processes, School of Chemical and Environmental Engineering, Technical University of Crete (TUC), 73100 Chania, Crete, Greece; yyentek@isc.tuc.gr; Tel.: +30-28210-37752
[2] Foundation for Research and Technology—Hellas/Institute of Geoenergy (FORTH/IG), Technical University of Crete, Building M1, University Campus, Acrotiri, 73100 Chania, Crete, Greece

1. Overview

As a result of the rapid growth of nanoscience and nanotechnology, including advanced methods of fabrication and characterization of nanostructured materials, great progress has been made in many fields of science, not least in environmental catalysis, energy production and sustainability [1–8].

Also known as "nano-catalysts" [1–8], they now play a leading role in environmental and energy science and engineering by providing innovative, cost-effective, and durable nanostructured materials with highly promising performance in the control of environmental pollutants, production of clean fuels and added-value chemicals, and circular-economy technologies [5–22].

Indeed, the rational design of materials, particularly at the nano-to-atom level, offers advantages and enables tailoring and fine-tuning of their critical points in catalysis textural, structural, physicochemical, and local surface chemistry properties, as well as the optimization of metal–metal and metal–support interactions, thus resulting in catalytic systems with outstanding activity and stability performance in numerous eco-friendly applications. These include, for example, emission-control catalysis, waste treatment, photocatalysis, bio-refinery, CO_2 utilization and fuel cell applications, as well as hydrocarbon processing for H_2, added-value chemicals and liquid fuels production [5–25].

Celebrating the 10th anniversary of *Nanomaterials*, the Editor-in-Chief, Prof. Dr. Shirley Chiang, and the members of the Editorial Office organized and led a series of Special Issues on key topics that cover and highlight the journal's aims and scope as well as future scientific and technical perspectives of the area. The present SI entitled "Advances in Environmental Nanoscience and Nanotechnology" was one of them, which aimed to host significant advances in the title.

In this context, this SI has succeeded in collecting five high-quality contributions, refs. [26–30], covering recent research progress in various sub-directions of the field. These contributions are briefly discussed below.

2. Special Issue's Contribution and Highlights

The capture, utilization, and recycling of CO_2 emissions pose major challenges today due to the urgent need for the mitigation of global warming through the greenhouse effect and the concomitant major climate changes. CO_2 hydrogenation to produce renewable fuels (e.g., methane or methanol) is one of the most attractive alternatives for this target, which will contribute to a cleaner and more sustainable future. Notably, CO_2 methanation, also known as the *Sabatier reaction*, becomes more attractive if hydrogen demands for CO_2 methanation reaction can be provided via a water-splitting process based on solar- or wind-powered systems. This further expands the eco-friendly and sustainability feature of the CO_2 methanation concept, which is described as the power-to-gas process and has

Citation: Yentekakis, I.V. The 10th Anniversary of Nanomaterials—Recent Advances in Environmental Nanoscience and Nanotechnology. *Nanomaterials* 2022, *12*, 915. https://doi.org/10.3390/nano12060915

Received: 25 February 2022
Accepted: 2 March 2022
Published: 10 March 2022

Publisher's Note: MDPI stays neutral with regard to jurisdictional claims in published maps and institutional affiliations.

attracted intense attention as a promising method for the storage of hydrogen, as well as overcoming the safety, storage, and transport difficulties in H_2 managing.

The above benefits of thermo-catalytic CO_2 methanation reaction are highlighted in the review paper of Tsiotsias et al. [26], in which the recent literature on the subject is overviewed and analyzed, giving particular attention on advanced bimetallic, Ni-M (M = Fe, Co, Cu, Ru, Rh, Pt, Pd, Re)-based catalyst formulations aiming to enhance the reaction activity and selectivity towards CH_4, especially in the low temperature region (ca. 200–350 °C). This thorough literature overview and analysis allowed the authors to conclude that although Ni is among the most active methanation catalysts, similarly to the very active Ru and Rh noble metals, it is being favored considering its low cost and high abundance in nature. However, there are some drawbacks of Ni as a methanation catalyst, such as insufficient low-temperature activity, low reducibility and a substantial propensity for nickel nanoparticle sintering. Literature results show that these inefficacies can be partly overcome via the incorporation of a second transition metal (e.g., Fe, Co) or a noble metal (e.g., Ru, Rh, Pt, Pd and Re) at low loading in Ni-based catalysts. In that way, the formation of several possible mixing nanoparticle (NP) structures, such as *"alloyed"* or "core-shell" NPs or even *"Janus"* heterostructures (closely located, and thus interacting, Ni and M individual particles), causes strong electronic metal-to-metal effects between the two active phases; thus, new high-performing and low-cost methanation catalysts can be obtained. For example, using Fe and Co as heteroatoms due to their similar size and electronic properties with the Ni atom allow for their easy dissolution into the Ni lattice, forming NiFe and NiCo alloyed particles, respectively. The specific composition of the formed alloys can lead to the optimization of CO_2 methanation performance, especially in the case of NiFe alloys. The combined bimetallic catalysts can also offer additional advantages, such as higher stability, and sulfur-poisoning tolerance. On the other hand, using noble metals as heteroatoms in the Ni-M combination, an increase in the reducibility and dispersion of the Ni primary phase can be obtained, useful for the CO_2 methanation reaction. Among noble metals, Rh and Pt can greatly enhance the catalytic activity for CO_2 methanation when dissolved or deposited in small quantities on Ni-based catalysts. Finally, the authors assume that a trade-off between cost and catalytic activity for CO_2 methanation catalysts can potentially be overcome via the development of bimetallic Ni-containing catalysts with an optimized Ni–dopant metal synergy.

In a highly comprehensive and interesting review, Al-Maqdi et al. [27] thoroughly analyze the recent research efforts on the development of enzyme-loaded flower-shaped nanomaterials, which are highly promising for biosensing, biocatalytic, and environmental applications. The authors explain how organic–inorganic hybrid nanoflowers (hNFs), a recently developed class of well-structured and well-oriented flower-like materials, through their unique structural and multifunctional properties, has gained intense interest and can be useful in several top technological applications. The structural attributes along with the surface-engineered functional entities of hNFs, such as size, shape, surface orientation, structural integrity, stability under reactive environments, enzyme stabilizing capability, and organic–inorganic ratio, all are key well-tailored properties and characteristics that can significantly contribute to and determine the hNFs applications. According to the authors, although the development of hNFs is still in its infancy, the rapid development of biotechnology in general and nanotechnology in particular makes hNFs a versatile platform for constructing enzyme-loaded/immobilized structures for a variety of applications, including detection and sensing-based, environmental- and sustainability-based, and biocatalytic and biotransformation applications. Readers of this work can find detailed information on many key issues related to the current advances in multifunctional NPCs that are particularly emphasized in the review, such as: (a) critical factors; (b) different metal/non-metal-based synthesizing processes (i.e., Cu-, Ca-, Mn-, Zn-, Co-, Fe-, multimetal-, and non-metal-based hNFs); (c) their applications; and finally, (d) the interfacial mechanism involved in hNF development considering the three critical points, which are the combination of metal ions and organic matter, the petal formation, and the generation of hNFs. Bearing in mind that the subject is new, the review of Al-Maqdi et al. [24] can undoubtedly be a valuable source of expert information and therefore a useful tool for hNF designers and hNF application engineers.

Aiming at a low carbon footprint energy future, H_2 production, storage and transport currently garners huge research interest. Besides the electrochemical water-splitting technology coupled with renewable energy sources for the high demand of electric power, hydrocarbons are the most important raw feedstock for hydrogen production. Although the primary focus is currently on the abundant-in-nature methane (as a key component of natural gas and biogas), the use of liquefied petroleum gas (LPG), mainly consisting of propane and butane, for hydrogen production has recently received increasing interest for various economic and availability reasons. Interestingly, two works on this theme were contributed to the present SI.

In the first paper, Ramantani et al. [28] systematically investigated the partial substitution of La by Sr and of Ni by Ru or Rh in the $LaNiO_3$ perovskite structure to explore the effects of the nature and composition of the A- and B-sites of such substituted perovskites on the propane steam reforming (PSR) reaction towards syngas (H_2+CO) production. Specifically, $LaNiO_3$ (LN) and $La_{0.8}Sr_{0.2}NiO_3$ (LSN) and noble metal-substituted LNM_x and $LSNM_x$ (M = Ru, Rh; $x = 0.01, 0.1$) perovskite-based catalysts were studied. The authors found that the incorporation of Ru and Rh foreign cations in the A and/or B sites of the perovskite structure resulted in an increase in the specific surface area, a shift of XRD lines toward lower diffraction angles, and a decrease in the mean primary crystallite size of the parent material, which was more pronounced for materials with higher NM loading due to distortion of the perovskite structure. Under PSR reaction conditions, the in situ development of new phases including metallic Ni and $La_2O_2CO_3$ on the LNM_x, and metallic Ni, $La_2O_2CO_3$, La_2SrO_x, La_2O_3, and $SrCO_3$ on the LSNMx were resulted, which enhance the PSR activity of these catalysts. Although the LN catalyst exhibited higher primary activity compared to LSN, its PSR catalytic performance did not appreciably change upon partial substitution of Ni by Ru. In striking contrast, the authors demonstrated that partial substitution of Ni by Ru and especially Rh in the LSN perovskite resulted in the significant promotion of catalytic performance on the title reaction, further improved upon increasing the noble metal content from $x = 0.01$ to 0.1 in the $LSNM_x$ (M = Ru, Rh) perovskite matrix. Thus, the $LSNRh_{0.1}$ catalyst found to perform extremely stable for at least 40 hours on stream due to the in situ formation of the $La_2O_2CO_3$ phase, which facilitates carbon oxidation, preventing its accumulation. This catalyst was the best over all the samples studied, offering propane conversion of ca. 92% accompanied by a selectivity towards H_2 as high as 97% at 600 °C; its pronounced catalytic performance was attributed to a synergy of well-dispersed Ru nanocrystallites with Ni^0 species on the perovskite surface [25]. The above results once again confirm the value of using such multifunctional materials in significant catalytic reactions. The specialized properties of perovskites, such as multiple types of active centers, including surface oxygen vacancies, as well as labile lattice oxygen and mobile O^{2-} ions—properties that are easily adapted and optimized on a case-by-case basis by partial substitution of A and B sites with A′ and B′ alternatives—are particularly useful in catalysis, as, in turn, these properties can play multiple roles as reaction promoters and stabilizers of the catalytic systems [23–25].

In the second paper concerning the same reaction (PSR), reported by Kokka et al. [29], the catalytic performance of supported Ni catalysts with respect to the nature of the oxide supports used for the dispersion of Ni particles was investigated. The authors found that Ni was much more active when supported on ZrO_2 or yttria-stabilized zirconia (YSZ: Y_2O_3-ZrO_2) compared to TiO_2, whereas Al_2O_3– and CeO_2-supported catalysts exhibited intermediate performance. The comparison of the PSR performance of the catalysts was based on the turnover frequency (TOF) kinetic measurements, which demonstrated that support-induced promotional effects on the TOF of C_3H_8 conversion can be more than *one order of magnitude* higher, following the order $Ni/TiO_2 < Ni/CeO_2 < Ni/Al_2O_3 < Ni/YSZ < Ni/ZrO_2$. These intrinsic rate increases were accompanied by a parallel increase in the selectivity toward the intermediate methane produced. Conducting in situ FTIR experiments, it was demonstrated that CH_x species produced via the dissociative adsorption of propane are the key reaction intermediates. Then, CH_x hydrogenation to CH_4 and/or conversion to formates, and eventually to CO, is favored over the most active Ni/ZrO_2 catalyst. In addition, the optimal Ni/ZrO_2 catalyst exhibited excellent stability for more than 30 h time-on-stream (TOS) and was therefore proposed by the authors as the most promising of the series for the PSR process for H_2 production.

Finally, Chalmpes et al. [30] reported a new synthetic approach towards carbon materials, which additionally provides a way to obtain useful materials out of waste or disposed rocket propellants. The method is based on the hypergolic ignition of furfuryl alcohol by fuming nitric acid at ambient conditions, which lead to the fast, spontaneous, and exothermic formation of carbon nanosheets in two steps: (i) polymerization of furfuryl alcohol to poly (furfuryl alcohol), and (ii) in situ carbonization of the polymer by an internal temperature increment near its decomposition point. A variety of advanced characterization techniques, such as XRD, infrared spectroscopy (Raman/IR), UV–vis, XPS, and SEM/TEM/AFM microscopies, were used to analyze the structure and morphology of the obtained carbon nanosheets. The method also provides for the direct conversion of the released energy into useful work by either heating acetone to boiling or spinning the Crookes radiometer. The authors argue that in a broader sense, the furfuryl alcohol-fuming nitric acid system could be the basis for a future "carbon from rocket fuel" perspective, especially given the wealth of available rocket bipropellants, as well as the growing progress for new hypergolic fuels. For such a perspective, the authors conclude that a technical upgrade of the method, as well as the search for new hypergolic pairs that will provide even higher carbon yields, remain future challenges for large-scale safe application.

Considering all of the above, I believe that the present SI has collected high-impact works that highlight valuable specific topics and current research interests in the field of environmental nanoscience and nanotechnology, as well as crucial perspectives and promising implementations. I would like to sincerely thank all the authors and reviewers who, with their valuable participation, have contributed to the production of these high-quality and high-impact papers that add value and can encourage researchers in the field.

Finally, I would like to take the opportunity to encourage researchers to submit their best works in *Nanomaterials*, a journal that has already been established as a high value, readability, and impact one, very suitable to host high-quality works in the field of nanomaterials/nanotechnology. As Editor-in-Chief of the highly active "Environmental Nanoscience and Nanotechnology" section of the journal, I would kindly appreciate your contributions to this area as well.

Acknowledgments: I would like to thank Nanomaterials MDPI journal for giving me the opportunity to edit this Special Issue, as well as the associated journal team and Assistant Editor, for their continued and close support during this effort.

Conflicts of Interest: The author declares no conflict of interest.

References

1. Yang, X.-F.; Wang, A.; Qiao, B.; Li, J.; Liu, J.; Zhang, T. Single-atom catalysts: A new frontier in heterogeneous catalysis. *Acc. Chem. Res.* **2013**, *46*, 1740–1748. [CrossRef] [PubMed]
2. Datye, A.; Wang, Y. Atom trapping: A novel approach to generate thermally stable and regenerable single-atom catalysts. *Natl. Sci. Rev.* **2018**, *5*, 630–632. [CrossRef]
3. Neagu, D.; Kyriakou, V.; Roiban, I.-L.; Aouine, M.; Tang, C.; Caravaca, A.; Kousi, K.; Schreur-Piet, I.; Metcalf, I.S.; Vernoux, P.; et al. In Situ Observation of Nanoparticle Exsolution from Perovskite Oxides: From Atomic Scale Mechanistic Insight to Nanostructure Tailoring. *ACS Nano* **2019**, *13*, 12996–13005. [CrossRef]
4. Wei, M.; Li, J.; Chu, W.; Wang, N. Phase control of 2D binary hydroxides nanosheets via controlling-release strategy for enhanced oxygen evolution reaction and supercapacitor performances. *J. Energy Chem.* **2019**, *38*, 26–33. [CrossRef]
5. Yentekakis, I.V.; Dong, F. Grand challenges for catalytic remediation in environmental and energy applications towards a cleaner and sustainable future. *Front. Environ. Chem.* **2020**, *1*, 5. [CrossRef]

6. Yentekakis, I.V.; Panagiotopoulou, P.; Artemakis, G. A review of recent efforts to promote dry reforming of methane (DRM) to syngas production via bimetallic catalyst formulations. *Appl. Catal. B Environ.* **2021**, *296*, 120210. [CrossRef]

7. Song, Y.; Ozdemir, E.; Ramesh, S.; Adishev, A.; Subramanian, S.; Harale, A.; Albuali, M.; Fadhel, B.A.; Jamal, A.; Moon, D.; et al. Dry reforming of methane by stable Ni-Mo nanocatalysts on single-crystalline MgO. *Science* **2020**, *367*, 777–781. Available online: https://science.sciencemag.org/content/367/6479/777/tab-pdf (accessed on 22 February 2022). [CrossRef]

8. Akri, M.; Zhao, S.; Li, X.; Zang, K.; Lee, A.F.; Isaacs, M.A.; Xi, W.; Gangarajula, Y.; Luo, J.; Ren, Y.; et al. Atomically dispersed nickel as coke resistant active sites for methane dry reforming. *Nat. Commun.* **2019**, *10*, 5181. [CrossRef]

9. Kousi, K.; Neagu, D.; Bekris, L.; Papaioannou, E.I.; Metcalfe, I.S. Endogenous Nanoparticles Strain Perovskite Host Lattice Providing Oxygen Capacity and Driving Oxygen Exchange and CH_4 Conversion to Syngas. *Angew. Chem. Int. Ed.* **2020**, *59*, 2510–2519. [CrossRef]

10. Kousi, K.; Neagu, D.; Metcalfe, I.S. Combining Exsolution and Infiltration for Redox, Low Temperature CH_4 Conversion to Syngas. *Catalysts* **2020**, *10*, 468. [CrossRef]

11. Nikolaraki, E.; Goula, G.; Panagiotopoulou, P.; Taylor, M.J.; Kousi, K.; Kyriakou, G.; Kondarides, D.I.; Lambert, R.M.; Yentekakis, I.V. Support Induced Effects on the Ir Nanoparticles Activity, Selectivity and Stability Performance under CO_2 Reforming of Methane. *Nanomaterials* **2021**, *11*, 2880. [CrossRef] [PubMed]

12. Botzolaki, G.; Goula, G.; Rontogianni, A.; Nikolaraki, E.; Chalmpes, N.; Zygouri, P.; Karakassides, M.; Gournis, D.; Charisiou, N.; Goula, M.; et al. CO_2 Methanation on supported Rh nanoparticles: The combined effect of support oxygen storage capacity and Rh particle size. *Catalysts* **2020**, *10*, 944. [CrossRef]

13. Goula, G.; Botzolaki, G.; Osatiashtiani, A.; Parlett, C.M.A.; Kyriakou, G.; Lambert, R.M.; Yentekakis, I.V. Oxidative Thermal Sintering and Redispersion of Rh Nanoparticles on Supports with High Oxygen Ion Lability. *Catalysts* **2019**, *9*, 541. [CrossRef]

14. Yentekakis, I.V.; Goula, G.; Panagiotopoulou, P.; Kampouri, S.; Taylor, M.J.; Kyriakou, G.; Lambert, R.M. Stabilization of catalyst particles against sintering on oxide supports with high oxygen ion lability exemplified by Ir-catalysed decomposition of N_2O. *Appl. Catal. B Environ.* **2016**, *192*, 357–364. [CrossRef]

15. Yentekakis, I.V.; Goula, G.; Hatzisymeon, M.; Betsi-Argyropoulou, I.; Botzolaki, G.; Kousi, K.; Kondarides, D.I.; Taylor, M.J.; Parlett, C.M.A.; Osatiashtiani, A.; et al. Effect of support oxygen storage capacity on the catalytic performance of Rh nanoparticles for CO_2 reforming of methane. *Appl. Catal. B Environ.* **2019**, *243*, 490–501. [CrossRef]

16. Yentekakis, I.V.; Goula, G.; Kampouri, S.; Betsi-Argyropoulou, I.; Panagiotopoulou, P.; Taylor, M.J.; Kyriakou, G.; Lambert, R.M. Ir-catalyzed Nitrous oxide (N_2O) decomposition: Effect of the Ir particle size and metal-support interactions. *Catal. Lett.* **2018**, *148*, 341–347. [CrossRef]

17. Yentekakis, I.V.; Vernoux, P.; Goula, G.; Caravaca, A. Electropositive Promotion by Alkalis or Alkaline Earths of Pt-Group Metals in Emissions Control Catalysis: A Status Report. *Catalysts* **2019**, *9*, 157. [CrossRef]

18. Li, X.; Li, D.; Tian, H.; Zeng, L.; Zhao, Z.-J.; Gong, J. Dry reforming of methane over Ni/La_2O_3 nanorod catalysts with stabilized Ni nanoparticles. *Appl. Catal. B Environ.* **2017**, *202*, 683–694. [CrossRef]

19. Marinho, A.L.A.; Rabelo-Neto, R.C.; Epron, F.; Bion, N.; Toniolo, F.S.; Noronha, F.B. Embedded Ni nanoparticles in $CeZrO_2$ as stable catalyst for dry reforming of methane. *Appl. Catal. B Environ.* **2020**, *268*, 118387. [CrossRef]

20. Helveg, S.; Lopez-Cartes, C.; Sehested, J.; Hansen, P.L.; Clausen, B.S.; Rostrup-Nielsen, J.R.; Abild-Pedersen, F.; Nørskov, J.K. Atomic-scale imaging of carbon nanofibre growth. *Nature* **2004**, *427*, 426–429. [CrossRef]

21. Pliangos, C.; Yentekakis, I.V.; Papadakis, V.G.; Vayenas, C.G.; Verykios, X.E. Support induced promotional effects on the activity of automotive exhaust catalysts: 1. The case of oxidation of light hydrocarbons (C_2H_4). *Appl. Catal. B Environ.* **1997**, *14*, 161–173. [CrossRef]

22. Vayenas, C.G. Promotion, electrochemical promotion and metal–support interactions: Their common features. *Catal. Lett.* **2013**, *143*, 1085–1097. [CrossRef]

23. Zhu, J.; Li, H.; Zhong, L.; Xiao, P.; Xu, X.; Yang, X.; Zhao, Z.; Li, J. Perovskite oxides: Preparation, characterizations, and applications in heterogeneous catalysis. *ACS Catal.* **2014**, *4*, 2917–2940. [CrossRef]

24. Royer, S.; Duprez, D.; Can, F.; Courtois, X.; Batiot-Dupeyrat, C.; Laassiri, S.; Alamdari, H. Perovskites as substitutes of noble metals for heterogeneous catalysis: Dream or reality. *Chem. Rev.* **2014**, *114*, 10292–10368. [CrossRef] [PubMed]

25. Yentekakis, I.V.; Georgiadis, A.G.; Drosou, K.; Charisiou, N.D.; Goula, M.A. Selective catalytic reduction of NO_x over perovskite-based catalysts using $C_xH_y(O_z)$, H_2 and CO as reducing agents–A review of the latest developments. *Nanomaterials* **2022**, *submitted for publication*.

26. Tsiotsias, A.I.; Charisiou, N.D.; Yentekakis, I.V.; Goula, M.A. Bimetallic Ni-Based Catalysts for CO_2 Methanation: A Review. *Nanomaterials* **2021**, *11*, 28. [CrossRef] [PubMed]

27. Al-Maqdi, K.A.; Bilal, M.; Alzamly, A.; Iqbal, H.M.N.; Shah, I.; Ashraf, S.S. Enzyme-Loaded Flower-Shaped Nanomaterials: A Versatile Platform with Biosensing, Biocatalytic, and Environmental Promise. *Nanomaterials* **2021**, *11*, 1460. [CrossRef]

28. Ramantani, T.; Bampos, G.; Vavatsikos, A.; Vatskalis, G.; Kondarides, D.I. Propane Steam Reforming over Catalysts Derived from Noble Metal (Ru, Rh)-Substituted $LaNiO_3$ and $La_{0.8}Sr_{0.2}NiO_3$ Perovskite Precursors. *Nanomaterials* **2021**, *11*, 1931. [CrossRef]

29. Kokka, A.; Petala, A.; Panagiotopoulou, P. Support Effects on the Activity of Ni Catalysts for the Propane Steam Reforming Reaction. *Nanomaterials* **2021**, *11*, 1948. [CrossRef]

30. Chalmpes, N.; Bourlinos, A.B.; Talande, S.; Bakandritsos, A.; Moschovas, D.; Avgeropoulos, A.; Karakassides, M.A.; Gournis, D. Nanocarbon from Rocket Fuel Waste: The Case of Furfuryl Alcohol-Fuming Nitric Acid Hypergolic Pair. *Nanomaterials* **2021**, *11*, 1. [CrossRef]

 *nanomaterials*

Review

Bimetallic Ni-Based Catalysts for CO$_2$ Methanation: A Review

Anastasios I. Tsiotsias [1], Nikolaos D. Charisiou [1], Ioannis V. Yentekakis [2] and Maria A. Goula [1,*]

[1] Laboratory of Alternative Fuels and Environmental Catalysis (LAFEC), Department of Chemical Engineering, University of Western Macedonia, GR-50100 Koila, Greece; antsiotsias@uowm.gr (A.I.T.); ncharisiou@uowm.gr (N.D.C.)

[2] Laboratory of Physical Chemistry & Chemical Processes, School of Environmental Engineering, Technical University of Crete, GR-73100 Chania, Greece; yyentek@isc.tuc.gr

* Correspondence: mgoula@uowm.gr; Tel.: +30-246-106-8296

Abstract: CO$_2$ methanation has recently emerged as a process that targets the reduction in anthropogenic CO$_2$ emissions, via the conversion of CO$_2$ captured from point and mobile sources, as well as H$_2$ produced from renewables into CH$_4$. Ni, among the early transition metals, as well as Ru and Rh, among the noble metals, have been known to be among the most active methanation catalysts, with Ni being favoured due to its low cost and high natural abundance. However, insufficient low-temperature activity, low dispersion and reducibility, as well as nanoparticle sintering are some of the main drawbacks when using Ni-based catalysts. Such problems can be partly overcome via the introduction of a second transition metal (e.g., Fe, Co) or a noble metal (e.g., Ru, Rh, Pt, Pd and Re) in Ni-based catalysts. Through Ni-M alloy formation, or the intricate synergy between two adjacent metallic phases, new high-performing and low-cost methanation catalysts can be obtained. This review summarizes and critically discusses recent progress made in the field of bimetallic Ni-M (M = Fe, Co, Cu, Ru, Rh, Pt, Pd, Re)-based catalyst development for the CO$_2$ methanation reaction.

Keywords: CO$_2$ methanation; bimetallic catalysts; Ni-based catalysts; promoters; alloy nanoparticles; bimetallic synergy

Citation: Tsiotsias, A.I.; Charisiou, N.D.; Yentekakis, I.V.; Goula, M.A. Bimetallic Ni-Based Catalysts for CO$_2$ Methanation: A Review. *Nanomaterials* **2021**, *11*, 28. https://dx.doi.org/10.3390/nano11010028

Received: 16 November 2020
Accepted: 22 December 2020
Published: 24 December 2020

Publisher's Note: MDPI stays neutral with regard to jurisdictional claims in published maps and institutional affiliations.

1. Introduction

During the last hundred years, rapid industrialization and the high energy demands of our society have disrupted the carbon cycle through ever increasing greenhouse gas emissions, and the ramp-up of renewable energy production has yet to offset the negative effects on our planet's climate and ecosystems [1,2]. However, progress made in hydrogen production technologies through water electrolysis has raised hopes for the utilization of this green fuel that produces no CO$_2$ emissions upon its combustion [3], despite the fact that its storage and transportation remain challenging when compared to other traditional energy carriers, such as natural gas [4]. In the last decade, research efforts have been focused on the development of catalysts that can utilize this excess renewable hydrogen in order to hydrogenate CO$_2$ released from industrial flue gases. This way, H$_2$ can be transformed into a reliable energy carrier, that is, CH$_4$ or synthetic natural gas (SNG), with a significantly higher energy density, all the while creating a closed carbon cycle [5]. The complete hydrogenation of CO$_2$ into CH$_4$, or CO$_2$ methanation, is also known as the Sabatier reaction and is an exothermic reaction with the following equation:

$$CO_2 + 4H_2 \rightarrow CH_4 + 2H_2O \quad \Delta H_{298\,K} = -165\ kJ/mol \tag{1}$$

Ni has become a favourite active metal for this reaction, since its high methanation activity, low cost and natural abundance render it attractive for industrial-scale applications [6]. Since CH$_4$ yield peaks at a relatively low temperature (300–400 °C, depending on the reaction conditions) [7], structural degradation of Ni-based catalysts, though not completely avoided, plays a minor role compared to other reactions (e.g., methane dry reforming) [8]. The choice of the metal oxide support also appears to be of great importance in the performance of Ni-based catalysts [9–11]. Ni/CeO$_2$ catalysts, for example, are much more active compared to Ni/Al$_2$O$_3$ or Ni/SiO$_2$ catalysts. This is mainly attributed to ceria's intricate redox and O^{2-}-defect chemistry, with it being able to transport oxygen species and oxygen ion vacancies throughout its lattice, having higher basicity compared to other metal oxides that favours CO$_2$ chemisorption and activation, as well as exhibiting a strong metal–support interaction that favours a higher Ni dispersion (Figure 1) [12].

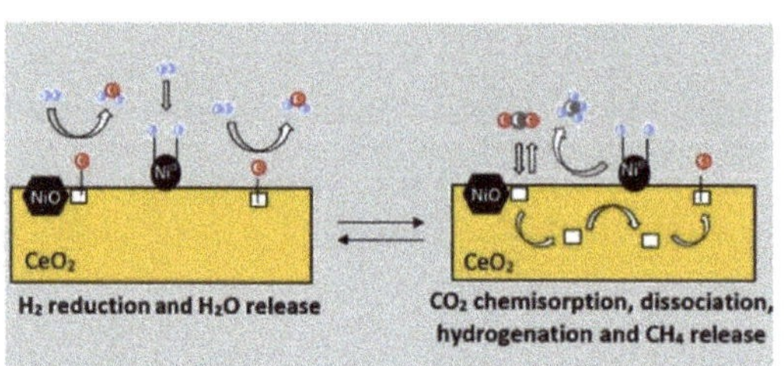

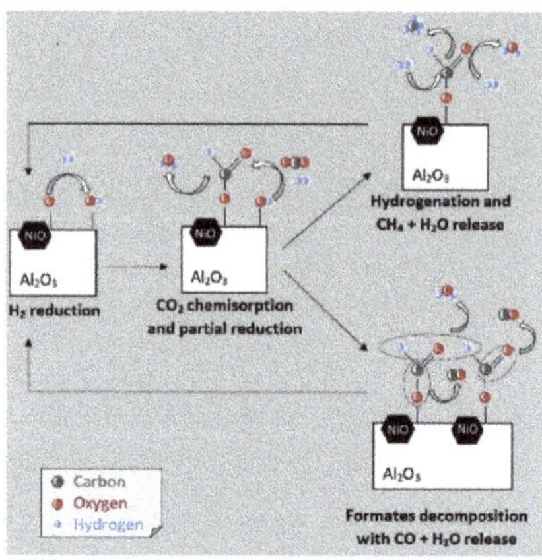

Figure 1. Scheme of the CO$_2$ + H$_2$ reaction mechanisms over Ni/CeO$_2$ and Ni/Al$_2$O$_3$ catalysts. Reproduced with permission from [12]. Copyright: Elsevier, Amsterdam, The Netherlands, 2020.

The activity of Ni-based catalysts can be further improved via modification of the metal oxide supports. For example, alkali and alkaline earth metals [13], transition metals [6] and rare-earth metals [14] can be used as promoters that modify the physicochemical properties of metal oxide supports. In some cases, these ions can enter the lattice of the metal oxide supports (e.g., Ca^{2+} ions in CeO$_2$ and ZrO$_2$ lattices) [15], or form segregated metal oxide phases supported on the support surface (e.g., La$_2$O$_3$, CeO$_2$ and MnO$_x$ in Al$_2$O$_3$) [16]. Such modifications can lead to an increase in support basicity, so that the initial step of CO$_2$ chemisorption step is accelerated, or to an increase in the active metal dispersion [17]. In most cases, the low-temperature activity and stability of Ni-based catalysts is enhanced following modification of the metal oxide supports [13,14,16].

Besides Ni, Ru and Rh noble metals have been extensively studied as active metallic phases in CO$_2$ methanation and they usually achieve a much higher activity at low temperatures [18,19]. Since CH$_4$ is thermodynamically favoured over other CO$_2$ hydrogenation products such as CO, at low temperatures, CH$_4$ selectivity can be significantly higher when using noble metal catalysts [7]. Among the two noble metals, Ru can achieve higher activity and its price is considerably lower compared to Rh, while it can also provide significant methanation activity when supported on cheap supports (e.g., Al$_2$O$_3$ or TiO$_2$) at a metal loading as low as 1% or even 0.5% [20]. Ru is also preferable to Ni for application in the combined capture and methanation of CO$_2$ derived from industrial flue-gases since the high reducibility of RuO$_x$ oxides allows for isothermal operation at low temperatures [21,22].

A popular method to counter some of the drawbacks of Ni-based catalysts is to use a second metal (e.g., Fe, Co or Ru) as a dopant, in order to create appropriate bimetallic CO$_2$ methanation catalysts [6]. Such an approach has been successfully employed in other reactions. For example, NiFe alloys are active and stable catalysts for dry reforming of methane, since Fe can promote carbon gasification and significantly reduce coking through an intricate dealloying and realloying mechanism [23]. The combination of Ni with other metals can either lead to the formation of Ni-M alloys, or monometallic heterostructures with closely located active metallic Ni-M phases [23,24]. There are two types of metals that are used in such Ni-M bimetallic catalysts, the one an early transition metal such as Fe and Co and the other a noble metal, namely Ru, Rh, Pt, Pd and Re.

Fe and Co can easily dissolve into the Ni lattice due to the similar crystallographic properties of the corresponding metallic phases. In the example of Fe, the dissolution of Fe atoms into the Ni lattice leads to the formation of NiFe alloys, with Ni_3Fe being the most thermodynamically stable [25,26]. The introduction of Fe causes an expansion of the Ni fcc lattice up to a specific Fe amount and a shift of the (111) Ni reflection in XRD towards lower 2θ values. At higher Fe contents, the lattice becomes Fe rich and switches to the more compact bcc structure of pure Fe [27]. The introduction of the dopant metal can be used to tailor the electronic properties of Ni, so that the new alloy phase can achieve superior activity compared to monometallic Ni. This can also lead to a higher dispersion, stability and/or resistance towards deactivation [24]. The application of computational methods has shown that specific alloys can lower the M-CO binding energy and lead to higher CO methanation activities [28].

Noble metals Ru, Rh, Pt, Pd and Re can increase the reaction activity by enhancing the reducibility of the primary Ni phase, by increasing the Ni dispersion, or by changing the reaction pathway [29]. Ru and Ni mostly form monometallic heterostructures that rely on the synergistic effect between the two separate metallic phases, while Pt and Pd mostly lead to the creation of NiPt and NiPd alloys [30–32]. It has been shown that an addition of only a miniscule amount of noble metal (e.g., 0.5% or 1%) can greatly enhance the reducibility and low-temperature activity of Ni-based catalysts without the need to use high loadings of precious metals [33].

In this review, we aim to summarize the most recent progress made in the field of Ni-based bimetallic catalyst development for the CO_2 methanation reaction. Due to the different nature of the reaction promotion, Ni-based catalysts combined with either early transition metals (Fe and Co), or noble metals (Ru, Rh, Pt, Pd and Re), will be discussed in separate chapters.

2. Promotion with Transition Metals

There are many works that use a transition metal additive to enhance the activity of Ni-based catalysts [24]. These additives may include: V, Cr, Mn, Fe, Co, Y and Zr. Y and Zr, for example, are mostly used as dopants to modify the lattice of the metal oxide support and enhance its defect chemistry. Zr is used to stabilize the CeO_2 structure and enhance its oxygen vacancies population (i.e., oxygen storage capacity, OSC) [19], while Y can generate oxygen vacancies in ZrO_2-based supports [34,35]. Mn mostly forms MnO_x phases that increase the catalyst basicity and favour CO_2 chemisorption. All these modifications on the support's properties can lead to an increase in CO_2 activation and, thus, Mn, Ce, Zr and Y are often regarded as efficient promoters in CO_2 methanation [19,34–36].

Fe and Co, when combined with Ni-based catalysts, allow the creation of NiFe and NiCo alloys [26,37]. The incorporation of these transition metal dopants into the lattice of the active Ni phase can directly interfere with nickel's electronic properties and methanation chemistry [38]. This can either lead to an increase in activity and stability or to a complete catalyst deactivation, depending on the Ni/dopant ratio, its degree of metal intermixing and the interaction of both metals with the support [39].

2.1. Promotion with Fe

Among all metals, Fe is by far the most studied element in bimetallic Ni-based catalysts for CO_2 methanation, since it is quite cheap and highly available and it exhibits a high solubility into the Ni lattice, favouring the formation of NiFe alloys [25]. It has been suggested, based on computational screening and catalytic experiments, that alumina supported and Ni-rich NiFe catalysts can improve the rate of CO_2 conversion, with the optimal Ni/(Ni + Fe) ratio being around 0.7–0.9 [40,41]. Many more works have focused on NiFe alloys prepared with different methods and supported on various metal oxides [42].

Generally, the Ni/Fe ratio in the alloy and the reducibility of the metal-oxide support appear to be the most critical parameters that determine whether Fe will promote or suppress the catalyst's CO_2 methanation activity.

Pandey et al. [43] were among the first to perform a systematic study about the Fe promotion in Ni-based methanation catalysts. The optimal active metal content of the catalysts consisted of 75% Ni and 25% Fe. When compared to their monometallic counterparts (Ni and Fe), the bimetallic NiFe alloy catalysts supported on alumina and silica were shown to exhibit higher CH_4 yields and this enhancement was more apparent in the alumina supported catalysts. This was attributed to the creation of a suitable alloy phase and to the increased CO_2 chemisorption at unreduced Fe_3O_4 sites. The authors then compared the activity of Ni_3Fe catalysts supported on different metal oxide supports, such as Al_2O_3, SiO_2, ZrO_2 and TiO_2, and noted that Al_2O_3 supported catalysts yielded the best results and provided the largest promotion due to Fe alloying [42]. Regarding the optimal alloy composition, a statistical technique, namely response surface methodology (RSM), was also used. The model equation predicted that a 32.78% Ni loading and 7.67% Fe loading would lead to the optimal methane yield, and the obtained experimental results confirmed this model prediction [44].

Ray et al. [38] from the same group managed to develop a descriptor for the methanation of CO_2 over Ni_3M bimetallic catalysts, with the second metal (M) being Fe and Cu. It was shown that the turnover frequency for methane production (TOF_{CH4}) could be linearly correlated with the number of d-density of states (d-DOS) at the Fermi level (N_{EF}) based on density functional theory (DFT) calculations, as shown in Figure 2. The N_{EF} descriptor successfully predicted the enhancement of CO_2 methanation performance over the Ni_3Fe catalyst, due to a favourable change in the electronic properties of the active Ni phase, while Ni_3Cu alloy formation was detrimental for the production of CH_4. In a follow-up study, Ray et al. [45] also included Ni_3Co catalysts in their calculations and concluded that Co alloying could also enhance the CO_2 methanation performance but by a smaller degree compared to Fe. They also showed that Co alloying led to the highest promotion regarding the dry methane reforming (DMR) reaction, with the location of the Ni d-band centre (ε_d) being the most appropriate descriptor to assess the catalytic activity for this reaction.

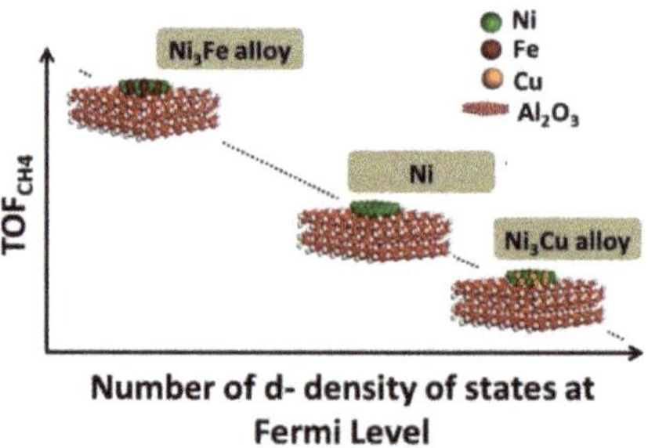

Figure 2. Linear correlation between the turnover frequency for CH_4 production (TOF_{CH4}) and the number of d-density of states (d-DOS) at Fermi level (N_{EF}) for Ni, Ni_3Fe and Ni_3Cu catalysts. Reproduced with permission from [38]. Copyright: Elsevier, Amsterdam, The Netherlands, 2017.

The Grunwaldt group have been amongst the pioneers in the development of NiFe-based methanation catalysts and the elucidation of the role of Fe in the overall catalytic performance. Mutz et al. [26] prepared Ni_3Fe catalysts supported on Al_2O_3 via deposition–precipitation. The formed alloy nanoparticles exhibited a small size and high dispersion, and the NiFe-based catalyst proved to be more active and stable at lower temperatures compared to the monometallic Ni-based catalyst. The alloy catalyst was also proven to have a quite stable and robust performance after a 45 h time-on-stream operation under industrially oriented conditions [26]. No carbon deposition could be observed under various gas feeds for such catalysts using operando Raman spectroscopy [46]. Farsi et al. [47] investigated the CO_2 methanation kinetics on such Ni_3Fe methanation catalysts under technical operation conditions. CO selectivity over CH_4 was found to increase over shorter residence times and higher temperatures, while water concentration was indicated as the main inhibiting factor.

Furthermore, Serrer et al. [48] focussed specifically on the role of Fe during CO_2 methanation, employing advanced operando spectroscopic methods. Fe was shown to act as a protective or "sacrificial" dopant upon cases of H_2-dropout during CO_2 methanation. While Ni-based catalysts are oxidized under such events and then fail to regain their initial activity upon the restoration of H_2 flow, NiFe-based catalysts retain the Ni^0 sites in their reduced state, due to the preferential oxidation of their Fe sites into FeO. Under normal operation, Fe was shown to increase the reducibility of Ni and result in the formation of small FeO_x clusters at the surface of the alloy nanoparticles, with Fe being found in various oxidation states [49]. It has been suggested that increased methanation performance of NiFe alloy catalysts could be due to these small FeO_x clusters on top of alloy nanoparticles that can potentially favour CO_2 chemisorption and activation, as shown in Figure 3.

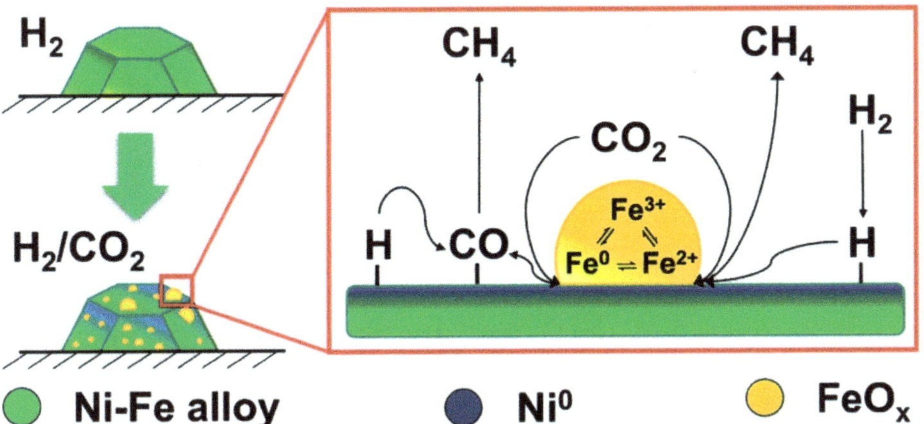

Figure 3. Scheme of the CO_2 activation mechanism on Ni–Fe alloy-based catalysts during methanation reaction under realistic conditions. Reproduced with permission from [49]. Copyright: Royal Society of Chemistry, London, UK, 2020.

Mebrahtu et al. [39] used hydrotalcite-precursors with a tailored Fe/Ni ratio in order to prepare $NiFe/(Mg,Al)O_x$ catalysts with high levels of metal intermixing and dispersion. The Fe/Ni ratio played a crucial role in the physicochemical properties and the catalytic performance of the prepared catalysts. An Fe/(Ni + Fe) ratio of 0.1 was shown to provide high metal dispersion, small nanoparticle sizes and an optimum amount of surface basic sites. Consequently, the catalyst with this specific ratio offered the best low-temperature catalytic performance in terms of CO_2 conversion and CH_4 selectivity, whereas catalysts with higher Fe contents experienced a significant drop for these values (Figure 4). Mebrahtu et al. [50] also indicated a possible deactivation pathway for monometallic Ni catalysts via the formation of Ni hydroxides caused by the water produced in situ upon methanation. It was found that the introduction of Fe prevented the formation of Ni-OH species, thus increasing the catalytic activity of such systems. It was also argued that Fe formed spinel phases on the alumina nanosheets and did not form alloys with Ni.

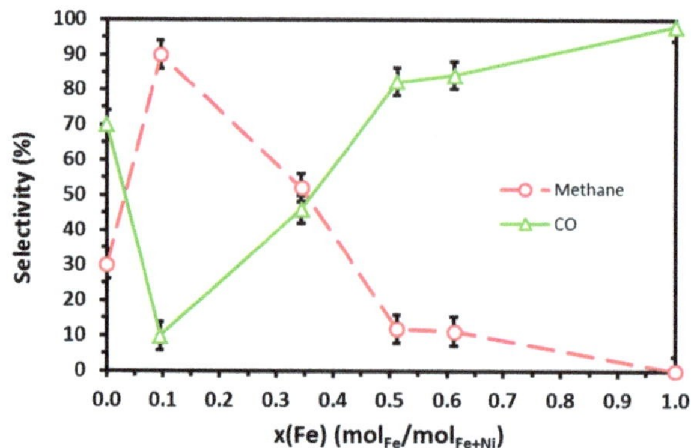

Figure 4. Effect of Fe content on CH_4 ($\bigcirc$) and CO ($\triangle$) selectivity in comparison to sole metals. The highest methane selectivity, with 90%, was achieved for the catalyst containing 10% Fe. Reproduced with permission from [39]. Copyright: Royal Society of Chemistry, London, UK, 2018.

Giorgianni et al. [51] attributed the beneficial role of Fe into such hydrotalcite-derived catalysts to the presence of Fe(II) species. These species could activate CO_2 molecules and adsorb H_2O produced in situ during the reaction, thus preventing the hydroxylation of nearby Ni^0 active sites. However, these Fe(II) species inevitably undergo oxidation into Fe(III) over time, reducing the catalytic performance.

In a similar fashion, Huynh et al. [52] observed high CO_2 methanation activity and stability for cheap hydrotalcite-derived $NiFe/(Mg,Al)O_x$ catalysts, which could achieve around 75% CO_2 conversion and 95% CH_4 selectivity at just 300 °C under high gas space velocities. They also showed that an Fe/(Ni + Fe) ratio of 0.2 (Ni_4Fe) could lead to the highest promotion in CH_4 yield at low temperatures, due to a considerable decrease in the activation energy for CH_4 formation [53]. The formate pathway was observed to be favoured over the direct dissociation of CO_2 into CO via in-situ diffuse reflectance infrared Fourier transform spectroscopy (DRIFTS) and density functional theory (DFT) [53]. Furthermore, NiFe-containing layered double hydroxides (LDHs) could be in situ grown over alumina and silica washcoated cordierite monoliths via an urea hydrolysis preparation method [54]. Using different washcoat materials and metal concentrations, a structured NiFe bimetallic catalyst with a thin catalytic layer on the cordierite substrate was prepared. This structured NiFe monolithic catalyst achieved high CH_4 yields under high gas flow rates, thus making it attractive for industrial-scale applications.

Burger et al. [55] prepared $NiAlO_x$ coprecipitated catalysts modified with Fe and Mn. Both metals improved the performance of the $NiAlO_x$ catalysts. Mn formed separate MnO_x phases that enhanced the CO_2 adsorption capacity and metal dispersion, while Fe formed NiFe alloys, promoting the electronic properties of Ni and enhancing the catalyst stability. The optimal Ni to promoter molar ratio was five. Such catalysts were also proven to be resistant upon sulphur poisoning, since H_2S was preferentially adsorbed at the metal promoter phases [56], thus preserving the Ni active sites [57].

Burger et al. [58,59] also studied the mechanism of Fe promotion in coprecipitated Ni/Al_2O_3 catalysts. In one study [58], Fe was deposited adjacent to Ni nanoparticles via a surface redox mechanism creating alloyed surface phases. Fe alloying also increased the activity of Ni/Al_2O_3 catalysts prepared via deposition–precipitation and the thermal stability of co-precipitated $NiAlO_x$ catalysts. (γFe,Ni) alloy nanoparticles were observed at co-precipitated $NiFeAlO_x$ catalysts and the presence of Fe(II) species upon aging provided additional CO_2 activation sites [59]. The generation of active Fe(II) species appeared to partially offset the negative effects caused by active metal sintering upon aging.

Other reports about Fe promotion in Ni/Al_2O_3 catalysts include the work of Li et al. [60]. It was found that adding 3% Fe on a 12% Ni/Al_2O_3 catalyst led to a mild improvement for CO_2 conversion and CH_4 selectivity, whereas the increase in the Fe content to 12% (Fe/Ni $\approx$ 1) caused a worse methanation performance. Liang et al. [61] studied the effect of various additives on Ni/Al_2O_3 catalysts and found the presence of an increased number of oxygen vacancies over the Fe-modified catalyst, as evidenced by electron paramagnetic resonance (EPR), that caused favourable changes to the reaction pathway. Finally, in contrast to other works, Daroughegi et al. [62] found that a 25% Ni/Al_2O_3 catalyst modified with 5% Fe exhibited a much worse methanation performance in terms of CO_2 conversion compared to the corresponding monometallic Ni catalyst.

Up until now, the NiFe-based catalysts studied were supported on Al_2O_3 or Al-based "inert" supports. The oxide support can, however, also play a major role in the methanation performance of alloy catalysts [63]. Ren et al. [64] showed that modification of a 30% Ni/ZrO_2 catalyst with 3% Fe enhanced its low temperature methanation performance. Modification with 5% Fe also yielded better results compared to the monometallic catalyst, but higher Fe contents led to a decline in methanation activity. The authors suggested that the majority of Fe in the catalysts was not fully reduced following pretreatment under H_2 flow, but remained at an Fe(II) oxidation state. These species could potentially improve the dispersion and reducibility of the Ni phase and also promote the reduction of ZrO_2, thus facilitating the presence of oxygen vacancies, that together with Fe(II) sites typically enhance CO_2 chemisorption and dissociation.

Yan et al. [65] employed low metal loadings of Ni and Fe (1.5% Ni and 0.5–4.5% Fe) supported ZrO_2 and assigned the various interfacial sites on the catalysts as selective for different CO_2 hydrogenation products. The Ni-ZrO_2 interface in the monometallic catalyst was characterized as active for the methanation reaction. Fe addition up to an equimolar amount to that of Ni led to a small improvement in CO_2 conversion and CH_4 selectivity and probably preserved the methane selective active metal–ZrO_2 interface. Only with the addition of a large amount of Fe (Fe/Ni ratio at 3) does the CO selectivity increase, via the creation of Ni-FeO_x interface that binds intermediate CO weakly and favours the reverse water gas shift (RWGS) reaction. Their experimental results are summarized in Figure 5. Furthermore, other studies showed that Fe promotion provided an enhancement of the methanation performance of Ni-based catalysts supported on Al_2O_3 and mesoporous clay modified with ZrO_2 [66,67].

It has been previously mentioned that CeO_2 support can offer significant advantages regarding the CO_2 methanation performance of Ni-based catalysts, when compared with other metal oxide supports [12]. However, Fe modification of Ni catalysts supported on CeO_2-based supports appears to exert a negative influence on CO_2 methanation. Winter et al. [68] studied CeO_2-supported NiFe catalysts with low metal loadings of Ni and Fe (1.5% Ni and 0.5–1.5% Fe). In contrast to the ZrO_2 supported catalysts reported by Yan et al. [65], Fe modification, even in small amounts, led to a significant drop in CO_2 conversion and rise in CO selectivity. The majority of Fe remained oxidized according to X-ray absorption near edge structure (XANES) analysis and these oxidized Fe species probably led to a weakened interaction between metal and intermediate CO, thus facilitating CO desorption and higher CO selectivity [68].

Pastor-Pérez et al. [69] prepared Fe and Co modified Ni catalysts supported on CeO_2-ZrO_2 and observed a negative effect of Fe addition, since the 3% Fe, 15% Ni/CeO_2-ZrO_2 catalyst exhibited lower values for CO_2 conversion and CH_4 selectivity compared to the monometallic Ni catalyst. Likewise, le Saché et al. [70] observed that Fe addition on a Ni/CeO_2-ZrO_2 catalyst led to a drop in CO_2 conversion and CH_4 selectivity. Furthermore, Frontera et al. [71] also reported a drop in CO_2 methanation activity upon Fe alloying over Ni catalysts supported on Gd-doped CeO_2 (GDC). In contrast to NiFe catalysts supported on inert supports such as Al_2O_3 and SiO_2 [26,43], Fe addition appeared to suppress the population of surface oxygen vacancies on the already defect-rich GDC support.

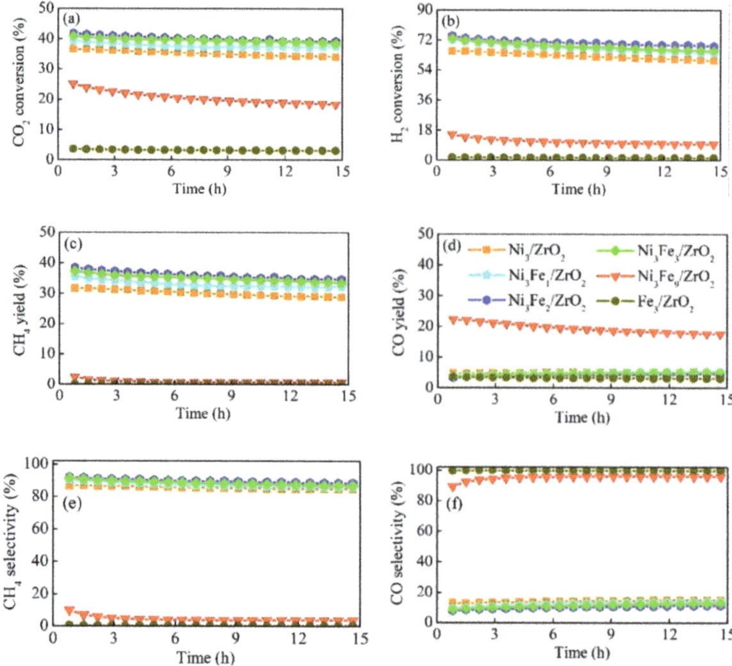

Figure 5. Conversions of (**a**) CO_2 and (**b**) H_2, yields of (**c**) CH_4 and (**d**) CO, and selectivities of (**e**) CH_4 and (**f**) CO on ZrO_2-supported catalysts plotted versus time on stream for reaction of CO_2 and H_2 (5 mL/min CO_2 + 10 mL/min H_2 + 25 mL/min Ar) at 673 K. Reproduced with permission from [65]. Copyright: Elsevier, Amsterdam, The Netherlands, 2019.

Carbon is another type of support in heterogeneous catalysis, which has been deemed as inactive regarding CO_2 methanation when using Ni-based catalysts [72]. However, Gonçalves et al. [73] managed to improve the performance of activated carbon (AC) supported Ni catalysts by modifying the surface chemistry of carbon and by introducing Fe as a second metal. They reported that 5% Fe addition on a 15% Ni catalyst supported on activated carbon with increased basicity (AC-R) improved the catalytic performance for CO_2 methanation. More specifically, NiFe alloy formation improved the low-temperature activity of the supported catalysts, increased the CH_4 selectivity and favoured catalyst stability, as shown in Figure 6.

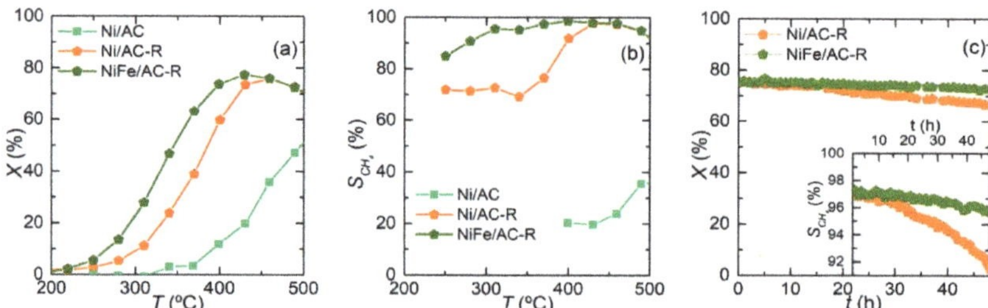

Figure 6. Comparison of the catalytic properties of the best Ni catalyst Ni/AC–R and the same promoted by Fe NiFe/AC–R: X_{CO2} and S_{CH4} as a function of temperature (**a,b**) as well as time-on-stream (TOS) at 450 °C (**c**). Reaction conditions: P = 1 bar; Weight Hourly Space Velocity (WHSV) = 60,000 mL g^{-1} h^{-1}; CO_2:H_2 = 1:4. Reproduced with permission from [73]. Copyright: Royal Society of Chemistry, London, UK, 2020.

The formation of NiFe alloy phase can sometimes take place via the migration of Fe atoms that are incorporated in the support lattice, towards the surface Ni particles, upon exposure to a reducing atmosphere. Wang et al. [74] impregnated Ni on an olivine support and observed the formation of NiFe alloyed phase from the Fe contained in the support. The good CO_2 methanation performance was ascribed to NiFe alloy formation and its favourable interaction with unreduced FeO_x segregated from the olivine support upon calcination and reduction treatments. Thalinger et al. [75], however, reported that NiFe alloy formation through Fe exsolution from a reducible $La_{0.6}Sr_{0.4}FeO_{3-\delta}$ perovskite with supported Ni nanoparticles could spoil the methanation chemistry of Ni through an unfavourable change in nickel's electronic properties. The suppression of CO_2 methanation was less pronounced on a $Ni/SrTi_{0.7}Fe_{0.3}O_{3-\delta}$ catalyst, due to the better structural stability of this perovskite support [76] and, thus, the lesser extent of Fe exsolution and alloying with Ni.

Lastly, Pandey et al. [77] performed a study in which they used unsupported NiO as a catalyst precursor and introduced various Fe amounts. Similarly to many of their supported counterparts [26,39], Ni catalysts modified with Fe in a stoichiometry of Ni/Fe > 1 exhibited an increase in catalytic activity for CO_2 methanation compared to metallic Ni catalyst. The catalysts that consisted of Ni-rich NiFe alloys with 10% Fe and 25% Fe yielded the best results. The one with 10% Fe was shown to be the most active and the higher activity was ascribed to the absence of unalloyed Fe in this catalyst [77].

By taking into consideration the large number of literature reports regarding Fe modified Ni catalysts [26,39,65], it can be concluded that, under specific circumstances, Fe can significantly promote the CO_2 methanation performance. The molar ratio between Ni and Fe appears to be the most important factor that determines whether Fe will promote or demote the methanation performance. An Fe/Ni ratio between 0.1 and 0.25 mostly yields the best results, while Fe-rich alloys significantly degrade the catalytic activity [39]. Furthermore, the type of support also influences the degree of promotion. Al_2O_3, ZrO_2, SiO_2 and carbon supported Ni catalysts are typically promoted upon Fe addition at a specific Fe/Ni ratio, but Ni catalysts supported on defect-rich CeO_2-based supports do not exhibit a similar behaviour [42,68]. Through spectroscopic techniques, it has been reported that partly reduced Fe(II) species play a major role in the promotion mechanism [49,50]. In general, Fe constitutes an ideal promoter for Ni-based catalysts, since it is also a cheap and earth-abundant metal and, through the favourable formation of NiFe alloy, it can be used to explicitly tailor the electronic properties of the catalytically active phase. Some of the most representative studies that include Ni-M (M = transition metal) bimetallic catalysts for CO_2 methanation are comparatively presented in Table 1.

2.2. Promotion with Co

Co is another transition metal commonly used to prepare bimetallic NiCo catalysts for CO_2 methanation. Co is right adjacent to Ni in the periodic table, it can dissolve into the lattice of metallic Ni similarly to Fe and its easy transition between the oxidation states of Co(III), Co(II) and Co^0 can induce modifications to the electronic properties of Ni-based catalysts [24,78]. Among the first works, Guo et al. [79] prepared several bimetallic NiCo catalysts supported on SiO_2 with different Co/Ni ratios. They found that higher Co loadings improved the activity of the catalysts and that a Co/Ni ratio of 0.4 led to the best performing catalyst. The formation of a homogeneous NiCo alloy supported on SiO_2 has also been shown to promote CO dissociation and hydrogen spillover during CO methanation, leading to higher activity [78].

Further works focussed on Al_2O_3 supported NiCo catalysts, with Xu et al. [80] and Liu et al. [37] introducing ordered mesoporosity into the catalyst structure via a one-pot evaporation induced self-assembly (EISA) synthesis method. Xu et al. [80] reported that a suitable Co/(Ni + Co) ratio of 20% could increase the catalytic activity and stability. The authors suggested that Ni and Co were supported as adjacent monometallic phases on the mesoporous Al_2O_3 structure and that they served as active sites for the chemisorption of H_2 and CO_2, respectively. The synergy between these two metals acted to decrease the activation energy for CO_2 methanation. Liu et al. [37] also prepared similar ordered mesoporous $NiCo/Al_2O_3$ composites. The authors claimed that NiCo alloy formation and the confinement effect of the mesoporous structure were responsible for the improved low-temperature activity and stability of the bimetallic catalysts. Their best performing 10N3COMA catalyst (3% Co_3O_4 and 10% NiO on ordered mesoporous Al_2O_3) exhibited 78% CO_2 conversion and 99% CH_4 selectivity at 400 °C, as well as great stability upon a 60 h time on stream test, as shown in Figure 7.

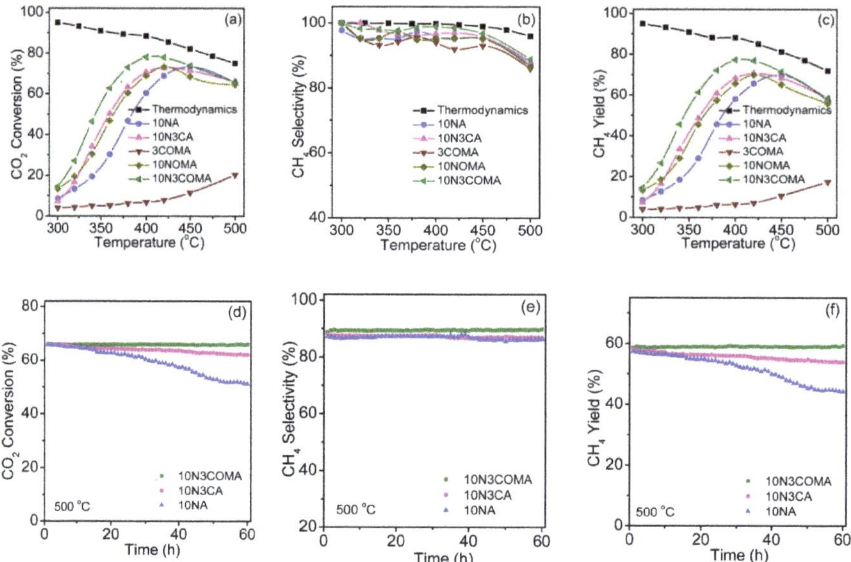

Figure 7. Catalytic properties of the catalysts: (**a,d**) CO_2 conversion, (**b,e**) CH_4 selectivity, and (**c,f**) CH_4 yield. Reproduced with permission from [37]. Copyright: Elsevier, Amsterdam, The Netherlands, 2018.

Alrafei et al. [81] prepared Ni and NiCo catalysts with various metal loadings supported on Al_2O_3 extrudates. According to their findings, 10% Ni and 10% Co were the most suitable metal loadings for the bimetallic catalysts. The bimetallic catalyst with these metal loadings outperformed the monometallic 10% Ni catalyst, as Co helped to increase the reducibility and dispersion of the Ni phase. However, the 20% monometallic Ni catalyst exhibited better performance in terms of CO_2 conversion and CH_4 selectivity when compared to the bimetallic catalyst, with a total Ni and Co metal loading of 20%. All catalysts presented a remarkable stability upon 200 h of operation. In contrast to other works, Fatah et al. [82] reported that 5% Co addition on a 5% Ni/Al_2O_3 catalyst caused roughly a 30% drop in CO_2 conversion and lower CH_4 selectivity compared to the Ni monometallic catalyst. The deactivation was attributed to the occurrence of larger particles in the NiCo bimetallic catalyst, while an increased formation of formate intermediate species was also identified via in situ FTIR.

Besides SiO_2 and Al_2O_3, a lot of works focus on NiCo catalysts supported on ZrO_2, CeO_2-ZrO_2 and other CeO_2-based supports. As discussed before, Ren et al. [64] studied 30% Ni/ZrO_2 catalysts modified with Fe, Co and Cu. The Co-modified catalyst exhibited higher CO_2 conversion, but slightly lower CH_4 selectivity, and Fe was shown to be more suitable as a promoter metal. Razzaq et al. [83], on the other hand, concluded that a Co promoted Ni/CeO_2-ZrO_2 catalyst was more suitable compared to other catalysts (modified with Mo and Fe) for the co-methanation of CO and CO_2 in CH_4-rich coke oven gas. Following that work, Zhu et al. [84] confirmed that 5% Co addition on a 15% Ni/CeO_2-ZrO_2 catalyst could promote catalytic activity and stability for CO_2 methanation. The optimum CeO_2/ZrO_2 molar ratio of 1/3 ($Ce_{0.25}Zr_{0.75}O_2$) was shown to also play a major role by providing a maximum amount of oxygen vacancies close to active metal sites that favour CO_2 activation and dissociation.

As previously discussed, Pastor-Pérez et al. [69] prepared Fe- and Co-modified Ni catalysts supported on CeO_2-ZrO_2. While Fe modification impeded the CO_2 methanation activity, the addition of 3% Co on a 15% Ni/CeO_2-ZrO_2 catalyst enhanced the CO_2 conversion, CH_4 selectivity and the long-term stability and it allowed for the use of higher space velocities. Following that work, Pastor-Pérez et al. [85] also showed that such Co modified Ni catalyst could prevent coke deposition during the reaction. The catalyst was quite robust upon long-term operation tests, while the presence of methane in the feed gas did not appear to impede the catalytic activity, but instead promoted it. In contrast to other works, the addition of Co in Ni catalysts supported on defect-rich GDC support did not offer any advantages to the catalytic performance for CO_2 methanation, as reported by Frontera et al. [86].

Apart from the typical catalyst supports, Jia et al. [87] supported the two metals of Ni and Co on TiO_2 coated silica spheres. The electronic effect of the reducible TiO_2 layer acted to promote metal dispersion and the adsorption of H_2 and CO_2, while the addition of Co second metal enhanced the intrinsic activity and long-term stability of the NiCo/TiO_2@SiO_2 catalyst in a fluidized bed reactor. Varun et al. [88] prepared NiO-MgO nanocomposites, via a sonochemical method, that were active for the CO_2 methanation reaction. The authors further modified the composites by impregnating 2% Co, Fe and Cu. Among the three dopants, only Co led to significant activity improvement by lowering the activation energy for the methanation pathway.

Table 1. Summary of some typical bimetallic Ni-based catalysts promoted with transition metals (Fe, Co and Cu) for the methanation of CO_2.

Second Metal	Catalyst Composition	Preparation Method	Conditions	Performance	Comments	Ref.
Fe	10% and 30% Ni_3Fe/Al_2O_3 and SiO_2	Incipient wetness impregnation	WHSV = 32,000 mL g^{-1} h^{-1} H_2/CO_2 = 24/1	X_{CO2} = 35%, $S_{CH4} \approx 100\%$ at 250 °C (30% Ni_3Fe/Al_2O_3)	A 25% Fe content in the NiFe bimetallic catalysts led to the highest methanation performance. Unreduced Fe_3O_4 sites may have contributed by increasing CO_2 chemisorption.	[43]
Fe	10% Ni_3Fe/Al_2O_3, SiO_2, ZrO_2, TiO_2 and Nb_2O_5	Incipient wetness impregnation	WHSV = 32,000 mL g^{-1} h^{-1} H_2/CO_2 = 24/1	X_{CO2} = 22.1%, at 250 °C (10% Ni_3Fe/Al_2O_3)	All Ni_3Fe bimetallic catalysts had a higher CH_4 yield compared to the monometallic ones. The largest activity enhancement was observed for the Al_2O_3-supported catalyst.	[42]
Fe, Co, Cu	15% Ni_3Fe, Ni_3Co and Ni_3Cu/Al_2O_3	Incipient wetness impregnation	WHSV = 60,000 mL g^{-1} h^{-1} H_2/CO_2 = 24/1	$TOF_{CH4} \times 10^3$ = 32.8 ± 2.3 s^{-1} at 250 °C (15% Ni_3Fe/Al_2O_3)	Ni_3Fe provided the highest TOF_{CH4} out of all bimetallic catalysts and the monometallic Ni one. Linear correlation was observed between the d-density of states at the Fermi level (N_{EF}) and TOF_{CH4} based on DFT calculations.	[45]
Fe	17% Ni_3Fe/Al_2O_3	Urea deposition–precipitation	WHSV = 60,000 ml_{CO2} $g^{-1} h^{-1}$ H_2/CO_2 = 4/1	X_{CO2} = 78%, S_{CH4} = 99.5% at 350 °C (17% Ni_3Fe/Al_2O_3)	Ni_3Fe bimetallic catalyst with small (4 nm) and highly dispersed NiFe alloy nanoparticles provided high activity and stability. Great stability upon 45 h experiments under industrially oriented conditions.	[26]
Fe	17% Ni_3Fe/Al_2O_3	Urea deposition–precipitation	H_2/CO_2 = 4/1 (Operando synchrotron studies)	X_{CO2} = 61%, S_{CH4} = 96% at 350 °C (17% Ni_3Fe/Al_2O_3)	FeO_x nanoclusters were formed at the surface of NiFe alloy nanoparticles, due to oxidation of Fe^0 in the alloy to Fe^{2+}. A redox cycle between Fe^0, Fe^{2+} and Fe^{3+} at the interface between FeO_x clusters and alloy nanoparticles promoted CO_2 activation.	[49]
Fe	12% Ni and 1.2–18% $Fe/MgO-Al_2O_3$	Co-precipitation (Hydrotalcite-derived catalysts)	GHSV = 12,020 h^{-1} H_2/CO_2 = 4/1	Rate of CO_2 conversion = 6.96 $mmol_{CO2}/mol_{Fe+Ni}/s$ S_{CH4} = 99.3% at 335 °C (12% Ni and 1.2% $Fe/MgO-Al_2O_3$)	An Fe/(Ni + Fe) ratio of 0.1 (Ni_9Fe_1) provided the highest metal dispersion, the smallest size of nanoparticles and an optimum amount of surface basic sites. The corresponding catalyst had the best performance at 335 °C with the highest CH_4 selectivity. Larger Fe loadings deactivated the catalyst for the CO_2 methanation reaction.	[39]
Fe	20% Ni and 2–10% $Fe/MgO-Al_2O_3$	Co-precipitation (Hydrotalcite-derived catalysts)	WHSV = 43,200 ml_{CO2} $g^{-1} h^{-1}$ H_2/CO_2 = 4/1	X_{CO2} = 53%, $S_{CH4} \approx 98\%$ at 270 °C (20% Ni and 5% $Fe/MgO-Al_2O_3$)	An Fe/(Ni + Fe) ratio of 0.25 (Ni_4Fe_1) provided the highest CH_4 yield at low temperatures by lowering the energy barrier for CH_4 formation. CO_2 hydrogenation via *HCOO (formate) intermediate was favoured over its direct dissociation to *CO	[53]

Table 1. Cont.

Fe	≈50% Ni and 8.5% Fe/Al$_2$O$_3$	Co-precipitation	WHSV = 600,000 mL g^{-1} h^{-1} H$_2$/CO$_2$ = 4/1	Weight time yield of CH$_4$ = 30 mol h^{-1} kg$_{cat}$$^{-1}$ at 230 °C after 70 h aging time	Fe promoted the electronic properties of Ni through the formation of (γFe,Ni) nanoparticles. Increase in catalytic activity upon exposure to aging conditions. Fe^{2+} sites on disordered FeO$_x$ were formed in-situ and promoted CO$_2$ activation.	[59]
Fe, Co, Cu	30% Ni and 3% Fe, Co, Cu/ZrO$_2$	Wet impregnation	GHSV = 10,000 h^{-1} H$_2$/CO$_2$ = 4/1 p = 0.5 MPa	X$_{CO2}$ = 82%, S$_{CH4}$ = 90% at 230 °C (30% Ni and 3% Fe/ZrO$_2$)	Addition of 3% Fe and Co enhanced the low-temperature CO$_2$ conversion. In addition, 3% Fe increased CH$_4$ selectivity, but higher Fe loadings negatively influenced the methanation performance. Partly reduced Fe^{2+} sites potentially increased the dispersion of Ni and the reducibility of ZrO$_2$, thus contributing to the formation of oxygen vacancies in the support.	[64]
Fe	1.51% Ni and 0.48–4.3% Fe/ZrO$_2$	Incipient wetness impregnation	WHSV = 30,000 mL g^{-1} h^{-1} H$_2$/CO$_2$ = 2/1	X$_{CO2}$ = 39.3%, S$_{CH4}$ = 88.5% at 400 °C (1.51% Ni and 0.96% Fe (Ni$_3$Fe$_2$)/ZrO$_2$)	Ni-ZrO$_2$ interface was considered active for the CO$_2$ methanation reaction. Fe addition up to a small amount could increase the methanation performance. Via high Fe/Ni ratios, the formed Ni-FeO$_x$ interface was selective for CO, due to the weak binding of intermediate *CO species.	[65]
Fe	15% Ni and 5% Fe/surface modified activated carbon (AC)	Incipient wetness impregnation	WHSV = 60,000 mL g^{-1} h^{-1} H$_2$/CO$_2$ = 4/1	X$_{CO2}$ = 77%, S$_{CH4}$ = 98% at 400 °C (15% Ni and 3% Fe/AC-R)	Increase in activity by enhancing the surface chemistry of AC and introducing Fe. NiFe alloy formation improved the low-temperature activity, CH$_4$ selectivity and stability due to the optimal CO dissociation energy.	[73]
Co	10% Ni and 3% Co/ordered mesoporous alumina (OMA)	Evaporaton-induced self-assembly (EISA)	WHSV = 10,000 mL g^{-1} h^{-1} H$_2$/CO$_2$ = 4/1	X$_{CO2}$ = 78%, S$_{CH4}$ = 99% at 400 °C (7.8% Ni and 2.2% Co/OMA)	NiCo bimetallic catalyst showed increased CO$_2$ conversion and CH$_4$ selectivity compared to the monometallic Ni one. Co increased H$_2$ uptake. Catalysts stable at 500 °C for 60 h. NiCo alloy formation and the confinement effect of the ordered mesostructure contributed to the high stability.	[37]
Co, Fe	15% Ni and 3% Co, Fe/CeO$_2$-ZrO$_2$	Wet impregnation	WHSV = 12,500 mL g^{-1} h^{-1} H$_2$/CO$_2$ = 4/1	X$_{CO2}$ = 83%, S$_{CH4}$ = 94% at 300 °C (15% Ni and 3% Co/CeO$_2$-ZrO$_2$)	Co improved the catalytic performance for CO$_2$ methanation above 250 °C. NiCo modified catalyst had great stability after many hours and could retain high activity under increased gas space velocities. Co increased Ni dispersion and could decrease coke deposition due to its redox properties.	[69]

Table 1. Cont.

Co, Fe, Cu	2% Co, Fe and Cu/NiO-MgO ($\approx$35% Ni)	Sonochemical synthesis and Wet impregnation	$GHSV = 47{,}760\ h^{-1}$ $H_2/CO_2 = 4/1$	$X_{CO2} = 90\%$, $S_{CH4} = 99\%$ at 400 °C (2% Co/NiO-MgO)	NiO-MgO nanocomposites prepared via the sonochemical method and promoted with 2% Co were highly active for CO_2 methanation. The activation energy for CO_2 methanation was lower for the Co impregnated catalysts, due to the reducing nature of Co.	[88]
Co	$LaNi_{1-x}Co_xO_3/MCF$, x = 0–0.2 ($\approx$13–16% Ni and 0–3% Co)	Citrate-assisted impregnation	$WHSV = 60{,}000\ mL\ g^{-1}\ h^{-1}$ $H_2/CO_2 = 4/1$	$X_{CO2} = 75\%$, $S_{CH4} = 98\%$ at 450 °C ($LaNi_{0.95}Co_{0.05}O_3/MCF$)	NiCo alloy nanoparticles adjacent to La_2O_3 and supported on mesostructured cellular foam silica (MCF) were formed after reduction. The catalyst was active and stable and 5% Co substitution in the perovskite B-site was optimal for the promotion of CO_2 methanation.	[89]
Cu	1% Cu and 9–10% Ni/SiO_2	Wet impregnation	$WHSV = 60{,}000\ mL\ g^{-1}\ h^{-1}$ $H_2/CO_2 = 4/1$	$X_{CO2} = 39.5\%$, $S_{CH4} = 44.4\%$ at 400 °C (1% Cu and 10% Ni/SiO_2)	Cu addition improved Ni dispersion and reducibility and NiCu alloys were formed. Cu introduction also decreased CO_2 conversion and CH_4 selectivity, favouring instead the production of CO via the reverse water–gas shift (RWGS) reaction.	[90]

Lastly, Zhang et al. [89] attempted to increase the metal intermixing between Ni and Co by incorporating Co into the lattice of LaNiO$_3$ perovskite and supporting the perovskite on a mesostructured cellular foam (MCF) silica support. Following reduction, NiCo alloy nanoparticles and adjacent La$_2$O$_3$ phase, highly dispersed on the MCF support, were generated. The promoting effect of Co alloying, the increased basicity and dispersion induced by the La$_2$O$_3$ phase and the confinement effect of the mesoporous support acted to yield a catalyst with increased activity and great stability. A low Co/(Ni + Co) ratio of 5% at a perovskite with nominal composition LaNi$_{0.95}$Co$_{0.05}$O$_3$ was found to be optimal by leading to NiCo alloy particles with the smallest size (Figure 8).

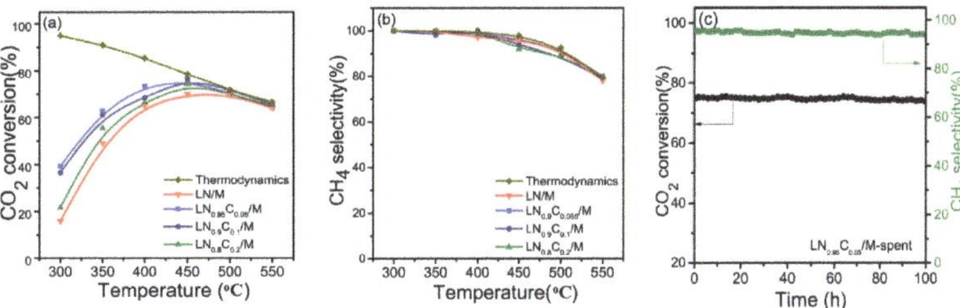

Figure 8. Catalytic properties of the catalysts at 60,000 mL g^{-1} h^{-1}: (**a**) CO$_2$ conversion, (**b**) CH$_4$ selectivity, and (**c**) lifetime test of LN$_{0.95}$C$_{0.05}$/M at 450 °C. Reproduced with permission from [89]. Copyright: Elsevier, Amsterdam, The Netherlands, 2020.

Overall, it can be concluded that Co mostly acts to promote the performance of CO$_2$ methanation catalysts. However, based on the reported literature, it appears that the type of support, either inert or reducible, does not affect the promotion mechanism [37,69]. The effect of the Co/(Ni + Co) ratio is also not as apparent when compared to similar NiFe bimetallic catalysts. However, it appears that Ni-rich NiCo catalysts lead to a more favourable activity [89].

2.3. Promotion with Cu

In all of the works discussed until now, whenever Cu is added in Ni-based catalysts, the CO$_2$ methanation reaction is hindered and the antagonistic reverse water–gas shift (RWGS) reaction is instead promoted. Thus, the Cu-containing bimetallic catalysts turned out to be inferior to monometallic Ni ones for CO$_2$ methanation [38,45,64,88]. Dias et al. [90] also showed that even 1% Cu addition on 10% Ni/SiO$_2$ could drastically decrease CO$_2$ conversion and increase the selectivity for CO. Although Cu enhanced the dispersion and reducibility of Ni, similarly to other transition metals (Fe and Co), it deactivated the catalyst for the CO$_2$ methanation reaction. This was probably a result of NiCu alloy formation and the inability of pure Cu and NiCu alloy to adsorb hydrogen. Despite that, the Cu-containing catalysts could be suitable for CO production, due to their high stability and CO selectivity. Table 1 below summarises typical bimetallic Ni-based catalysts promoted with transition metals (Fe, Co and Cu) for the methanation of CO$_2$.

3. Promotion with Noble Metals

Although noble metals are significantly pricier than transition metals, they can provide significant advantages regarding the CO$_2$ methanation reaction, due to their excellent low-temperature activity, as well as their high reducibility and stability once oxidized [18,19,91–93]. The combination of Ni with a noble metal aims to transfer some of these properties into Ni-based catalysts, without the need for the use of a high noble metal loading [33]. Combined Ni and Ru catalysts exist mostly in the form of heterostructures, rather than alloys, and Ru can provide additional methanation sites, as well as spillover hydrogen into nearby Ni sites [94]. On the other hand, Pt and Pd mostly modify the electronic properties of Ni via alloy formation [31,32].

3.1. Promotion with Ru

Ru is the cheapest among the discussed noble metals and, at the same time, the most active metallic phase in CO_2 methanation. The combination of the metals Ru and Ni has been described as "privileged" due to the great benefits that can arise from the NiRu bimetallic synergy [94]. Ru can drastically improve the reducibility of Ni catalysts, as well as improve the Ni metallic dispersion and provide additional methanation sites.

In an early work, Zhen et al. [30] impregnated Ni and Ru on γ-Al_2O_3. Ru was found to be segregated at the outer surface and not forming an alloy with the dominant Ni metallic phase. The catalyst with 1% Ru and 10% Ni presented high activity, CH_4 selectivity and stable performance upon 100 h of operation. It was suggested that CO_2 molecules dissociated on Ru particles and H_2 molecules on Ni ones, followed by hydrogen spillover to hydrogenate the adsorbed carbon species (Figure 9). Lange et al. [95] also found that part of Ru noble metal could be substituted by Ni in Ru/ZrO_2 catalysts without compromising the catalytic activity. They claimed that low metal loadings led to the formation of alloys, but higher loadings led to the separation of the two metallic phases.

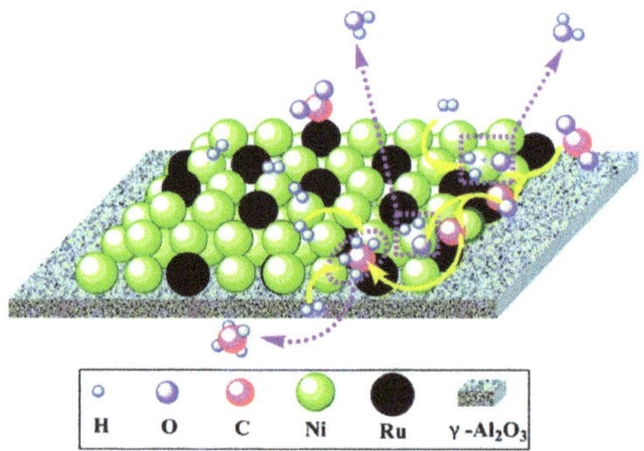

Figure 9. The proposed a possible reaction mechanism of CO_2 methanation over 10Ni–1.0Ru catalyst. Reproduced with permission from [30]. Copyright: Royal Society of Chemistry, London, UK, 2014.

Hwang et al. [96] modified the Ru content on a 35% Ni and 5% Fe/Al_2O_3 xerogel. A volcano-shaped trend was observed regarding the Ru content, with the small loading of 0.6% yielding the best results. Liu et al. [97] synthesized Ni and Ru-doped ordered mesoporous CaO-Al_2O_3 nanocomposites. The confinement of the active nanoparticles due to the ordered mesoporous structure of the catalyst prevented their sintering, while the CaO component increased the basicity and favoured CO_2 chemisorption. The addition of a small quantity of Ru as a second metal enhanced the activation of H_2 and CO_2, due to its synergy with the Ni primary phase, leading to a final nanocomposite catalyst with superior catalytic activity and stability. In a more recent study, Chein et al. [98] also reported that a 1% Ru and 10% Ni/Al_2O_3 catalyst had a better low-temperature activity compared to the respective monometallic catalysts, also agreeing with the fact that a small amount of added Ru can induce significant changes to the catalytic activity of Ni-based catalysts.

NiRu bimetallic catalysts are also viable candidates to be studied under more industrial-like conditions. Bustinza et al. [99] prepared bimetallic NiRu structured monolithic catalysts with low Ru contents by homogeneously dispersing Ni and Ru precursors over alumina washcoated monoliths. Through an appropriate preparation method, Ni was supported in the form of small nanoparticles (2–4 nm), while Ru was atomically dispersed over the structured support. The bimetallic catalyst with these characteristics provided stable CO_2 methanation performance for many hours under high space velocities. Navarro et al. [100] structured a catalyst consisting of $MgAl_2O_4$ supported 0.5% Ru and 15% Ni washcoated on metal micromonoliths. The structured catalyst was stable upon 100 h of continuous operation. The effect of CH_4 presence in the initial gas stream (simulated biogas) was also addressed by Stangeland et al. [101]. CO_2 conversion was increased upon promotion of a 20% Ni/Al_2O_3 catalyst with 0.5% Ru and the catalyst was not greatly affected by the CH_4 presence in the gas stream, exhibiting a stable methanation activity.

Another practical implementation of the NiRu bimetallic combination is in dual-function materials (DFMs), used for the integrated capture and methanation of CO_2 [102]. Having Ni active methanation phase in DFMs presents a major problem, since Ni is deactivated during the CO_2 capture step from flue gases and cannot be reactivated under H_2 flow at the usual operation temperature of 320 °C [21,103]. Adding just 1% Ru in a DFM consisting of 10% Ni increased the reducibility of the Ni phase by 70% and thus enabled CH_4 production under isothermal cyclic operation [104]. Employing in-situ DRIFTS, it was shown that both Ru and Ni are active sites for the methanation reaction, with bicarbonate and bidentate carbonate intermediate species spilling over to both metals before being hydrogenated into CH_4 [105].

Besides Al_2O_3-based supports, ZrO_2, CeO_2-ZrO_2 and other CeO_2-based defect-rich supports are also commonly used in NiRu-based catalysts and will be discussed thereafter. Ocampo et al. [106] modified 5% Ni catalysts supported on CeO_2-ZrO_2 by changing the CeO_2/ZrO_2 ratio and by adding a noble metal phase (Ru and Rh). A mass ratio between CeO_2 and ZrO_2 of 1.5 in the support led to the best results and 0.5% noble metal addition increased the catalytic activity and stability by enhancing Ni dispersion. In a later study, Shang et al. [107] prepared Ru-doped 30% Ni catalysts supported on CeO_2-ZrO_2 via an one-pot hydrolysis method. They found that Ru addition could enhance Ni dispersion and promote the basicity of the catalysts, thereby achieving a quite increased low-temperature activity. They reported that their best performing catalyst 3% Ru, 30% $Ni/Ce_{0.9}Zr_{0.1}O_2$ could achieve 98.2% CO_2 conversion and 100% CH_4 selectivity under the reaction conditions tested.

Renda et al. [33] performed an extensive study on the effect of noble metal loading on the catalytic activity of 10% Ni catalysts supported on CeO_2. At first, CeO_2 was identified as the most suitable catalyst support over $CeZrO_4$ and CeO_2/SiO_2, which is in agreement with other studies [12]. In the case of Ru, a volcano-shaped trend could be observed between the Ru loading and the yield for methane at low-temperatures, with 1% Ru leading to the best performance, as shown in Figure 10. The yield for CH_4 at 300 °C was almost double for the 1% Ru promoted 10% Ni/CeO_2 catalyst since Ru offered additional active sites for CO_2 methanation in synergy with Ni. The worse performance at higher Ru loadings was attributed to the worsening of active metal dispersion. Renda et al. [108] also showed that using ruthenium acetylacetonate instead of ruthenium chloride as the Ru precursor salt could improve the dispersion of the active metals, due to the templating effect of the precursor salt, leading to an enhanced methanation performance.

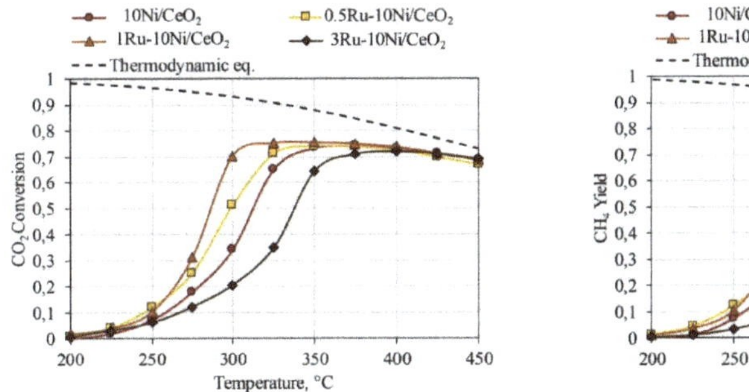

Figure 10. CO_2 conversion and CH_4 yield trends for the samples doped with 0.5%, 1% and 3% ruthenium (H_2:Ar:CO_2 = 4:5:1; WHSV = 60 1/(h·g_{cat})). Reproduced with permission from [33]. Copyright: Elsevier, Amsterdam, The Netherlands, 2020.

In another interesting study, le Saché et al. [70] reported that Ru addition on a Ni/CeO$_2$-ZrO$_2$ catalyst could improve the dispersion of Ni and increase the overall intrinsic activity for the reduction of CO$_2$. The bimetallic NiRu catalyst exhibited enhanced CO$_2$ methanation performance in terms of CO$_2$ conversion and CH$_4$ selectivity at lower temperatures compared to the monometallic Ni catalyst. Furthermore, the NiRu-based catalyst was also active for CO production via the reverse water–gas shift (RWGS) reaction upon temperature increase. It was shown that the catalyst could be stable and high-performing over long time for both reactions (CO$_2$ methanation and RWGS) at the respective reaction temperatures (350 and 700 °C). Their experimental results are summarized in Figure 11. It should also be noted that Fe addition on the Ni/CeO$_2$-ZrO$_2$ catalyst led to inferior CO$_2$ methanation performance, as discussed previously.

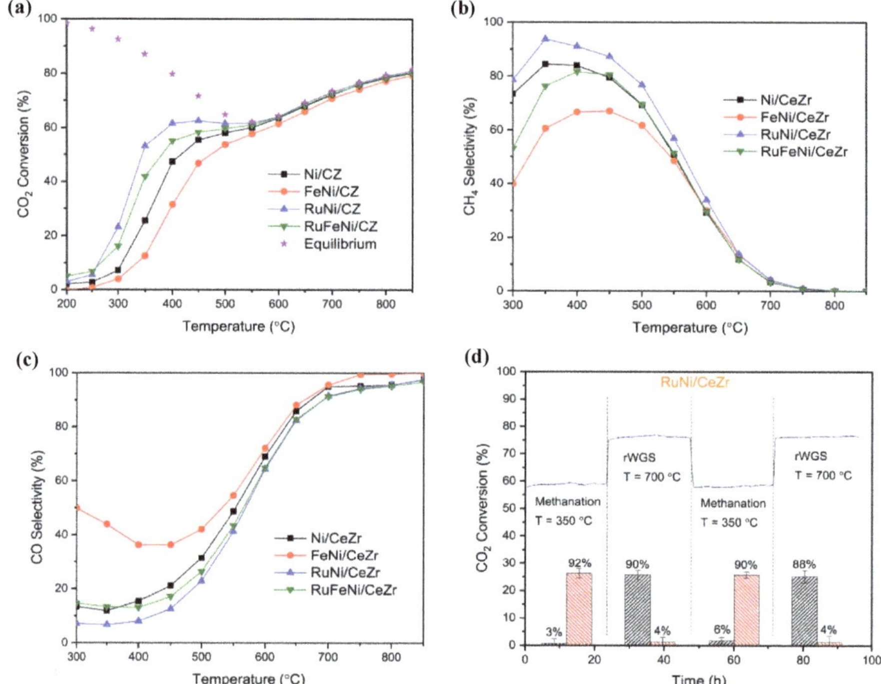

Figure 11. (**a**) CO$_2$ conversion, (**b**) CH$_4$ selectivity and (**c**) CO selectivity for all catalysts as a function of temperature. (**d**) Stability test for the RuNi/CeZr catalyst, varying the temperature between 350 and 700 °C and starting with the methanation cycle. Product selectivity is represented as columns, black for CO and red for CH$_4$. Reproduced with permission from [70]. Copyright: American Society of Chemistry, Washington, DC, USA, 2020.

Through a novel preparation method, Polanski et al. [109] prepared Ru nanoparticles supported on metallic Ni grains via first preparing Ru nanoparticles supported on an intermediate silica carrier. Ru and Ni were then deposited on the carrier and the silica was etched upon suspension in a strong basic solution to yield a 1.5% Ru/Ni catalyst with an oxide passivation layer. The prepared catalyst exhibited a greatly enhanced low-temperature activity with about 100% CO$_2$ conversion being achieved at a temperature as low as 200 °C. Carbon deposition was shown to deactivate the catalyst over time, with the initial activity being regained upon treatment under H$_2$ atmosphere. In a follow-up work, Siudyga et al. [110] synthesized 1.5% Ru catalysts supported on Ni nanowires via a similar preparation method. The higher surface area of the 1D nanostructure of the Ni support provided an additional advantage, with around 100% CO$_2$ conversion being reached at 179 °C and the onset temperature for the reaction being observed at just 130 °C. Siudyga et al. [111] also observed that Ru/Ni catalysts are highly active for CO methanation from syngas, with 100% CO conversion being reached at 178 °C and observable CO conversion being detected even at −7 °C.

The choice of Ru as a promoter for the CO_2 methanation reaction in Ni-based catalysts is a sensible option, since the combination of these two metals appears to yield favourable results, as reported by numerous works [30,33,70]. The synergistic effect through the presence of Ru and Ni adjacent phases can be observed via the increased metal dispersion, the higher reducibility, the improved low-temperature activity and the superior stability. Ni and Ru are both very active CO_2 methanation sites on their own and their combination can lead to an enhancement of the intrinsic catalytic activity, and to a decrease in the activation energy, through the favourable activation of H_2 and CO_2 reactant molecules [33,94]. The only drawback for the use of Ru is its higher price compared to transition metals like Ni, Fe and Co, even though it is significantly cheaper when compared to other noble metals such as Rh or Pt. On rare occasions, Ru addition does lead to a drop in CO_2 methanation activity, like in the work of Wei et al. [112] regarding Ni catalysts supported on zeolites. Some of the most representative studies that include Ni-M (M = noble metal) bimetallic catalysts for CO_2 methanation are comparatively presented in Table 2.

3.2. Promotion with Rh

Rhodium (Rh) is one of the most expensive noble metals and it is known to be highly active for CO_2 methanation at low temperatures, even when used under very low loadings [19,113]. However, its presence in bimetallic Ni-Rh catalysts does not always lead to a desirable activity increase, especially when taking into consideration the increased cost of the catalytically active material. One of the earlier works that does report a promoting effect of Rh in Ni-based catalysts was conducted by Ocampo et al. [106]. The addition of just 0.5% Rh on a 5% Ni/CeO_2-ZrO_2 catalyst greatly increased the catalytic activity of the catalyst, and also by a larger degree compared to the modification with 0.5% Ru. Higher stability for the noble metal modified catalysts was also observed and the better performance was attributed to the enhancement of the dispersion of nickel nanoparticles. Swalus et al. [114] followed a different path by mechanically mixing 1% Rh/Al_2O_3 and 1% Ni/activated carbon catalysts. The combined catalyst presented increased methane production through a synergistic collaboration between Ni and Rh metallic sites. More specifically, Ni sites were active for H_2 adsorption and Rh sites for CO_2 adsorption, respectively. Active hydrogen species could then spill over to Rh sites to facilitate the hydrogenation of chemisorbed CO_2 molecules.

Via a complex material architecture, Arandiyan et al. [115] prepared highly active NiRh CO_2 methanation catalysts. They first synthesized a 3-dimensionally ordered macroporous (3DOM) perovskite with nominal stoichiometry $LaAl_{0.92}Ni_{0.08}O_3$ via a poly(methyl methacrylate) (PMMA) colloidal crystal-templating preparation method. Rh was then introduced via wet impregnation of a Rh precursor salt. Upon a high-temperature reduction treatment, exsolution of Ni nanoparticles occurred, followed by NiRh alloy formation with the surface Rh atoms. The Ni exsolution procedure facilitated a high dispersion of Ni on the entire perovskite surface and a strong metal–support interaction that could otherwise not be achieved by traditional impregnation techniques [116]. The Rh-Ni/3DOM LAO catalyst with highly dispersed NiRh alloy nanoparticles on a macroporous perovskite support exhibited a significantly enhanced CO_2 methanation activity, when compared to similar monometallic Ni and Rh catalysts [115].

The architecture of bimetallic NiRh catalysts supported on Al_2O_3 was also modified by Wang et al. [117]. On the one hand, the method of galvanic replacement (GR) led to the formation of a Ni@Rh core-shell catalyst, where Ni nanoparticles were encapsulated by an atomically thin RhO_x shell. On the other hand, the method of chemical reduction (CR) led to a higher degree of intermixing between Ni and Rh. The catalyst prepared by galvanic replacement presented a superior CO_2 methanation performance, leading to the conclusion that Rh atoms highly dispersed as a thin shell over Ni nanoparticles were more active compared to large Rh nanoparticles or NiRh alloys (Figure 12). Dissociative adsorption of CO_2 over Rh was observed as the key intermediate step. Reducing the loading of Rh could also further enhance the specific activity per noble metal atom. Thus, the galvanic replacement method, which rests on the reduction potential difference between the various redox metal pairs, was shown to be an effective strategy for the preparation of bimetallic catalysts [118].

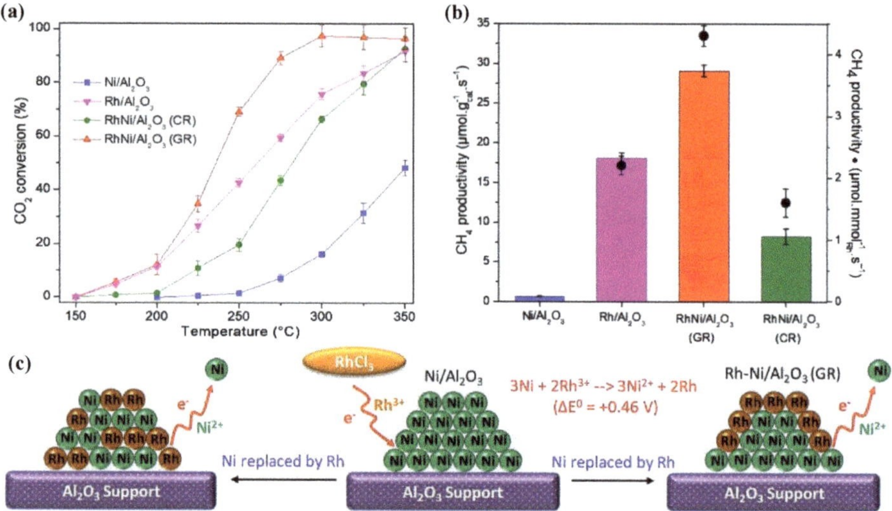

Figure 12. Catalytic activity of Ni/Al_2O_3, Rh/Al_2O_3, $RhNi/Al_2O_3$ (CR), and $RhNi/Al_2O_3$ (GR) for CO_2 methanation: (a) CO_2 conversion versus reaction temperature; (b) Rh normalised CH_4 productivity at 250 °C. (c) Possible structures of as-prepared bimetallic $RhNi/Al_2O_3$ catalysts synthesised by galvanic replacement. Reproduced with permission from [117]. Copyright: Elsevier, Amsterdam, The Netherlands, 2020.

Contrasting the results discussed until now [115,117], there are a number of works that observe a deactivating effect of Rh presence in Ni-based catalysts. Mutz et al. [46] reported that a $NiRh_{0.1}$ catalyst supported on Al_2O_3 showed inferior CO_2 conversion to a similar monometallic Ni catalyst over the entire temperature range tested. Moreover, Mihet et al. [119] and Renda et al. [33] both showed that 0.5% Rh addition on a 10% Ni catalyst supported on Al_2O_3 and CeO_2, respectively, led to a drop in the CO_2 conversion and the overall CH_4 yield. A NiRh combined catalyst supported on Al_2O_3 and prepared Heyl et al. [120] was also less active for CO_2 methanation compared to a monometallic Rh catalyst.

Therefore, Rh insertion in Ni-based catalysts does not always lead to an activity enhancement and it appears that certain requirements need to be followed, such as the existence of highly dispersed NiRh alloys with strong metal–support interaction, or the fine dispersion of Rh on top of Ni particles [115,117]. In any case, the very high cost of Rh renders it unattractive for use as promoter in Ni-based catalysts, especially when comparable activity promotion can be achieved via the use of cheaper metals, like Ru.

3.3. Promotion with Pt

Monometallic Pt catalysts used in the reduction of CO_2 favour the formation of CO rather than CH_4 and are attractive catalysts for the reverse water–gas shift reaction (RWGS) [121]. However, NiPt alloys can have a different performance compared to monometallic Ni and Pt. Mihet et al. [119] observed a promoting effect for CO_2 methanation by 0.5% Pt addition on 10% Ni/Al_2O_3, through the enhancement of Ni dispersion and NiO reducibility. Renda et al. [33] observed a similar behaviour when 0.5% Pt was added on a 10% Ni/CeO_2 catalyst. Interestingly, as shown in Figure 13, the increase in Pt loading above 0.5% decreased the quantity of produced CH_4. This behaviour was attributed to the fact that Pt sites could easily dissociate CO_2 into CO, but the further conversion of intermediate CO species took place at the nearby Ni sites. When Pt was finely dispersed, CH_4 production was high, whereas higher Pt loadings could lead to increased CO production that blocked the active Ni methanation sites.

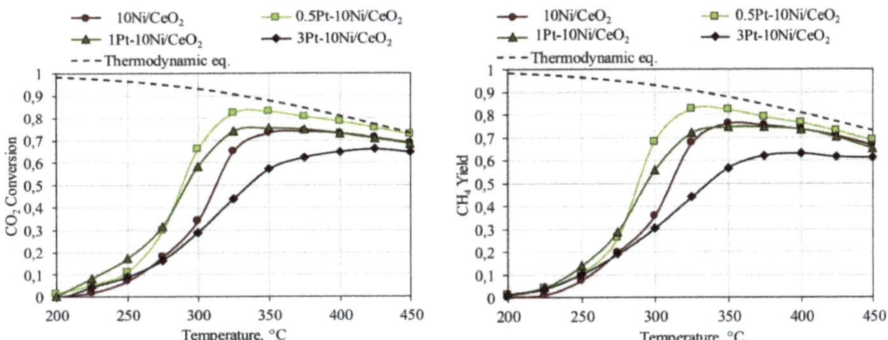

Figure 13. CO_2 conversion and CH_4 yield trends for the samples doped with 0.5%, 1% and 3% of platinum (H_2:Ar:CO_2 = 4:5:1; WHSV = 60 L/(h·g_{cat})). Reproduced with permission from [33]. Copyright: Elsevier, Amsterdam, The Netherlands, 2020.

Kikkawa et al. [31] elucidated the promoting role of isolated Pt atoms that are finely dispersed in Ni-rich NiPt alloy nanoparticles supported on Al_2O_3 for the CO_2 methanation reaction. Through Ni and Pt precursor salt impregnation and reduction, Pt atoms could dissolve into the lattice of Ni nanoparticles and were slightly negatively charged as $Pt^{\delta-}$, surrounded by $Ni^{\delta+}$ atoms. It was shown that isolated $Pt^{\delta-}$ enhanced the adsorption of intermediate CO species, while also weakening the C-O bond energy. These chemisorbed CO species on $Pt^{\delta-}$ could then be easily hydrogenated into CH_4. As a result, $Ni_{95}Pt_5$ supported on Al_2O_3 with 5% Pt atoms in the NiPt alloy led to a higher CH_4 formation rate and CH_4 yield compared to the monometallic Ni catalyst and other NiPt alloys with a higher Pt content. Overall, the promoting role of isolated Pt atoms on $Ni_{95}Pt_5/Al_2O_3$ rested on them acting both as CO adsorption sites, as well as H_2 dissociation sites.

In a later work [122], it was shown, based on in-situ FTIR observations, that, on such NiPt alloy catalysts (*iso*-Pt), the adsorbed intermediate CO species that form CH_4 were rather bridging CO between isolated Pt atoms and nearby Ni atoms. On the monometallic Pt catalyst, strongly anchored CO species were mainly found as on-top CO (terminal mode, Pt-C≡O), which were difficult to be hydrogenated into CH_4. These on-top CO species were then preferentially desorbed as CO rather than being converted into CH_4 and, hence, the monometallic Pt catalyst exhibited a high CO selectivity and reverse water–gas shift (RWGS) reactivity. However, the initially formed on-top CO species over isolated Pt atoms in NiPt alloys were quickly transformed into bridging CO between these isolated Pt atoms and their neighbouring Ni atoms (μ_2-bridging mode, Pt-(C=O)-Ni). These bridging intermediate CO species were also similar to those found in the monometallic Ni catalyst between Ni-Ni neighbours and were easily further converted into CH_4. Therefore, it was concluded that CH_4 (and CO) selectivity could be determined by the type of adsorbed intermediate CO species. High CH_4 selectivity for the monometallic Ni catalyst and the NiPt alloy catalyst (*iso*-Pt) that contained isolated Pt atoms was due to the occurrence of bridging CO, while CO selectivity was drastically increased in the monometallic Pt catalyst where CO was adsorbed as on-top CO.

Furthermore, the rate determining step (RDS) for CO_2 methanation over such NiPt alloy catalysts with isolated Pt atoms was suggested to be the chemisorption of H_2, based on kinetic studies [122]. That could be explained by the fact that although Pt sites favoured the dissociation of H_2 at low temperatures, isolated surface Pt atoms were preferentially covered by CO rather than H_2, due to the enhanced Pt-CO bonding. As a result, this limited availability of H_2 dissociation sites on isolated Pt atoms became the factor that restricted the formation of CH_4 over NiPt alloys. Therefore, the RDS for CO_2 methanation over NiPt alloys was suggested to be H_2 dissociation rather than H-assisted CO dissociation or hydrogenation of surface carbon species, which are commonly indicated as the RDS over monometallic Ni methanation catalysts [7,122]. The experimental results and reaction mechanism for these two works are summarized in Figure 14 [31,122].

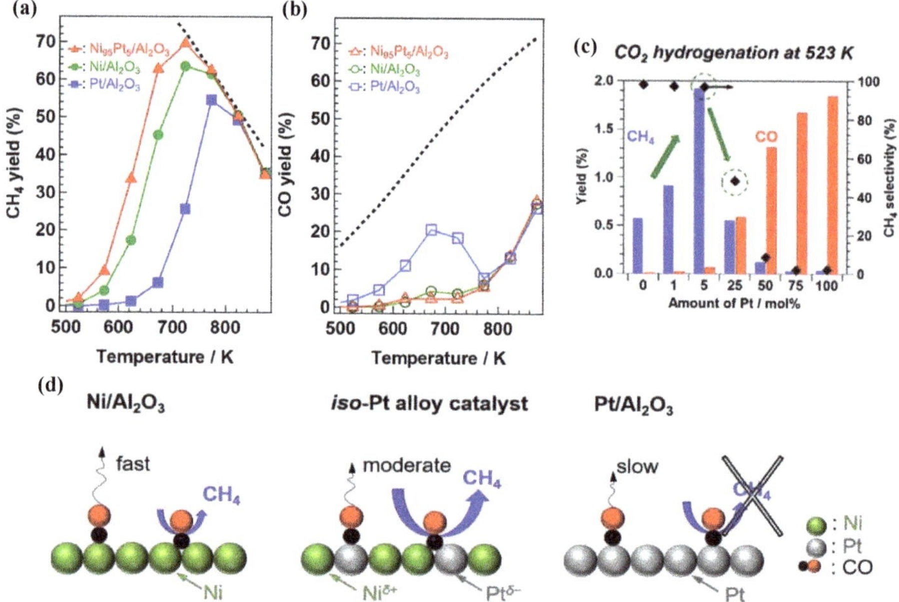

Figure 14. Yields of (**a**) CH_4 and (**b**) CO following hydrogenation of CO_2 over Ni−Pt/Al$_2$O$_3$, as well as (**c**) CH_4 yield and selectivity as a function of molar Pt percentage in the NiPt alloys; m_{cat} = 100 mg, 10% CO_2, and 40% H_2 in He (total flow rate = 50 mL min^{-1}). (**d**) Illustration of the adsorbed species and the reactions taking place over Ni/Al$_2$O$_3$, the *iso*-Pt catalyst, and Pt/Al$_2$O$_3$ during the hydrogenation of CO_2. Reproduced with permission from [31,122]. Copyright: American Chemical Society, Washington, DC, USA, 2019, 2020.

As discussed previously with regard to Ru doping, adding Pt in Ni-based DFMs could also increase the low-temperature reducibility of NiO. The addition of 1% Pt increased the NiO reducibility by 50%, compared to the 70% increase caused by the addition of 1% Ru [104]. Employing in-situ DRIFTS, it was shown that Pt acted to further promote CO_2 chemisorption and dissociation by forming Pt-CO species [105]. However, since Pt is not active for CH_4 formation, these chemisorbed carbonyl intermediates were found to spill over towards Ni sites upon H_2 inflow, where they were further hydrogenated into CH_4. Overall, the NiRu bimetallic DFMs yielded more CH_4 compared to the NiPt ones, since Ru sites could also participate in the CO_2 methanation reaction [104].

In short, despite monometallic Pt catalysts being very active for the RWGS reaction, Ni-rich NiPt alloys appear to promote CH_4 formation [31,33]. The dilution of Pt atoms over Ni metallic nanoparticles changes the CO_2 methanation pathway, since Pt sites accommodate adsorbed CO intermediates due to their high affinity with carbonyl [31]. In general, it seems that a very low amount of Pt added in Ni catalysts, even compared to other noble metals, can lead to a significant promotion for CO_2 methanation [33]. It should, however, be argued whether the extent of such promotion can be achieved via less expensive metals, like Ru.

3.4. Promotion with Pd

Pd is another noble metal with a high number of applications in heterogeneous catalysis [123,124]. Stand-alone Pd catalysts are not commonly used for CO_2 methanation, but NiPd combinations can sometimes have a superior activity for this reaction. An example of such promotion is given by the work of Mihet et al. [119]. The addition of 0.5% Pd on 10% Ni/Al_2O_3 catalysts remarkably increased the low-temperature activity and the NiPd bimetallic catalyst was quite stable, outperforming other bimetallic Ni-based catalysts promoted with Pt and Rh. The addition of Pd, as well as that of other noble metals, increased the reducibility of NiO and favoured higher metal dispersion and H_2 chemisorption. Arellano-Treviño et al. [104] also showed that Pd addition could increase NiO reducibility in DFMs, though its effectiveness for CH_4 production is not so high compared to Ru and Pt addition.

Li et al. [32] synthesized an effective CO_2 methanation catalyst composed of NiPd alloys supported on an SBA-15 mesoporous siliceous support. There was a strong synergetic effect between the two metals, as evidenced by H_2-TPR, and this synergy led to the bimetallic NiPd catalysts exhibiting enhanced CO_2 conversion compared to the monometallic Pd and Ni catalysts. The content of Ni and Pd in the alloy was also varied and it was shown that the $Ni_{0.75}Pd_{0.25}$ alloy with a Ni/Pd molar ratio of three led to the best results.

Furthermore, Yan et al. [125] prepared a composite hierarchically structured bimetallic nanocatalyst that consisted of metallic Pd nanoclusters adjacent to NiO with local tetrahedral symmetry. The support consisted of acid-treated carbon nanotubes (CNTs) and the catalysts were further decorated with a thin tetramethyl orthosilicate (TMOS) layer. This composite NiPd bimetallic catalyst produced a maximum amount of CH_4 and CO at 300 °C, compared to the monometallic Pd catalyst and other active catalysts with similar metal loadings that are reported in the literature. As shown in Figure 15, the higher performance of NiO_TPd-T catalyst was attributed to the local synergy at the interface between the Ni phase and the adjacent Pd phase, with H and CO being, respectively, chemisorbed over Ni and Pd interfacial sites. Adsorbed H atoms then helped reduce nearby NiO sites and thus increase the number of metallic Ni sites that are available for the methanation reaction.

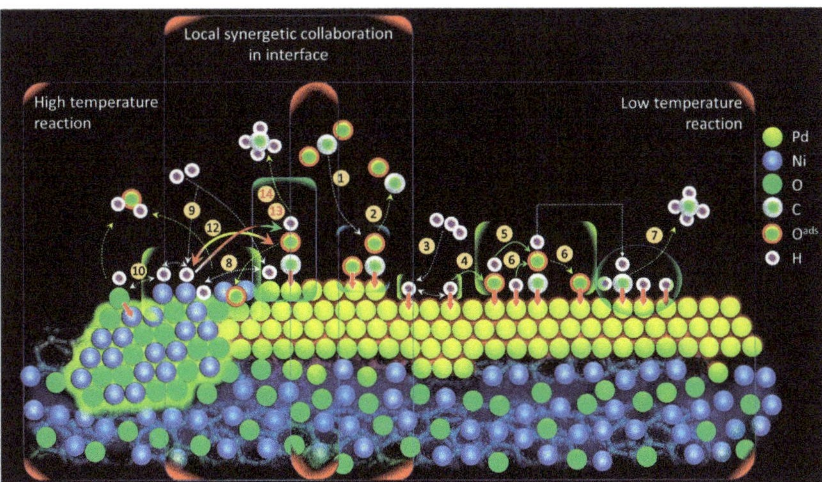

Figure 15. Schematic representation of the reaction coordinates in the NiO_TPd-T. Reproduced with permission from [125]. Copyright: Royal Society of Chemistry, London, UK, 2020.

3.5. Promotion with Re

Rhenium (Re) is not a very common element in heterogeneous catalysis, and it is only sometimes considered to be a noble metal. An example of its application in heterogeneous catalysis is the deoxydehydration (DODH) of biomass-derived polyols to produce adipic acid over ReO_x supported on ZrO_2 [126]. There has, however, recently been a handful of reports where Re is studied as a promoter in Ni-based catalysts for methanation.

Initially, Han et al. [127] screened 63 different elements as potential promoters in Ni/Al_2O_3 catalysts for the methanation of CO, with the help of data mining technology. Their model predicted that promoting a Ni/Al_2O_3 catalyst with Re could potentially lead to the largest drop in the temperature for 50% conversion (T_{50}) by 78 °C, compared to the unpromoted Ni/Al_2O_3. In another computational study, Yuan et al. [128] showed that the high oxygen affinity of Re can render it a favourable element to facilitate C-O bond scission during CO_2 methanation. Their calculations indicated that isolated Re atoms on top of Ni(111) facets can accommodate O adatoms cleaved from COOH*/HCOO* and CHO*/CO* intermediates, whereas CO* and H* can be found adsorbed on the Ni(111) surface. The synergistic collaboration between the Ni surface and Re dopant atoms could promote the scission of C-O bonds, increasing CH_4 selectivity and favouring the overall catalytic activity for CO_2 methanation.

Dong et al. [129] experimentally tested the promoting effect of Re on Ni catalysts supported on H_2O_2-modified manganese sand for CO methanation coupled with water–gas shift. Re was reported to decrease the activation energy for CO methanation and lead to a faster CH_4 formation rate. Subsequently, Dong et al. [130] studied Re-promoted Ni catalysts supported on coal combustion fly ash (CCFA) derived from industrial wastes for the CO_2 methanation reaction. The bimetallic Ni-Re/CCFA catalyst exhibited higher CO_2 conversion and CH_4 selectivity (99.5% and 70.3%, respectively), compared to the monometallic Ni/CCFA catalyst (Figure 16). The promoting effect of Re was attributed to the increase in Ni dispersion, as well as to the resistance of the bimetallic catalyst towards sintering and coke formation. Bicarbonate, bidentate formate and methoxyl species were detected as the reaction intermediates via in-situ DRIFTS. Table 2, below, summarises typical bimetallic Ni-based catalysts promoted with noble metals (Ru, Rh, Pt, Pd and Re) for the methanation of CO_2.

Table 2. Summary of some typical bimetallic Ni-based catalysts promoted with noble metals (Ru, Rh, Pt, Pd and Re) for the methanation of CO_2.

Second Metal	Catalyst Composition	Preparation Method	Conditions	Performance	Comments	Ref.
Ru	10% Ni and 0.5–5% Ru/Al$_2$O$_3$	Wet impregnation (sequential and co-impregnation)	GHSV = 9000 h^{-1} H_2/CO_2 = 4/1	X_{CO2} = 82.7%, S_{CH4} = 100% at 400 °C (10% Ni and 1% Ru/Al$_2$O$_3$)	Co-impregnation of Ni and Ru precursor salts could provide more active sites in the catalyst. The best results were achieved using 1% Ru loading. Ru and Ni were found to form separate phases and it was suggested that CO_2 and H_2 activation took place at Ru and Ni sites, respectively.	[30]
Ru	10% Ni and 1% Ru/2% CaO—ordered mesoporous alumina (OMA)	Evaporaton-induced self-assembly (EISA)	WHSV = 30,000 mL g^{-1} h^{-1} H_2/CO_2 = 4/1	X_{CO2} = 83.8%, S_{CH4} = 100% at 380 °C (10% Ni and 1% Ru/2% CaO—OMA)	A step increase in the CO_2 conversion and CH_4 selectivity was observed after the addition of the CaO basic promoter and the Ru second metal. Catalysts stable at 550 °C for 109 h due to the confinement effect. Ni and Ru synergy could reduce the activation energy for CO_2 methanation.	[97]
Ru	11.1–12.7% Ni and 0.9–4.8% Ru/Al$_2$O$_3$-washcoated cordierite	Equilibrium adsorption (Ni) and wet impregnation (Ru)	GHSV = 104,000 h^{-1} H_2/CO_2 = 4/1	X_{CO2} = 55%, S_{CH4} ≈ 100% at 350 °C (12.7% Ni and 0.9% Ru/Al$_2$O$_3$-washcoated cordierite)	Ni was homogeneously dispersed over the structured support as small nanoparticles (2–4 nm) via equilibrium adsorption, while Ru was atomically dispersed via wet impregnation. The structured catalyst on Al$_2$O$_3$-washcoated monolith provided stable performance with low pressure drop under high space velocities.	[99]
Ru	30% Ni and 0–5% Ru/CeO$_2$-ZrO$_2$	One-pot hydrolysis of metal nitrates with (NH$_4$)$_2$CO$_3$	GHSV = 2400 h^{-1} H_2/CO_2 = 4/1	X_{CO2} = 98.2%, S_{CH4} = 100% at 230 °C (30% Ni and 3% Ru/Ce$_{0.9}$Zr$_{0.1}$O$_2$)	Ru addition on Ni/CeO$_2$-ZrO$_2$ could increase surface basicity and promote Ni dispersion. Thus, the formation of a NiRu bimetallic catalyst enhanced the low-temperature catalytic activity for CO_2 methanation.	[107]
Ru	10% Ni 0.5–3% Ru/CeO$_2$-ZrO$_2$	Wet impregnation with different Ru precursor salts	WHSV = 60,000 mL g^{-1} h^{-1} H_2/CO_2 = 4/1	X_{CO2} ≈ 80%, S_{CH4} ≈ 100% at 350 °C (10% Ni and 0.5% Ru/CeO$_2$-ZrO$_2$)	Adding Ru in 10% Ni/CeO$_2$-ZrO$_2$ increased the CO_2 methanation performance. When ruthenium acetylacetonate was used instead of ruthenium chloride as the precursor salt, the metal dispersion and catalytic activity were enhanced due to the templating effect of the precursor salt molecule.	[108]
Ru	15% Ni and 1% Ru/CeO$_2$-ZrO$_2$	Wet impregnation	WHSV = 24,000 mL g^{-1} h^{-1} H_2/CO_2 = 4/1	X_{CO2} = 53%, S_{CH4} = 93% at 350 °C (10% Ni and 0.5% Ru/CeO$_2$-ZrO$_2$)	The introduction of 1% Ru in 15% Ni/CeO$_2$-ZrO$_2$ improved the dispersion of Ni and the intrinsic activity for CO_2 reduction. The catalyst was active for both CO_2 methanation at 350 °C and reverse water–gas shift (RWGS) at 700 °C.	[70]

Table 2. em Cont.

Ru	1.5% Ru/Ni	Ni deposition on Ru/SiO$_2$ intermediate and then silica etching	Flow rate = 3000 mL h^{-1} H$_2$/CO$_2$ = 4/1	$X_{CO2} \approx 100\%$, $S_{CH4} \approx 100\%$ at 200 °C (1.5% Ru/Ni)	This novel synthesis method, using an intermediate silica carrier to disperse Ru, yielded fine Ru nanoparticles supported on Ni grains. The 1.5% Ru/Ni catalyst had an oxide passivation layer and a very high low-temperature catalytic activity.	[109]
Ru	1.5% Ru/Ni nanowires (NWs)	Ni NW deposition on Ru/SiO$_2$ intermediate and then silica etching	Flow rate = 3000 mL h^{-1} H$_2$/CO$_2$ = 4/1	$X_{CO2} \approx 100\%$, $S_{CH4} \approx 100\%$ at 179 °C (1.5% Ru/Ni NWs)	When Ni nanowires were used instead of Ni powder, the higher surface area of the 1D nanostructure led to a higher CO$_2$ methanation catalytic activity, with 100% CO$_2$ conversion being reached at just 179 °C.	[110]
Rh, Ru	5% Ni and 0.5% Rh, Ru/CeO$_2$-ZrO$_2$	Pseudo sol-gel in propionic acid	GHSV = 43,000 h^{-1} H$_2$/CO$_2$ = 4/1	$X_{CO2} = 77.8\%$, $S_{CH4} = 99.2\%$ at 350 °C (0.5% Rh and 5% Ni/Ce$_{0.72}$Zr$_{0.28}$O$_2$)	Noble metal addition (Rh and Ru) increased the dispersion of Ni and the catalytic activity for CO$_2$ methanation. Rh addition led to slightly better results compared to Ru.	[106]
Rh	Rh-Ni/3DOM-LaAlO$_3$ ($\approx$2% Ni and 1% Rh)	Rh wet impregnation and Ni exsolution from LaAl$_{0.92}$Ni$_{0.08}$O$_3$	WHSV = 48,000 mL g^{-1} h^{-1} H$_2$/CO$_2$ = 4/1	$X_{CO2} \approx 93\%$, S_{CH4} high at 308 °C (Rh-Ni/3DOM-LaAlO$_3$)	Rh-Ni/3DOM-LaAlO$_3$ with bimetallic NiRh alloy nanoparticles was highly efficient for CO$_2$ methanation. 3DOM LaAl$_{0.92}$Ni$_{0.08}$O$_3$ perovskite was prepared via PMMA colloidal crystal templating and Rh was added via wet impregnation. Ni exsolution and NiRh alloy formation followed after reduction treatment.	[115]
Rh	1.56–1.9% Ni 0.69–1.18% Rh/Al$_2$O$_3$	Galvanic replacement (GR) and chemical reduction (CR)	WHSV = 48,000 mL g^{-1} h^{-1} H$_2$/CO$_2$ = 4/1	$X_{CO2} \approx 97\%$, $S_{CH4} \approx 100\%$, at 300 °C (1.56% Ni and 1.08% Rh/Al$_2$O$_3$ (GR))	RhNi/Al$_2$O$_3$ catalysts prepared by galvanic replacement exhibited superior CO$_2$ methanation performance at low temperatures. Galvanic replacement led to Ni nanoparticles encapsulated by an atomically thin RhO$_x$ shell, while chemical reduction led to a higher degree of Ni and Rh intermixing.	[117]
Pt, Ru, Rh	10% Ni and 0.5–3% Pt, Ru and Rh/CeO$_2$	Wet impregnation (sequential)	WHSV = 60,000 mL g^{-1} h^{-1} H$_2$/CO$_2$ = 4/1	$X_{CO2} = 82\%$, $S_{CH4} \approx 100\%$, at 325 °C (10% Ni and 0.5% Pt/CeO$_2$)	Promotion of 10% Ni/CeO$_2$ with Pt and Ru further increased the catalytic activity, while Rh led to the worst performance. The optimal Pt loading was 0.5%, while for Ru it was 1%. Pt enhanced the dissociation of CO$_2$ into intermediate CO, while Ru provided additional methanation sites.	[33]
Pt	Ni$_{100-x}$Pt$_x$/Al$_2$O$_3$, (1 mmol Ni + Pt metal g$_{cat}$$^{-1}$)	Wet impregnation	WHSV = 30,000 mL g^{-1} h^{-1} H$_2$/CO$_2$ = 4/1	$X_{CO2} = 70\%$, $S_{CH4} = 97\%$, at 427 °C (Ni$_{95}$Pt$_5$/Al$_2$O$_3$)	Single atom alloy catalysts (SAAC) with Pt atoms dissolved into the lattice of Ni nanoparticles supported on Al$_2$O$_3$ were highly active for CO$_2$ methanation. A Pt/(Ni + Pt) molar ratio of 5% was optimal. Isolated Pt atoms adjacent to Ni enhanced the adsorption of intermediate CO, while weakening the C-O bond energy and thus favoured the further conversion to CH$_4$.	[31]

Table 2. em Cont.

Pd, Pt and Rh	10% Ni and 0.5% Pd, Pt and Rh/Al_2O_3	Incipient wetness impregnation	GHSV = 5700 h^{-1} H_2/CO_2 = 4/1	X_{CO2} = 90.6%, S_{CH4} = 97%, at 300 °C (10% Ni and 0.5% Pd/Al_2O_3)	Pd and Pt addition improved the catalytic activity of 10% Ni/Al_2O_3, while Rh addition led to worse performance. Pd and Pt increased the dispersion and reducibility of NiO and provided active sites for H_2 chemisorption and activation. The Pd-promoted catalyst was slightly better than the Pt-promoted one.	[119]
Pd	$Ni_{1-x}Pd_x$/SBA-15	NiPd nanoparticle synthesis in oil amine and impregnation on SBA-15 with hexane	WHSV = 6000 mL g^{-1} h^{-1} H_2/CO_2 = 4/1	X_{CO2} = 96.1%, S_{CH4} = 97.5%, at 430 °C ($Ni_{0.75}Pd_{0.25}$/SBA-15)	The synergy between Ni and Pd metals led to active NiPd alloy catalysts for CO_2 methanation supported on mesoporous SBA-15 support. All bimetallic catalysts were better than the monometallic ones. A Ni/(Ni + Pd) ratio of 3 ($Ni_{0.75}Pd_{0.25}$) led to the best performing catalyst.	[32]
Pd	30% NiO_TPd-TMOS/acid-treated CNTs (Pd/Ni molar ratio = 1.5)	Ni wet impregnation on acid-treated CNTs, Pd addition with $NaBH_4$ and TMOS decoration	WHSV = 100,000 mL g^{-1} h^{-1} H_2/CO_2 = 3/1	Methane yield: Y_{CH4} = 1905.1 μmol CH_4 g_{cat}^{-1}, at 300 °C (NiO_TPd-TMOS/acid-treated CNTs)	The catalyst consisted of Pd nanoclusters adjacent to NiO with tetrahedral symmetry, supported on acid-treated CNTs and decorated with a layer of tetramethyl orthosilicate (TMOS). The maximum amount of CH_4 was produced due to the Ni-Pd synergy at their interface, with H and CO being adsorbed over interfacial Ni and Pd sites, respectively.	[125]
Re	15% Ni and 1% Re/CCFA (coal combustion fly ash)	Wet impregnation (co-impregnation)	GHSV = 2000 h^{-1} H_2/CO_2 = 4/1	X_{CO2} = 99.6%, S_{CH4} = 70.3%, at 400 °C (NiRe/CCFA)	NiRe bimetallic catalysts were supported on coal combustion fly ash (CCFA) from industrial waste. Re addition improved Ni dispersion and the catalyst resistance towards sintering and coking, leading to better performance for CO_2 methanation.	[130]

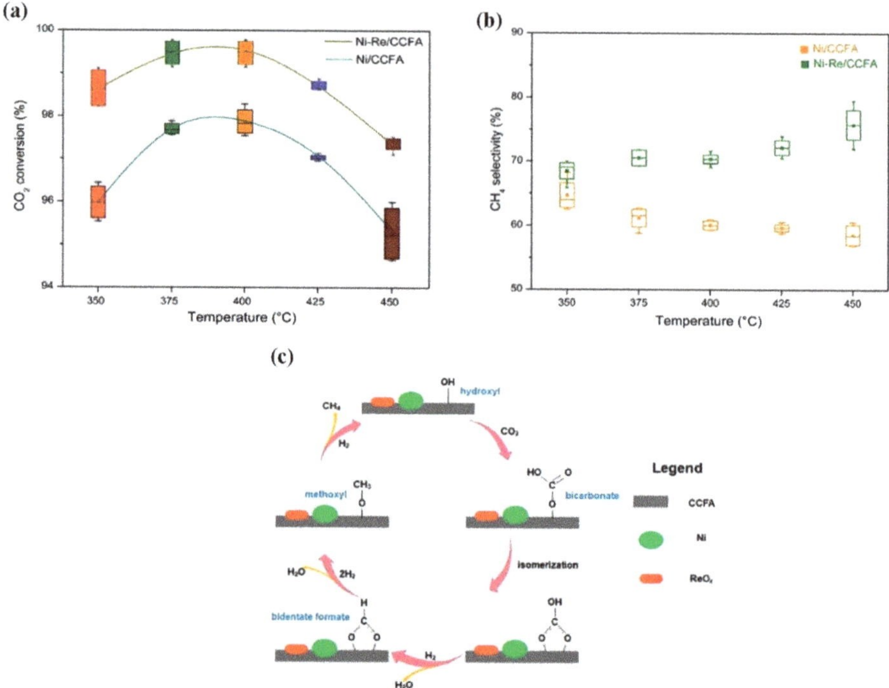

Figure 16. The effect of reaction temperature (**a**,**b**) on the CO$_2$ conversion and CH$_4$ selectivity. The conditions of reaction temperature test are H$_2$:CO$_2$:N$_2$ = 4:1:0.5, GHSV = 2000 h^{-1} and 1 atm. (**c**) Proposed reaction mechanism of CO$_2$ adsorption and methanation over reduced Ni-Re/CCFA. Reproduced with permission from [130]. Copyright: Elsevier, Amsterdam, The Netherlands, 2020.

4. Conclusions

The race for the development of low-cost and high-performing CO$_2$ methanation catalysts stems from the need to efficiently convert excess electricity and H$_2$ generated from renewables, as well as CO$_2$ captured from flue gases, into a reliable energy carrier. Ni is the standard option to be used in CO$_2$ methanation catalysts, due to its high activity and low cost. However, insufficient low-temperature activity and the degradation of Ni catalysts over time due to oxidation and sintering creates the need for the employment of specific metal additives to counter such drawbacks. These additives can fall in two generalized categories: other transition metals (including Fe and Co) and noble metals (including Ru, Rh, Pt, Pd and Re).

The transition metals Fe and Co offer the obvious advantage of being cheap like Ni and their similar size and electronic properties allow for their intricate interaction with the Ni primary phase and their easy dissolution into the Ni lattice, forming NiFe and NiCo alloys, respectively. The composition of the formed alloy is of great importance, since only specific bimetallic combinations can lead to an optimal CO$_2$ methanation performance, especially in the case of NiFe alloys. The combined bimetallic catalysts can also offer additional advantages, such as higher stability, as well as resistance towards oxidation and sulphur poisoning.

Noble metals generally increase the reducibility and dispersion of the Ni primary phase and they can also participate in the reaction as active CO$_2$ methanation phases.

Stand-alone Ru catalysts are highly active for low-temperature CO_2 methanation and the presence of Ru in bimetallic Ni catalysts as a separate monometallic phase also boosts catalytic activity. Additionally, the cost-effectiveness of Ru compared to other noble metals renders the bimetallic NiRu combinations quite popular in the field of heterogeneous catalysis. Rh and Pt can also greatly enhance the catalytic activity for CO_2 methanation when dissolved or deposited upon Ni in small quantities. Lastly, Pd and Re have been also tested as potential promoters in Ni-based catalysts.

The assumed trade-off between cost and catalytic activity for CO_2 methanation catalysts can be potentially overcome via the development of bimetallic Ni-containing catalysts with an optimised Ni–dopant metal synergy. Recent advances in operando spectroscopic techniques can shed light on how the reaction mechanism differs between Ni-based alloys or Ni–dopant metal interfaces and monometallic Ni, allowing for the development of catalysts with the lowest possible cost and highest possible performance.

Author Contributions: Conceptualization, A.I.T.; Data curation, A.I.T.; Formal analysis, A.I.T.; Funding acquisition, I.V.Y., M.A.G.; Investigation, Methodology, M.A.G.; Project administration, M.A.G., I.V.Y.; Project coordination, I.V.Y.; Resources, M.A.G.; Supervision, N.D.C.; Writing—original draft, A.I.T., N.D.C.; Writing—review and editing, N.D.C., I.V.Y. and M.A.G. All authors have read and agreed to the published version of the manuscript.

Funding: This research has been co-financed by the European Union and Greek national funds through the operational program "Regional Excellence" and the operational program Competitiveness, Entrepreneurship and Innovation, under the call Research—Create—Innovate (Project code: T1EDK-00782).

Institutional Review Board Statement: Not applicable.

Informed Consent Statement: Not applicable.

Data Availability Statement: Data sharing is not applicable to this article as no new data were created or analyzed in this study.

Conflicts of Interest: The authors declare no conflict of interest.

References

1. Mardani, A.; Štreimikienė, D.; Cavallaro, F.; Loganathan, N.; Khoshnoudi, M. Carbon dioxide (CO_2) emissions and economic growth: A systematic review of two decades of research from 1995 to 2017. *Sci. Total. Environ.* **2019**, *649*, 31–49. [CrossRef]
2. Gielen, D.; Boshell, F.; Saygin, D.; Bazilian, M.D.; Wagner, N.; Gorini, R. The role of renewable energy in the global energy transformation. *Energy Strategy Rev.* **2019**, *24*, 38–50. [CrossRef]
3. Yentekakis, I.V.; Dong, F. Grand Challenges for Catalytic Remediation in Environmental and Energy Applications toward a Cleaner and Sustainable Future. *Front. Environ. Chem.* **2020**, *1*, 5. [CrossRef]
4. Ren, J.; Musyoka, N.M.; Langmi, H.W.; Mathe, M.; Liao, S. Current research trends and perspectives on materials-based hydrogen storage solutions: A critical review. *Int. J. Hydrogen Energy* **2017**, *42*, 289–311. [CrossRef]
5. Thema, M.; Bauer, F.; Sterner, M. Power-to-Gas: Electrolysis and methanation status review. *Renew. Sustain. Energy Rev.* **2019**, *112*, 775–787. [CrossRef]
6. Lv, C.; Xu, L.; Chen, M.; Cui, Y.; Wen, X.; Li, Y.; Wu, C.E.; Yang, B.; Miao, Z.; Hu, X.; et al. Recent Progresses in Constructing the Highly Efficient Ni Based Catalysts with Advanced Low-Temperature Activity Toward CO_2 Methanation. *Front. Chem.* **2020**, *8*, 269. [CrossRef]
7. Lee, W.J.; Li, C.; Prajitno, H.; Yoo, J.; Patel, J.; Yang, Y.; Lim, S. Recent trend in thermal catalytic low temperature CO_2 methanation: A critical review. *Catal. Today* **2020**. [CrossRef]
8. Charisiou, N.D.; Siakavelas, G.; Tzounis, L.; Sebastian, V.; Monzon, A.; Baker, M.A.; Hinder, S.J.; Polychronopoulou, K.; Yentekakis, I.V.; Goula, M.A. An in depth investigation of deactivation through carbon formation during the biogas dry reforming reaction for Ni supported on modified with CeO_2 and La_2O_3 zirconia catalysts. *Int. J. Hydrogen Energy* **2018**, *43*, 18955–18976. [CrossRef]
9. Charisiou, N.D.; Papageridis, K.N.; Tzounis, L.; Sebastian, V.; Hinder, S.J.; Baker, M.A.; AlKetbi, M.; Polychronopoulou, K.; Goula, M.A. Ni supported on $CaO-MgO-Al_2O_3$ as a highly selective and stable catalyst for H_2 production via the glycerol steam reforming reaction. *Int. J. Hydrogen Energy* **2019**, *44*, 256–273. [CrossRef]
10. Charisiou, N.D.; Tzounis, L.; Sebastian, V.; Hinder, S.J.; Baker, M.A.; Polychronopoulou, K.; Goula, M.A. Investigating the correlation between deactivation and the carbon deposited on the surface of Ni/Al_2O_3 and Ni/La_2O_3 -Al_2O_3 catalysts during the biogas reforming reaction. *Appl. Surf. Sci.* **2019**, *474*, 42–56. [CrossRef]

11. Papageridis, K.N.; Charisiou, N.D.; Douvartzides, S.; Sebastian, V.; Hinder, S.J.; Baker, M.A.; AlKhoori, S.; Polychronopoulou, K.; Goula, M.A. Promoting effect of CaO-MgO mixed oxide on $Ni/\gamma-Al_2O_3$ catalyst for selective catalytic deoxygenation of palm oil. *Renew. Energy* **2020**, *162*, 1793–1810. [CrossRef]

12. Cárdenas-Arenas, A.; Quindimil, A.; Davó-Quiñonero, A.; Bailón-García, E.; Lozano-Castelló, D.; De-La-Torre, U.; Pereda-Ayo, B.; González-Marcos, J.A.; González-Velasco, J.R.; Bueno-López, A. Isotopic and in situ DRIFTS study of the CO_2 methanation mechanism using Ni/CeO_2 and Ni/Al_2O_3 catalysts. *Appl. Catal. B Environ.* **2020**, *265*, 118538. [CrossRef]

13. Tsiotsias, A.I.; Charisiou, N.D.; Yentekakis, I.V.; Goula, M.A. The Role of Alkali and Alkaline Earth Metals in the CO_2 Methanation Reaction and the Combined Capture and Methanation of CO_2. *Catalysts* **2020**, *10*, 812. [CrossRef]

14. Siakavelas, G.I.; Charisiou, N.D.; AlKhoori, S.; AlKhoori, A.A.; Sebastian, V.; Hinder, S.J.; Baker, M.A.; Yentekakis, I.V.; Polychronopoulou, K.; Goula, M.A. Highly selective and stable nickel catalysts supported on ceria promoted with Sm_2O_3, Pr_2O_3 and MgO for the CO_2 methanation reaction. *Appl. Catal. B Environ.* **2021**, *282*, 119562. [CrossRef]

15. Everett, O.E.; Zonetti, P.C.; Alves, O.C.; de Avillez, R.R.; Appel, L.G. The role of oxygen vacancies in the CO_2 methanation employing Ni/ZrO_2 doped with Ca. *Int. J. Hydrogen Energy* **2020**, *45*, 6352–6359. [CrossRef]

16. Garbarino, G.; Wang, C.; Cavattoni, T.; Finocchio, E.; Riani, P.; Flytzani-Stephanopoulos, M.; Busca, G. A study of $Ni/La-Al_2O_3$ catalysts: A competitive system for CO_2 methanation. *Appl. Catal. B Environ.* **2019**, *248*, 286–297. [CrossRef]

17. Liu, K.; Xu, X.; Xu, J.; Fang, X.; Liu, L.; Wang, X. The distributions of alkaline earth metal oxides and their promotional effects on Ni/CeO_2 for CO_2 methanation. *J. CO_2 Util.* **2020**, *38*, 113–124. [CrossRef]

18. Garbarino, G.; Bellotti, D.; Finocchio, E.; Magistri, L.; Busca, G. Methanation of carbon dioxide on Ru/Al_2O_3: Catalytic activity and infrared study. *Catal. Today* **2016**, *277*, 21–28. [CrossRef]

19. Botzolaki, G.; Goula, G.; Rontogianni, A.; Nikolaraki, E.; Chalmpes, N.; Zygouri, P.; Karakassides, M.; Gournis, D.; Charisiou, N.D.; Goula, M.A.; et al. CO_2 Methanation on Supported Rh Nanoparticles: The combined Effect of Support Oxygen Storage Capacity and Rh Particle Size. *Catalysts* **2020**, *10*, 944. [CrossRef]

20. Falbo, L.; Visconti, C.G.; Lietti, L.; Szanyi, J. The effect of CO on CO_2 methanation over Ru/Al_2O_3 catalysts: A combined steady-state reactivity and transient DRIFT spectroscopy study. *Appl. Catal. B Environ.* **2019**, *256*, 117791. [CrossRef]

21. Arellano-Treviño, M.A.; He, Z.; Libby, M.C.; Farrauto, R.J. Catalysts and adsorbents for CO_2 capture and conversion with dual function materials: Limitations of Ni-containing DFMs for flue gas applications. *J. CO_2 Util.* **2019**, *31*, 143–151. [CrossRef]

22. Porta, A.; Visconti, C.G.; Castoldi, L.; Matarrese, R.; Jeong-Potter, C.; Farrauto, R.; Lietti, L. Ru-Ba synergistic effect in dual functioning materials for cyclic CO_2 capture and methanation. *Appl. Catal. B Environ.* **2021**, *283*, 119654. [CrossRef]

23. Bian, Z.; Das, S.; Wai, M.H.; Hongmanorom, P.; Kawi, S. A Review on Bimetallic Nickel-Based Catalysts for CO_2 Reforming of Methane. *ChemPhysChem* **2017**, *18*, 3117–3134. [CrossRef] [PubMed]

24. De, S.; Zhang, J.; Luque, R.; Yan, N. Ni-based bimetallic heterogeneous catalysts for energy and environmental applications. *Energy Environ. Sci.* **2016**, *9*, 3314–3347. [CrossRef]

25. Mangla, A.; Deo, G.; Apte, P.A. NiFe local ordering in segregated Ni_3Fe alloys: A simulation study using angular dependent potential. *Comput. Mater. Sci.* **2018**, *153*, 449–460. [CrossRef]

26. Mutz, B.; Belimov, M.; Wang, W.; Sprenger, P.; Serrer, M.A.; Wang, D.; Pfeifer, P.; Kleist, W.; Grunwaldt, J.D. Potential of an alumina-supported Ni_3Fe catalyst in the methanation of CO_2: Impact of alloy formation on activity and stability. *ACS Catal.* **2017**, *7*, 6802–6814. [CrossRef]

27. Bieniek, B.; Pohl, D.; Schultz, L.; Rellinghaus, B. The effect of oxidation on the surface-near lattice relaxation in FeNi nanoparticles. *J. Nanoparticle Res.* **2011**, *13*, 5935–5946. [CrossRef]

28. Andersson, M.P.; Bligaard, T.; Kustov, A.; Larsen, K.E.; Greeley, J.; Johannessen, T.; Christensen, C.H.; Nørskov, J.K. Toward computational screening in heterogeneous catalysis: Pareto-optimal methanation catalysts. *J. Catal.* **2006**, *239*, 501–506. [CrossRef]

29. Álvarez, A.M.; Bobadilla, L.F.; Garcilaso, V.; Centeno, M.A.; Odriozola, J.A. CO_2 reforming of methane over Ni-Ru supported catalysts: On the nature of active sites by operando DRIFTS study. *J. CO_2 Util.* **2018**, *24*, 509–515.

30. Zhen, W.; Li, B.; Lu, G.; Ma, J. Enhancing catalytic activity and stability for CO_2 methanation on $Ni-Ru/\gamma-Al_2O_3$ via modulating impregnation sequence and controlling surface active species. *RSC Adv.* **2014**, *4*, 16472–16479. [CrossRef]

31. Kikkawa, S.; Teramura, K.; Asakura, H.; Hosokawa, S.; Tanaka, T. Isolated Platinum Atoms in $Ni/\gamma-Al_2O_3$ for Selective Hydrogenation of CO_2 toward CH_4. *J. Phys. Chem. C* **2019**, *123*, 23446–23454. [CrossRef]

32. Li, Y.; Zhang, H.; Zhang, L.; Zhang, H. Bimetallic Ni-Pd/SBA-15 alloy as an effective catalyst for selective hydrogenation of CO_2 to methane. *Int. J. Hydrogen Energy* **2019**, *44*, 13354–13363. [CrossRef]

33. Renda, S.; Ricca, A.; Palma, V. Study of the effect of noble metal promotion in Ni-based catalyst for the Sabatier reaction. *Int. J. Hydrogen Energy* **2020**, in press. [CrossRef]

34. Ashok, J.; Ang, M.L.; Kawi, S. Enhanced activity of CO_2 methanation over Ni/CeO_2-ZrO_2 catalysts: Influence of preparation methods. *Catal. Today* **2017**, *281*, 304–311. [CrossRef]

35. Kesavan, J.K.; Luisetto, I.; Tuti, S.; Meneghini, C.; Iucci, G.; Battocchio, C.; Mobilio, S.; Casciardi, S.; Sisto, R. Nickel supported on YSZ: The effect of Ni particle size on the catalytic activity for CO_2 methanation. *J. CO_2 Util.* **2018**, *23*, 200–211. [CrossRef]

36. Vrijburg, W.L.; Garbarino, G.; Chen, W.; Parastaev, A.; Longo, A.; Pidko, E.A.; Hensen, E.J.M. Ni-Mn catalysts on silica-modified alumina for CO_2 methanation. *J. Catal.* **2020**, *382*, 358–371. [CrossRef]

37. Liu, Q.; Bian, B.; Fan, J.; Yang, J. Cobalt doped Ni based ordered mesoporous catalysts for CO_2 methanation with enhanced catalytic performance. *Int. J. Hydrogen Energy* **2018**, *43*, 4893–4901. [CrossRef]

38. Ray, K.; Deo, G. A potential descriptor for the CO_2 hydrogenation to CH_4 over Al_2O_3 supported Ni and Ni-based alloy catalysts. *Appl. Catal. B Environ.* **2017**, *218*, 525–537. [CrossRef]

39. Mebrahtu, C.; Krebs, F.; Perathoner, S.; Abate, S.; Centi, G.; Palkovits, R. Hydrotalcite based Ni-Fe/(Mg,Al)O_X catalysts for CO_2 methanation-tailoring Fe content for improved CO dissociation, basicity, and particle size. *Catal. Sci. Technol.* **2018**, *8*, 1016–1027. [CrossRef]

40. Sehested, J.; Larsen, K.E.; Kustov, A.L.; Frey, A.M.; Johannessen, T.; Bligaard, T.; Andersson, M.P.; Nørskov, J.K.; Christensen, C.H. Discovery of technical methanation catalysts based on computational screening. *Top. Catal.* **2007**, *45*, 9–13. [CrossRef]

41. Kang, S.H.; Ryu, J.H.; Kim, J.H.; Seo, S.J.; Yoo, Y.D.; Sai Prasad, P.S.; Lim, H.J.; Byun, C.D. Co-methanation of CO and CO_2 on the Ni_X-Fe_{1-X}/Al_2O_3 catalysts; effect of Fe contents. *Korean J. Chem. Eng.* **2011**, *28*, 2282–2286. [CrossRef]

42. Pandey, D.; Deo, G. Effect of support on the catalytic activity of supported Ni-Fe catalysts for the CO_2 methanation reaction. *J. Ind. Eng. Chem.* **2016**, *33*, 99–107. [CrossRef]

43. Pandey, D.; Deo, G. Promotional effects in alumina and silica supported bimetallic Ni-Fe catalysts during CO_2 hydrogenation. *J. Mol. Catal. A Chem.* **2014**, *382*, 23–30. [CrossRef]

44. Pandey, D.; Deo, G. Determining the Best Composition of a Ni–Fe/Al_2O_3 Catalyst used for the CO_2 Hydrogenation Reaction by Applying Response Surface Methodology. *Chem. Eng. Commun.* **2016**, *203*, 372–380. [CrossRef]

45. Ray, K.; Bhardwaj, R.; Singh, B.; Deo, G. Developing descriptors for CO_2 methanation and CO_2 reforming of CH_4 over Al_2O_3 supported Ni and low-cost Ni based alloy catalysts. *Phys. Chem. Chem. Phys.* **2018**, *20*, 15939–15950. [CrossRef]

46. Mutz, B.; Sprenger, P.; Wang, W.; Wang, D.; Kleist, W.; Grunwaldt, J.D. Operando Raman spectroscopy on CO_2 methanation over alumina-supported Ni, Ni_3Fe and $NiRh_{0.1}$ catalysts: Role of carbon formation as possible deactivation pathway. *Appl. Catal. A Gen.* **2018**, *556*, 160–171. [CrossRef]

47. Farsi, S.; Olbrich, W.; Pfeifer, P.; Dittmeyer, R. A consecutive methanation scheme for conversion of CO_2—A study on Ni_3Fe catalyst in a short-contact time micro packed bed reactor. *Chem. Eng. J.* **2020**, *388*, 124233. [CrossRef]

48. Serrer, M.A.; Kalz, K.F.; Saraci, E.; Lichtenberg, H.; Grunwaldt, J.D. Role of Iron on the Structure and Stability of $Ni_{3.2}Fe$/Al_2O_3 during Dynamic CO_2 Methanation for P2X Applications. *ChemCatChem* **2019**, *11*, 5018–5021. [CrossRef]

49. Serrer, M.-A.; Gaur, A.; Jelic, J.; Weber, S.; Fritsch, C.; Clark, A.H.; Saraçi, E.; Studt, F.; Grunwaldt, J.-D. Structural dynamics in Ni–Fe catalysts during CO_2 methanation—Role of iron oxide clusters. *Catal. Sci. Technol.* **2020**, *10*, 7542–7554. [CrossRef]

50. Mebrahtu, C.; Perathoner, S.; Giorgianni, G.; Chen, S.; Centi, G.; Krebs, F.; Palkovits, R.; Abate, S. Deactivation mechanism of hydrotalcite-derived Ni-AlO_X catalysts during low-temperature CO_2 methanation via Ni-hydroxide formation and the role of Fe in limiting this effect. *Catal. Sci. Technol.* **2019**, *9*, 4023–4035. [CrossRef]

51. Giorgianni, G.; Mebrahtu, C.; Schuster, M.E.; Large, A.I.; Held, G.; Ferrer, P.; Venturini, F.; Grinter, D.; Palkovits, R.; Perathoner, S.; et al. Elucidating the mechanism of the CO_2 methanation reaction over Ni-Fe hydrotalcite-derived catalysts via surface-sensitive in situ XPS and NEXAFS. *Phys. Chem. Chem. Phys.* **2020**, *22*, 18788–18797. [CrossRef] [PubMed]

52. Huynh, H.L.; Tucho, W.M.; Yu, X.; Yu, Z. Synthetic natural gas production from CO_2 and renewable H_2: Towards large-scale production of Ni–Fe alloy catalysts for commercialization. *J. Clean. Prod.* **2020**, *264*, 121720. [CrossRef]

53. Huynh, H.L.; Zhu, J.; Zhang, G.; Shen, Y.; Tucho, W.M.; Ding, Y.; Yu, Z. Promoting effect of Fe on supported Ni catalysts in CO_2 methanation by in situ DRIFTS and DFT study. *J. Catal.* **2020**, *392*, 266–277. [CrossRef]

54. Huynh, H.L.; Tucho, W.M.; Yu, Z. Structured NiFe catalysts derived from in-situ grown layered double hydroxides on ceramic monolith for CO_2 methanation. *Green Energy Environ.* **2020**, in press. [CrossRef]

55. Burger, T.; Koschany, F.; Thomys, O.; Köhler, K.; Hinrichsen, O. CO_2 methanation over Fe- and Mn-promoted co-precipitated Ni-Al catalysts: Synthesis, characterization and catalysis study. *Appl. Catal. A Gen.* **2018**, *558*, 44–54. [CrossRef]

56. Georgiadis, A.G.; Charisiou, N.D.; Goula, M.A. Removal of hydrogen sulfide from various industrial gases: A review of the most promising adsorbing materials. *Catalysts* **2020**, *10*, 521. [CrossRef]

57. Wolf, M.; Wong, L.H.; Schüler, C.; Hinrichsen, O. CO_2 methanation on transition-metal-promoted Ni-Al catalysts: Sulfur poisoning and the role of CO_2 adsorption capacity for catalyst activity. *J. CO_2 Util.* **2020**, *36*, 276–287. [CrossRef]

58. Burger, T.; Augenstein, H.M.S.; Hnyk, F.; Döblinger, M.; Köhler, K.; Hinrichsen, O. Targeted Fe-Doping of Ni−Al Catalysts via the Surface Redox Reaction Technique for Unravelling its Promoter Effect in the CO_2 Methanation Reaction. *ChemCatChem* **2020**, *12*, 649–662. [CrossRef]

59. Burger, T.; Ewald, S.; Niederdränk, A.; Wagner, F.E.; Köhler, K.; Hinrichsen, O. Enhanced activity of co-precipitated NiFeAlO_x in CO_2 methanation by segregation and oxidation of Fe. *Appl. Catal. A Gen.* **2020**, *604*, 117778. [CrossRef]

60. Li, Z.; Zhao, T.; Zhang, L. Promotion effect of additive Fe on Al_2O_3 supported Ni catalyst for CO_2 methanation. *Appl. Organomet. Chem.* **2018**, *32*, 1–7. [CrossRef]

61. Liang, C.; Ye, Z.; Dong, D.; Zhang, S.; Liu, Q.; Chen, G.; Li, C.; Wang, Y.; Hu, X. Methanation of CO_2: Impacts of modifying nickel catalysts with variable-valence additives on reaction mechanism. *Fuel* **2019**, *254*, 115654. [CrossRef]

62. Daroughegi, R.; Meshkani, F.; Rezaei, M. Characterization and evaluation of mesoporous high surface area promoted Ni-Al_2O_3 catalysts in CO_2 methanation. *J. Energy Inst.* **2020**, *93*, 482–495. [CrossRef]

63. Aziz, M.A.A.; Jalil, A.A.; Triwahyono, S.; Ahmad, A. CO_2 methanation over heterogeneous catalysts: Recent progress and future prospects. *Green Chem.* **2015**, *17*, 2647–2663. [CrossRef]

64. Ren, J.; Qin, X.; Yang, J.Z.; Qin, Z.F.; Guo, H.L.; Lin, J.Y.; Li, Z. Methanation of carbon dioxide over Ni-M/ZrO_2 (M = Fe, Co, Cu) catalysts: Effect of addition of a second metal. *Fuel Process. Technol.* **2015**, *137*, 204–211. [CrossRef]

65. Yan, B.; Zhao, B.; Kattel, S.; Wu, Q.; Yao, S.; Su, D.; Chen, J.G. Tuning CO_2 hydrogenation selectivity via metal-oxide interfacial sites. *J. Catal.* **2019**, *374*, 60–71. [CrossRef]
66. Lu, H.; Yang, X.; Gao, G.; Wang, J.; Han, C.; Liang, X.; Li, C.; Li, Y.; Zhang, W.; Chen, X. Metal (Fe, Co, Ce or La) doped nickel catalyst supported on ZrO_2 modified mesoporous clays for CO and CO_2 methanation. *Fuel* **2016**, *183*, 335–344. [CrossRef]
67. Wu, Y.; Lin, J.; Xu, Y.; Ma, G.; Wang, J.; Ding, M. Transition metals modified Ni-M (M = Fe, Co, Cr and Mn) catalysts supported on Al_2O_3-ZrO_2 for low-temperature CO_2 methanation. *ChemCatChem* **2020**, *12*, 3553–3559. [CrossRef]
68. Winter, L.R.; Gomez, E.; Yan, B.; Yao, S.; Chen, J.G. Tuning Ni-catalyzed CO_2 hydrogenation selectivity via Ni-ceria support interactions and Ni-Fe bimetallic formation. *Appl. Catal. B Environ.* **2018**, *224*, 442–450. [CrossRef]
69. Pastor-Pérez, L.; Le Saché, E.; Jones, C.; Gu, S.; Arellano-Garcia, H.; Reina, T.R. Synthetic natural gas production from CO_2 over Ni-x/CeO_2-ZrO_2 (x = Fe, Co) catalysts: Influence of promoters and space velocity. *Catal. Today* **2018**, *317*, 108–113. [CrossRef]
70. Le Saché, E.; Pastor-Pérez, L.; Haycock, B.J.; Villora-Picó, J.J.; Sepúlveda-Escribano, A.; Reina, T.R. Switchable Catalysts for Chemical CO_2 Recycling: A Step Forward in the Methanation and Reverse Water-Gas Shift Reactions. *ACS Sustain. Chem. Eng.* **2020**, *8*, 4614–4622. [CrossRef]
71. Frontera, P.; Macario, A.; Monforte, G.; Bonura, G.; Ferraro, M.; Dispenza, G.; Antonucci, V.; Aricò, A.S.; Antonucci, P.L. The role of Gadolinia Doped Ceria support on the promotion of CO_2 methanation over Ni and Ni–Fe catalysts. *Int. J. Hydrogen Energy* **2017**, *42*, 26828–26842. [CrossRef]
72. Liu, J.; Bing, W.; Xue, X.; Wang, F.; Wang, B.; He, S.; Zhang, Y.; Wei, M. Alkaline-assisted Ni nanocatalysts with largely enhanced low-temperature activity toward CO_2 methanation. *Catal. Sci. Technol.* **2016**, *6*, 3976–3983. [CrossRef]
73. Gonçalves, L.P.L.; Sousa, J.P.S.; Soares, O.S.G.P.; Bondarchuk, O.; Lebedev, O.I.; Kolen'ko, Y.V.; Pereira, M.F.R. The role of surface properties in CO_2 methanation over carbon-supported Ni catalysts and their promotion by Fe. *Catal. Sci. Technol.* **2020**, *10*, 7217–7225. [CrossRef]
74. Wang, G.; Xu, S.; Jiang, L.; Wang, C. Nickel supported on iron-bearing olivine for CO_2 methanation. *Int. J. Hydrogen Energy* **2016**, *41*, 12910–12919. [CrossRef]
75. Thalinger, R.; Gocyla, M.; Heggen, M.; Dunin-Borkowski, R.; Grünbacher, M.; Stöger-Pollach, M.; Schmidmair, D.; Klötzer, B.; Penner, S. Ni-perovskite interaction and its structural and catalytic consequences in methane steam reforming and methanation reactions. *J. Catal.* **2016**, *337*, 26–35. [CrossRef]
76. Kayaalp, B.; Lee, S.; Klauke, K.; Seo, J.; Nodari, L.; Kornowski, A.; Jung, W.; Mascotto, S. Template-free mesoporous $La_{0.3}Sr_{0.7}Ti_{1-x}Fe_xO_{3\pm\delta}$ for CH_4 and CO oxidation catalysis. *Appl. Catal. B Environ.* **2019**, *245*, 536–545. [CrossRef]
77. Pandey, D.; Ray, K.; Bhardwaj, R.; Bojja, S.; Chary, K.V.R.; Deo, G. Promotion of unsupported nickel catalyst using iron for CO_2 methanation. *Int. J. Hydrogen Energy* **2018**, *43*, 4987–5000. [CrossRef]
78. Zhao, B.; Liu, P.; Li, S.; Shi, H.; Jia, X.; Wang, Q.; Yang, F.; Song, Z.; Guo, C.; Hu, J.; et al. Bimetallic Ni-Co nanoparticles on SiO_2 as robust catalyst for CO methanation: Effect of homogeneity of Ni-Co alloy. *Appl. Catal. B Environ.* **2020**, *278*, 119307. [CrossRef]
79. Guo, M.; Lu, G. The regulating effects of cobalt addition on the catalytic properties of silica-supported Ni–Co bimetallic catalysts for CO_2 methanation. *React. Kinet. Mech. Catal.* **2014**, *113*, 101–113. [CrossRef]
80. Xu, L.; Lian, X.; Chen, M.; Cui, Y.; Wang, F.; Li, W.; Huang, B. CO_2 methanation over Co-Ni bimetal-doped ordered mesoporous Al_2O_3 catalysts with enhanced low-temperature activities. *Int. J. Hydrogen Energy* **2018**, *43*, 17172–17184. [CrossRef]
81. Alrafei, B.; Polaert, I.; Ledoux, A.; Azzolina-Jury, F. Remarkably stable and efficient Ni and Ni-Co catalysts for CO_2 methanation. *Catal. Today* **2020**, *346*, 23–33. [CrossRef]
82. Fatah, N.A.A.; Jalil, A.A.; Salleh, N.F.M.; Hamid, M.Y.S.; Hassan, Z.H.; Nawawi, M.G.M. Elucidation of cobalt disturbance on Ni/Al_2O_3 in dissociating hydrogen towards improved CO_2 methanation and optimization by response surface methodology (RSM). *Int. J. Hydrogen Energy* **2020**, *45*, 18562–18573. [CrossRef]
83. Razzaq, R.; Zhu, H.; Jiang, L.; Muhammad, U.; Li, C.; Zhang, S. Catalytic methanation of CO and CO_2 in coke oven gas over Ni-Co/ZrO_2-CeO_2. *Ind. Eng. Chem. Res.* **2013**, *52*, 2247–2256. [CrossRef]
84. Zhu, H. Catalytic Methanation of Carbon Dioxide by Active Oxygen Material $Ce_xZr_{1-x}O_2$ Supported Ni-Co Bimetallic Nanocatalysts. *AIChE J.* **2012**, *59*, 215–228.
85. Pastor-Pérez, L.; Patel, V.; Le Saché, E.; Reina, T.R. CO_2 methanation in the presence of methane: Catalysts design and effect of methane concentration in the reaction mixture. *J. Energy Inst.* **2020**, *93*, 415–424. [CrossRef]
86. Frontera, P.; Malara, A.; Modafferi, V.; Antonucci, V.; Antonucci, P.; Macario, A. Catalytic activity of Ni–Co supported metals in carbon dioxides methanation. *Can. J. Chem. Eng.* **2020**, *98*, 1924–1934. [CrossRef]
87. Jia, C.; Dai, Y.; Yang, Y.; Chew, J.W. Nickel–cobalt catalyst supported on TiO_2-coated SiO_2 spheres for CO_2 methanation in a fluidized bed. *Int. J. Hydrogen Energy* **2019**, *44*, 13443–13455. [CrossRef]
88. Varun, Y.; Sreedhar, I.; Singh, S.A. Highly stable M/NiO–MgO (M = Co, Cu and Fe) catalysts towards CO_2 methanation. *Int. J. Hydrogen Energy* **2020**, *45*, 28716–28731. [CrossRef]
89. Zhang, T.; Liu, Q. Mesostructured cellular foam silica supported bimetallic $LaNi_{1-x}Co_xO_3$ catalyst for CO_2 methanation. *Int. J. Hydrogen Energy* **2020**, *45*, 4417–4426. [CrossRef]
90. Dias, Y.R.; Perez-Lopez, O.W. Carbon dioxide methanation over Ni-Cu/SiO_2 catalysts. *Energy Convers. Manag.* **2020**, *203*, 112214. [CrossRef]
91. Goula, G.; Botzolaki, G.; Osatiashtiani, A.; Parlett, C.M.A.; Kyriakou, G.; Lambert, R.M.; Yentekakis, I.V. Oxidative thermal sintering and redispersion of Rh nanoparticles on supports with high oxygen ion lability. *Catalysts* **2019**, *9*, 541. [CrossRef]

92. Yentekakis, I.V.; Goula, G.; Panagiotopoulou, P.; Katsoni, A.; Diamadopoulos, E.; Mantzavinos, D.; Delimitis, A. Dry reforming of methane: Catalytic performance and stability of Ir catalysts supported on γ-Al_2O_3, $Zr_{0.92}Y_{0.08}O_{2-\delta}$ (YSZ) or $Ce_{0.9}Gd_{0.1}O_{2-\delta}$ (GDC) supports. *Top. Catal.* **2015**, *58*, 1228–1241. [CrossRef]

93. Yentekakis, I.V.; Goula, G.; Panagiotopoulou, P.; Kampouri, S.; Taylor, M.J.; Kyriakou, G.; Lambert, R.M. Stabilization of catalyst particles against sintering on oxide supports with high oxygen ion lability exemplified by Ir-catalysed decomposition of N_2O. *Appl. Catal. B* **2016**, *192*, 357–364. [CrossRef]

94. Polanski, J.; Lach, D.; Kapkowski, M.; Bartczak, P.; Siudyga, T.; Smolinski, A. Ru and Ni—Privileged Metal Combination for Environmental Nanocatalysis. *Catalysts* **2020**, *10*, 992. [CrossRef]

95. Lange, F.; Armbruster, U.; Martin, A. Heterogeneously-Catalyzed Hydrogenation of Carbon Dioxide to Methane using RuNi Bimetallic Catalysts. *Energy Technol.* **2015**, *3*, 55–62. [CrossRef]

96. Hwang, S.; Lee, J.; Hong, U.G.; Baik, J.H.; Koh, D.J.; Lim, H.; Song, I.K. Methanation of carbon dioxide over mesoporous Ni-Fe-Ru-Al_2O_3 xerogel catalysts: Effect of ruthenium content. *J. Ind. Eng. Chem.* **2013**, *19*, 698–703. [CrossRef]

97. Liu, Q.; Wang, S.; Zhao, G.; Yang, H.; Yuan, M.; An, X.; Zhou, H.; Qiao, Y.; Tian, Y. CO_2 methanation over ordered mesoporous NiRu-doped CaO-Al_2O_3 nanocomposites with enhanced catalytic performance. *Int. J. Hydrogen Energy* **2018**, *43*, 239–250. [CrossRef]

98. Chein, R.-Y.; Wang, C.-C. Experimental Study on CO_2 Methanation over Ni/Al_2O_3, Ru/Al_2O_3, and Ru-Ni/Al_2O_3 Catalysts. *Catalysts* **2020**, *10*, 1112. [CrossRef]

99. Bustinza, A.; Frías, M.; Liu, Y.; García-Bordejé, E. Mono- and bimetallic metal catalysts based on Ni and Ru supported on alumina-coated monoliths for CO_2 methanation. *Catal. Sci. Technol.* **2020**, *10*, 4061–4071. [CrossRef]

100. Navarro, J.C.; Centeno, M.A.; Laguna, O.H.; Odriozola, J.A. Ru–Ni/$MgAl_2O_4$ structured catalyst for CO_2 methanation. *Renew. Energy* **2020**, *161*, 120–132. [CrossRef]

101. Stangeland, K.; Kalai, D.Y.; Li, H.; Yu, Z. Active and stable Ni based catalysts and processes for biogas upgrading: The effect of temperature and initial methane concentration on CO_2 methanation. *Appl. Energy* **2018**, *227*, 206–212. [CrossRef]

102. Duyar, M.S.; Treviño, M.A.A.; Farrauto, R.J. Dual function materials for CO_2 capture and conversion using renewable H_2. *Appl. Catal. B Environ.* **2015**, *168–169*, 370–376. [CrossRef]

103. Wang, S.; Schrunk, E.T.; Mahajan, H.; Farrauto, R.J. The role of ruthenium in CO_2 capture and catalytic conversion to fuel by dual function materials (DFM). *Catalysts* **2017**, *7*, 88. [CrossRef]

104. Arellano-Treviño, M.A.; Kanani, N.; Jeong-Potter, C.W.; Farrauto, R.J. Bimetallic catalysts for CO_2 capture and hydrogenation at simulated flue gas conditions. *Chem. Eng. J.* **2019**, *375*, 121953. [CrossRef]

105. Proaño, L.; Arellano-Treviño, M.A.; Farrauto, R.J.; Figueredo, M.; Jeong-Potter, C.; Cobo, M. Mechanistic assessment of dual function materials, composed of Ru-Ni, Na_2O/Al_2O_3 and Pt-Ni, Na_2O/Al_2O_3, for CO_2 capture and methanation by in-situ DRIFTS. *Appl. Surf. Sci.* **2020**, *533*, 147469. [CrossRef]

106. Ocampo, F.; Louis, B.; Kiwi-Minsker, L.; Roger, A.C. Effect of Ce/Zr composition and noble metal promotion on nickel based $Ce_xZr_{1-x}O_2$ catalysts for carbon dioxide methanation. *Appl. Catal. A Gen.* **2011**, *392*, 36–44. [CrossRef]

107. Shang, X.; Deng, D.; Wang, X.; Xuan, W.; Zou, X.; Ding, W.; Lu, X. Enhanced low-temperature activity for CO_2 methanation over Ru doped the Ni/$Ce_xZr_{(1-x)}O_2$ catalysts prepared by one-pot hydrolysis method. *Int. J. Hydrogen Energy* **2018**, *43*, 7179–7189. [CrossRef]

108. Renda, S.; Ricca, A.; Palma, V. Precursor salts influence in Ruthenium catalysts for CO_2 hydrogenation to methane. *Appl. Energy* **2020**, *279*, 115767. [CrossRef]

109. Polanski, J.; Siudyga, T.; Bartczak, P.; Kapkowski, M.; Ambrozkiewicz, W.; Nobis, A.; Sitko, R.; Klimontko, J.; Szade, J.; Lelątko, J. Oxide passivated Ni-supported Ru nanoparticles in silica: A new catalyst for low-temperature carbon dioxide methanation. *Appl. Catal. B Environ.* **2017**, *206*, 16–23. [CrossRef]

110. Siudyga, T.; Kapkowski, M.; Janas, D.; Wasiak, T.; Sitko, R.; Zubko, M.; Szade, J.; Balin, K.; Klimontko, J.; Lach, D.; et al. Nano-Ru supported on Ni nanowires for low-temperature carbon dioxide methanation. *Catalysts* **2020**, *10*, 513. [CrossRef]

111. Siudyga, T.; Kapkowski, M.; Bartczak, P.; Zubko, M.; Szade, J.; Balin, K.; Antoniotti, S.; Polanski, J. Ultra-low temperature carbon (di)oxide hydrogenation catalyzed by hybrid ruthenium-nickel nanocatalysts: Towards sustainable methane production. *Green Chem.* **2020**, *22*, 5143–5150. [CrossRef]

112. Wei, L.; Kumar, N.; Haije, W.; Peltonen, J.; Peurla, M.; Grénman, H.; de Jong, W. Can bi-functional nickel modified 13X and 5A zeolite catalysts for CO_2 methanation be improved by introducing ruthenium? *Mol. Catal.* **2020**, *494*, 111115. [CrossRef]

113. Karelovic, A.; Ruiz, P. Mechanistic study of low temperature CO_2 methanation over Rh/TiO_2 catalysts. *J. Catal.* **2013**, *301*, 141–153. [CrossRef]

114. Swalus, C.; Jacquemin, M.; Poleunis, C.; Bertrand, P.; Ruiz, P. CO_2 methanation on Rh/γ-Al_2O_3 catalyst at low temperature: "In situ" supply of hydrogen by Ni/activated carbon catalyst. *Appl. Catal. B Environ.* **2012**, *125*, 41–50. [CrossRef]

115. Arandiyan, H.; Wang, Y.; Scott, J.; Mesgari, S.; Dai, H.; Amal, R. In Situ Exsolution of Bimetallic Rh-Ni Nanoalloys: A Highly Efficient Catalyst for CO_2 Methanation. *ACS Appl. Mater. Interfaces* **2018**, *10*, 16352–16357. [CrossRef]

116. Burnat, D.; Kontic, R.; Holzer, L.; Steiger, P.; Ferri, D.; Heel, A. Smart material concept: Reversible microstructural self-regeneration for catalytic applications. *J. Mater. Chem. A* **2016**, *4*, 11939–11948. [CrossRef]

117. Wang, Y.; Arandiyan, H.; Bartlett, S.A.; Trunschke, A.; Sun, H.; Scott, J.; Lee, A.F.; Wilson, K.; Maschmeyer, T.; Schlögl, R.; et al. Inducing synergy in bimetallic RhNi catalysts for CO_2 methanation by galvanic replacement. *Appl. Catal. B Environ.* **2020**, *277*, 119029. [CrossRef]
118. Touni, A.; Papaderakis, A.; Karfaridis, D.; Vourlias, G.; Sotiropoulos, S. Oxygen evolution reaction at IrO_2/Ir(Ni) film electrodes prepared by galvanic replacement and anodization: Effect of precursor Ni film thickness. *Molecules* **2019**, *24*, 2095. [CrossRef]
119. Mihet, M.; Lazar, M.D. Methanation of CO_2 on Ni/γ-Al_2O_3: Influence of Pt, Pd or Rh promotion. *Catal. Today* **2018**, *306*, 294–299. [CrossRef]
120. Heyl, D.; Rodemerck, U.; Bentrup, U. Mechanistic Study of Low-Temperature CO_2 Hydrogenation over Modified Rh/Al_2O_3 Catalysts. *ACS Catal.* **2016**, *6*, 6275–6284. [CrossRef]
121. Yang, X.; Su, X.; Chen, X.; Duan, H.; Liang, B.; Liu, Q.; Liu, X.; Ren, Y.; Huang, Y.; Zhang, T. Promotion effects of potassium on the activity and selectivity of Pt/zeolite catalysts for reverse water gas shift reaction. *Appl. Catal. B Environ.* **2017**, *216*, 95–105. [CrossRef]
122. Kikkawa, S.; Teramura, K.; Asakura, H.; Hosokawa, S.; Tanaka, T. Ni–Pt Alloy Nanoparticles with Isolated Pt Atoms and Their Cooperative Neighboring Ni Atoms for Selective Hydrogenation of CO_2 Toward CH_4 Evolution: In Situ and Transient Fourier Transform Infrared Studies. *ACS Appl. Nano Mater.* **2020**, *3*, 9633–9644. [CrossRef]
123. Jiang, H.; Gao, Q.; Wang, S.; Chen, Y.; Zhang, M. The synergistic effect of Pd NPs and UiO-66 for enhanced activity of carbon dioxide methanation. *J. CO_2 Util.* **2019**, *31*, 167–172. [CrossRef]
124. Goula, M.A.; Charisiou, N.D.; Papageridis, K.N.; Delimitis, A.; Papista, E.; Pachatouridou, E.; Iliopoulou, E.F.; Marnellos, G.; Konsolakis, M.; Yentekakis, I.V. A comparative study of the H_2-assisted selective catalytic reduction of nitric oxide by propene over noble metal (Pt, Pd, Ir)/γ-Al_2O_3 catalysts. *J. Environ. Chem. Eng.* **2016**, *4*, 1629–1641. [CrossRef]
125. Yan, C.; Wang, C.H.; Lin, M.; Bhalothia, D.; Yang, S.S.; Fan, G.J.; Wang, J.L.; Chan, T.S.; Wang, Y.L.; Tu, X.; et al. Local synergetic collaboration between Pd and local tetrahedral symmetric Ni oxide enables ultra-high-performance CO_2 thermal methanation. *J. Mater. Chem. A* **2020**, *8*, 12744–12756. [CrossRef]
126. Lin, J.; Song, H.; Shen, X.; Wang, B.; Xie, S.; Deng, W.; Wu, D.; Zhang, Q.; Wang, Y. Zirconia-supported rhenium oxide as an efficient catalyst for the synthesis of biomass-based adipic acid ester. *Chem. Commun.* **2019**, *55*, 11017–11020. [CrossRef]
127. Han, X.; Zhao, C.; Li, H.; Liu, S.; Han, Y.; Zhang, Z.; Ren, J. Using data mining technology in screening potential additives to Ni/Al_2O_3 catalysts for methanation. *Catal. Sci. Technol.* **2017**, *7*, 6042–6049. [CrossRef]
128. Yuan, H.; Zhu, X.; Han, J.; Wang, H.; Ge, Q. Rhenium-promoted selective CO_2 methanation on Ni-based catalyst. *J. CO_2 Util.* **2018**, *26*, 8–18. [CrossRef]
129. Dong, X.; Jin, B.; Kong, Z.; Sun, Y. Promotion effect of Re additive on the bifunctional Ni catalysts for methanation coupling with water gas shift of biogas: Insights from activation energy. *Chin. J. Chem. Eng.* **2020**, *28*, 1628–1636. [CrossRef]
130. Dong, X.; Jin, B.; Cao, S.; Meng, F.; Chen, T.; Ding, Q.; Tong, C. Facile use of coal combustion fly ash (CCFA) as Ni-Re bimetallic catalyst support for high-performance CO_2 methanation. *Waste Manag.* **2020**, *107*, 244–251. [CrossRef]

 *nanomaterials*

Review

Enzyme-Loaded Flower-Shaped Nanomaterials: A Versatile Platform with Biosensing, Biocatalytic, and Environmental Promise

Khadega A. Al-Maqdi [1], Muhammad Bilal [2], Ahmed Alzamly [1], Hafiz M. N. Iqbal [3], Iltaf Shah [1,*] and Syed Salman Ashraf [4,*]

[1] Department of Chemistry, College of Science, UAE University, Al Ain P. O. Box 15551, United Arab Emirates; 200935138@uaeu.ac.ae (K.A.A.-M.); ahmed.alzamly@uaeu.ac.ae (A.A.)

[2] School of Life Science and Food Engineering, Huaiyin Institute of Technology, Huaian 223003, China; bilaluaf@hotmail.com

[3] Tecnologico de Monterrey, School of Engineering and Sciences, Monterrey 64849, Mexico; hafiz.iqbal@tec.mx

[4] Department of Chemistry, College of Arts and Sciences, Khalifa University, Abu Dhabi P. O. Box 127788, United Arab Emirates

* Correspondence: altafshah@uaeu.ac.ae (I.S.); syed.ashraf@ku.ac.ae (S.S.A.); Tel.: +971-569-111-552 (I.S.); +971-503-126-075 (S.S.A.)

Citation: Al-Maqdi, K.A.; Bilal, M.; Alzamly, A.; Iqbal, H.M.N.; Shah, I.; Ashraf, S.S. Enzyme-Loaded Flower-Shaped Nanomaterials: A Versatile Platform with Biosensing, Biocatalytic, and Environmental Promise. *Nanomaterials* **2021**, *11*, 1460. https://doi.org/10.3390/nano11061460

Academic Editor: Daniela Lannazzo

Received: 29 March 2021
Accepted: 28 May 2021
Published: 31 May 2021

Abstract: As a result of their unique structural and multifunctional characteristics, organic–inorganic hybrid nanoflowers (hNFs), a newly developed class of flower-like, well-structured and well-oriented materials has gained significant attention. The structural attributes along with the surface-engineered functional entities of hNFs, e.g., their size, shape, surface orientation, structural integrity, stability under reactive environments, enzyme stabilizing capability, and organic–inorganic ratio, all significantly contribute to and determine their applications. Although hNFs are still in their infancy and in the early stage of robust development, the recent hike in biotechnology at large and nanotechnology in particular is making hNFs a versatile platform for constructing enzyme-loaded/immobilized structures for different applications. For instance, detection- and sensing-based applications, environmental- and sustainability-based applications, and biocatalytic and biotransformation applications are of supreme interest. Considering the above points, herein we reviewed current advances in multifunctional hNFs, with particular emphasis on (1) critical factors, (2) different metal/non-metal-based synthesizing processes (i.e., (i) copper-based hNFs, (ii) calcium-based hNFs, (iii) manganese-based hNFs, (iv) zinc-based hNFs, (v) cobalt-based hNFs, (vi) iron-based hNFs, (vii) multi-metal-based hNFs, and (viii) non-metal-based hNFs), and (3) their applications. Moreover, the interfacial mechanism involved in hNF development is also discussed considering the following three critical points: (1) the combination of metal ions and organic matter, (2) petal formation, and (3) the generation of hNFs. In summary, the literature given herein could be used to engineer hNFs for multipurpose applications in the biosensing, biocatalysis, and other environmental sectors.

Keywords: hybrid nanoflowers; biosynthesis; influencing factors; biosensing cues; bio-catalysis

1. Introduction

Enzymes are proteins (or ribonucleic acids) that act as catalysts to quicken chemical reactions by decreasing the activation energy. They are environmentally friendly and do not alter or get consumed during chemical reactions. Enzymes as biocatalysts have recently become an intensive area of research. Currently, enzymes are used in many industrial applications, including food, drugs, and water remediation [1–7], and have numerous benefits, including a high catalytic efficiency, a high selectivity, and biodegradability. Despite all of these benefits, the use of enzymes in industrial applications has some limitations, such as a low operational stability, difficult recovery, low reproducibility, and a high cost. Immobilization of the enzyme on an insoluble solid support has been found to be a useful way to overcome some of these limitations. The solid support must be inert, insoluble, nontoxic, environmentally safe, easily accessible, affordable, highly resistant to decay and microbial attack, and have an affinity to the enzyme used. Enzyme immobilization has many advantages, including functional stability, stability against extreme reaction conditions (changes in the pH and temperature), reusability, easy separation and recovery of enzymes, and increased catalytic performance. Many researchers have shown that immobilized enzymes are more stable than free enzymes. However, in some instances, immobilization can limit the enzyme performance and lower its catalytic activity. The reasons for this are the blocking of the active site on the enzyme and the conformational changes that happen to the enzyme after immobilization, as well as the limitations of mass transfer [8–14]. Therefore, there is a need to develop new and unique methods and materials to overcome these shortcomings caused by traditional immobilization methods. There is currently growing interest in the use of nanoscale materials, such as nanoparticles and nanocrystals, for enzyme immobilization [15–19].

More recently, Jun Ge et al. developed a new method of immobilizing enzymes on solid supports, known as organic–inorganic hybrid nanoflowers or hybrid nanoflowers, (hNFs), which are flower-like hybrid nanomaterials produced between a metal node and a protein through coordination interactions [20]. This review will focus on enzyme immobilization using organic–inorganic hybrid nanoflowers (hNFs), and will cover their synthesis, advantages, different types of nanoflowers, and applications.

2. Organic–Inorganic Hybrid Nanoflowers

Organic–inorganic hybrid nanoflowers were accidentally discovered in 2012. hNFs were first detected when 0.8 mM CuSO4 was added to phosphate-buffered saline (PBS) with 0.1 mgml^{-1} bovine serum albumin (BSA). The reaction pH was 7.4 at 25 °C. After 3 days, a blue precipitate was formed at the bottom of the reaction tube, resembling a flower structure. The formation of the organic–inorganic hNFs was confirmed using scanning electron microscopy (SEM) and transmission electron microscopy (TEM) [20]. The formation mechanism (self-assembly) of organic–inorganic hybrid nanoflowers occurs through the following three stages (Figure 1): nucleation, growth, and completion. In the first step, the nucleation of primary crystals is formed from protein molecules and Cu^{2+} ions. This occurs through binding to the amide group onto the protein backbone. The second step is the growth of seed particles of the Cu^{2+} binding site, which leads to the production of nano petals. This continues with protein nanoparticles and primary crystals. In the last step, the formation of organic–inorganic hybrid nanoflowers is completed. Figure 2 shows the SEM images of the organic–inorganic hybrid nanoflower formation after 2 h, 12 h, and 3 days [20].

Additional research on organic–inorganic hybrid nanoflowers shows that the presence of a protein (enzyme) is essential for the formation of nanoflowers, and without an enzyme, only a crystal structure is formed. A research group formed hNFs between *Burkholderia cepacia* lipase (BCL) and calcium phosphate, and a part of their results (Figure 3) shows that the morphology transforms from sheet stacking to flower-like after the addition of lipase [21]. Another research group used α-chymotrypsin (ChT) and calcium phosphate to form hNFs, testing the effect of different enzyme concentrations on the formation of hNFs (Figure 4). It is clear from the SEM image that for samples a1 and a2, where ChT was not added, there is no formation of nanoflowers, and only the presence of large crystals are observed. However, when 0.05 mg/mL ChT was added, small buds were observed (b1 and b2), and as the enzyme concentration increased, nanoflower formation was more evident [22]. Lin et al. immobilized trypsin on hNFs and studied its application as a reactor for highly efficient protein digestion. One of the experiments they ran shows that for hNFs to be formed, the trypsin enzyme must be added, and without trypsin, large crystals are formed (Figure 5). The figure also shows that with an increasing trypsin concentration, a flower-like structure appears [23]. Researchers in another study produced hNFs from soybean peroxidase (SBP) and Cu^{2+}. Their outcomes showed the same result as previous work, where the absence of a protein (SBP) produced a crystal-like structure, but not nanoflowers (Figure 6) [24]. Moreover, their research shows that the presence of proteins in the formation of hybrid nanoflowers is crucial.

Currently the majority of enzyme immobilization techniques use preexisting carriers. In such a case, the enzyme can be immobilized by physical adsorption or attachment through covalent bonds between the carrier and the enzyme. These methods usually consist of the following two stages: one is the synthesis of the carriers, and the second is the immobilization process of the enzyme on the carrier. These two steps can cause a reduction in the reaction efficiency. Moreover, they require a lengthier process, which results in a higher cost. One factor that distinguishes hNFs from other immobilization techniques is that it is a one-step reaction, i.e., the carrier synthesis and enzyme immobilization processes occur in one step. This leads to a simpler procedure, and thus a lower cost [25].

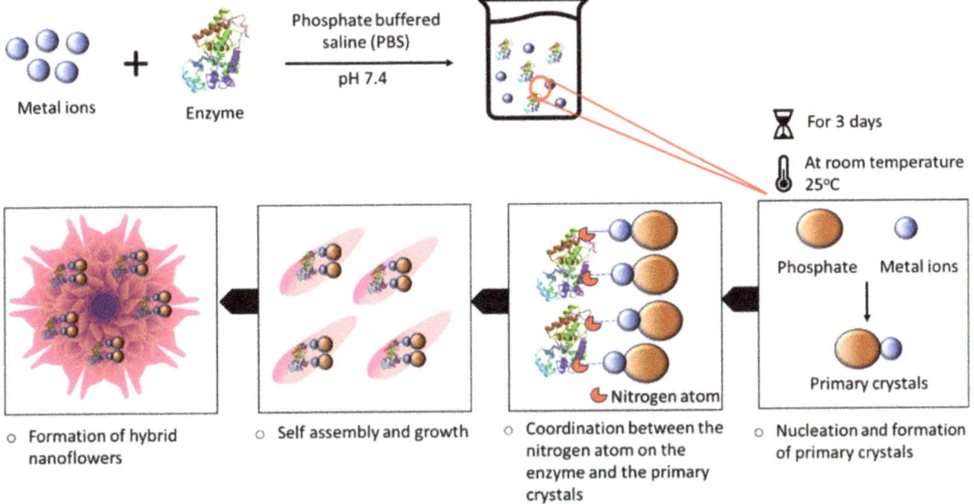

Figure 1. Schematic illustration of the hybrid nanoflower (hNF) formation process.

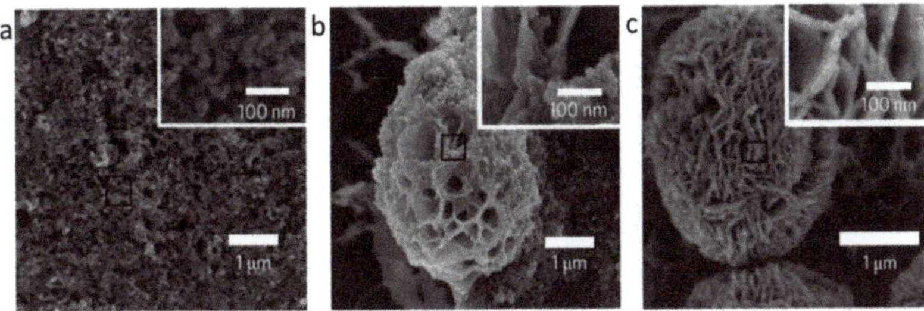

Figure 2. Formation of BSA-Cu$_3$(PO$_4$)$_2$.3H$_2$O nanoflowers. The SEM images at different times: (**a**) 2 h, (**b**) 12 h, and (**c**) 3 days. Reprinted from [20] with permission from Springer Nature. Copyright © 2021, Nature Publishing Group. License Number: 5031780508604.

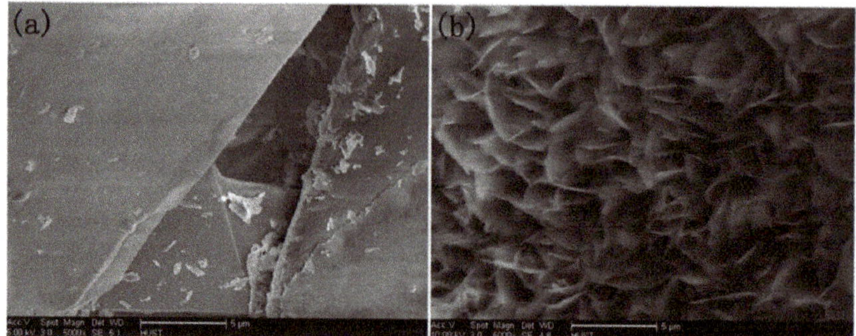

Figure 3. SEM image (**a**) without the addition of lipase enzyme, (**b**) with the addition of lipase enzyme. Reprinted from [21] with permission from the Royal Society of Chemistry. Copyright © The Royal Society of Chemistry. License Number: 1105113-1.

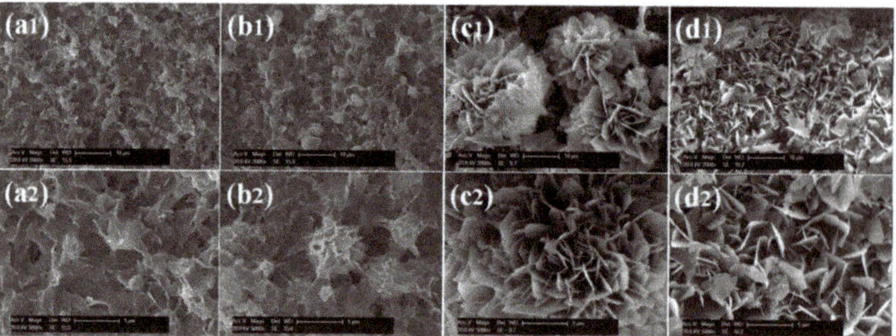

Figure 4. SEM images of different ChT concentrations on the formation of nanoflowers: (**a1,a2**) 0.0 mg/mL, (**b1,b2**) 0.05 mg/mL, (**c1,c2**) 0.1 mg/mL, and (**d1,d2**) 0.5 mg/mL. Reprinted from [22] with permission from the Royal Society of Chemistry. Copyright © The Royal Society of Chemistry. License Number: 1105121-1.

Figure 5. SEM image of different trypsin concentrations on the formation of nanoflowers: (**A1,A2**) 0.0 mg/mL, (**B1,B2**) 0.02 mg/mL, (**C1,C2**) 1.0 mg/mL, and (**D1,D2**) 5.0 mg/mL. Reprinted from [23] with permission from the Royal Society of Chemistry. Copyright © The Royal Society of Chemistry. License Number: 1105124-1.

Figure 6. SEM image of different soybean peroxidase (SBP) concentrations on the formation of nanoflowers: (**A**) 0.0 mg/mL, (**B**) 0.5 mg/mL, (**C**) 1.0 mg/mL, and (**D**) 2.0 mg. Reprinted from [24] with permission from Elsevier. Copyright © 2021 Elsevier B.V. License Number: 5031790214453.

3. Advantages or (Properties) of Hybrid Nanoflowers

3.1. Catalytic Activity

Several enzymes, including carbonic anhydrase, lipase, trypsin, laccase, and different types of peroxidases, have been used for the formation of hNFs through immobilization techniques [23,24,26–30]. A significant advantage of hNF formation over other immobilization methods is the increase in the immobilized enzyme's catalytic activity [24]. Studies have shown that immobilization only improves the stability, not the catalytic activity of the enzyme (with rare exceptions). This could be as a result of the mass transfer limitations and conformational changes in the enzyme [31,32]. For hNFs, the enhancement of the catalytic activity of the immobilized enzyme possibly arises from several reasons, namely: (i) the high surface area of hNFs, (ii) less mass transfer limitation, (iii) cooperative effect of the nanoscale-entrapped enzyme, and (iv) favorable enzyme conformation in hNFs [31,33,34].

A study showed that the enzymatic activities of soybean peroxidase in hNFs formed from different concentrations of crude SBP (0.5, 1, and 2 mg/mL) were 787, 1857, and 2500 U/mg, respectively. These results showed a ~137%, ~325%, and ~446% increase in activity compared with free crude SBP, respectively, which has an activity of 572 U/mg [24]. Another study produced hNFs and magnetic hNFs with laccase enzyme and copper(II) sulfate pentahydrate ($CuSO_4.5H_2O$). Laccase showed a higher activity in both hNFs than free laccase. The laccase hNF activity was 3.3 times greater than that of free laccase, and the laccase magnetic hNF activity was 2.7 times greater than that of free laccase. The reduction in activity in the magnetic hNFs was attributed to the shielding of active sites on laccase by the magnetic nanoparticles on the surface of the hNFs [34]. Another research group working with lipase enzymes found that lipase/$Zn_3(PO_4)_2$ hNFs had a higher enzyme activity (855 ± 13 U/g) than free lipase (328 ± 6 U/g). The increase in the enzyme activity was 147% of the pure enzyme [35]. Yin et al. synthesized α-chymotrypsin (ChT) hNFs and studied their use as immobilized ChT reactors for successful protein digestion. They determined the ChT activity in hNFs to be 3410 U/mg compared with free ChT, which has an activity of 1123 U/mg. The results show that the enhancement in the ChT activity was approximately 266% higher [22]. Lin et al., in another study, synthesized hNFs using horseradish peroxidase enzyme (HRP) and copper phosphate ($Cu_3(PO_4).3H_2O$) to use as a colorimetric platform for the visual identification of phenol and hydrogen peroxide. The results obtained showed a considerable improvement in the activity of the embedded HRP enzyme in the nanoflowers. The free HRP activity was 2970.5 U/mg, whereas the embedded HRP activity was 15,040.5 U/mg. This led to a 506% increase in the inactivity of the HRP-embedded nanoflowers [36]. All of these studies solidified that organic–inorganic hybrid nanoflowers significantly boost the catalytical activity of the embedded enzyme, which can be ascribed to the four previously-mentioned reasons.

3.2. Stability

3.2.1. Thermal Stability

Another advantage of hNFs is the stability they provide to the enzyme. Regarding thermal stability, a research group, Yu et al., studied the temperature effect on hNFs formed using calcium phosphate and six different enzymes, namely: papain, bromelain, trypsin, lipase from porcine pancreas (PPL), lipase from *Thermomyces lanuginosus* (TLL), and lipase B from *Candida antarctica* (CALB). The results were tested in different temperature ranges of 50, 60, and 70 °C, and showed that all enzyme-hNFs were more thermally stable than their corresponding free enzymes. For example, after heating at 70 °C for 6 h, the residual activity of the enzymes in hNFs was as follows: TLL-hNFs (78.3%), PPL-hNFs (72.9%), and CALB-hNFs (84.3%), counting for a 4.57, 2.61, and 2.35 times higher activity than the corresponding free enzymes, respectively. The authors attributed this stability to the strong interaction between the Ca^{2+} ions and the functional groups on the enzymes in the hNFs and the rigidity in the inorganic hNFs that enclosed the enzyme and stopped the peptide chains from unfolding, thus improving the thermal stability of the enzyme [37]. Another study performed on magnetic hNFs embedded with the laccase enzyme showed that hNFs had a significantly better thermal stability than free laccase. One example is that at 55 °C (incubation time of 1 h), the magnetic laccase-hNFs retained 80% of their activity, which was more than that of free laccase. The results showed that a temperature of 35 °C had no effect on the activity, and a temperature of 85 °C made laccase inactive for both laccase-hNFs and free laccase [34].

3.2.2. Storage Stability

The storage stability of enzymes in hNFs is another essential influencing factor that has been intensively studied. The study mentioned above also examined the storage stability of the laccase enzyme in magnetic hNFs. The findings showed that, at room temperature, free laccase lost 77% of its activity after 30 days and 90% after 60 days. On the other hand, laccase-magnetic hNFs sustained 60% of their activity after 30 days and 45% after 60 days. At 4 °C and higher, over 60 days, both laccase-magnetic hNFs and free laccase maintained high activities. The authors suggested that the loss of activity for the free laccase was a result of conformational changes in the enzyme. Additionally, the morphology of the magnetic hNFs was examined using SEM over a 60-day storage period, and there was no visible difference in the nanoflower size and hierarchical structure. In addition, leaching of hNFs at room temperature and at 4 °C over 60 days of storage was studied by examining the protein content in the supernatant of hNFs. The results showed that there was nearly no detectable protein in the supernatant at the two different temperatures, which indicates the stability of the formed hNF complex [34].

Another study synthesized hNFs using copper ions and urease enzyme at different pH values, and examined the enzyme activity after 30 days at 4 °C and room temperature. Both experiments showed that hNFs had a better storage stability over time than the free enzyme. At 4 °C, the hNFs produced at different pH values of 6, 7.4, 8, and 9 lost 22.55%, 3.7%, 10%, and 15% of their initial activities, respectively. On the other hand, free urease lost 73.55% of its initial activity. At room temperature, the hNFs lost 35%, 9.28%, 13.22%, and

22.34% of their initial activities at a pH of 6, 7.4, 8, and 9, respectively, whereas free urease lost 90.25% of its initial activity. These findings not only show that enzyme-embedded hNFs are more stable than their corresponding free enzyme, but also indicate that hNFs synthesized at pH 7.4 have the best storage stability [38].

Nadar et al. prepared hNFs from glucoamylase enzymes and copper ions, and studied their storage stability over 25 days with five-day intervals at 30 °C. Their results showed that the free glucoamylase activity slowly decreased to 68% of its original activity, whereas the hNFs were able to maintain 91% of their original activity [39]. Similarly, Patel et al. found that laccase-hNFs and cross-linked-laccase-hNFs were more stable than free laccase when stored at 4 °C for 60 days. The laccase residual activity was 53.3%, 91.5%, and 3.8% for laccase-hNFs, cross-linked-laccase-hNFs, and free laccase, respectively. Hence, laccase-hNFs and cross-linked-laccase-hNFs had 14 and 24 times, respectively, the free enzyme's residual activity [30].

3.3. Reusability

Research shows that hNFs remain active for multiple reaction cycles before they lose activity regarding reusability. In addition, research has shown that they can be recycled by adding fresh enzymes. Yn et al. rebloomed the hNFs they produced using the following method (Figure 7). First, 0.2 mL of acetic acid or phosphoric acid was added to dissolve the original hNFs. After that, the reaction mixture was heated for 10 min at 100 °C to denature all of the enzymes. The denatured enzymes were then removed via filtration or centrifugation. Then, the solution pH was adjusted to 6.7 using $Ca(OH)_2$. A rebloom of the hNFs occurred when fresh enzymes were added to the solution, and co-crystallization occurred with $Ca(PO_4)_2$. The reaction was kept at 4 °C for 24 h. Then, the nanoflowers were separated and used as the original hNFs. The researchers examined both the activity of the dual cycle hNFs and the recovery of $Ca(PO_4)_2$ for six different enzyme models. The results showed no noticeable differences between the original hNFs and the dual-cycle hNFs, which suggested that certain molecules, such as amino acids, do not affect the catalysis of the dual-cycle hybrid nanoflowers. While examining the recovery rate percentage of $Ca(PO_4)_2$ by checking their weight while dry and after recrystallization, the result showed up to a 99% recovery of $Ca(PO_4)_2$ for six enzymes-hNFs. In summary, the activity of the enzymes and the recovery of $Ca(PO_4)_2$ before and after the dual cycle were nearly constant for all of the tested enzyme model hNFs [37].

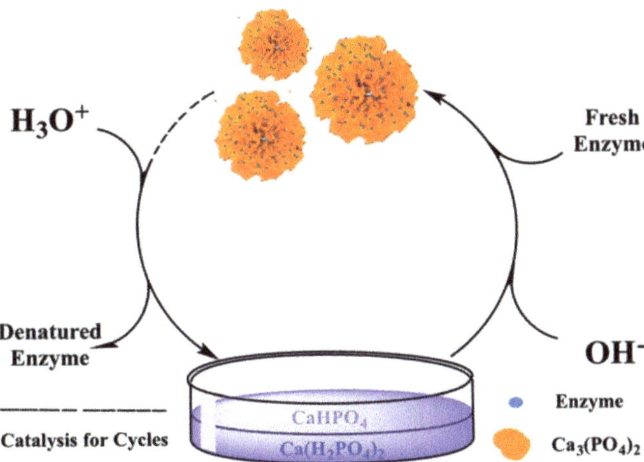

Figure 7. Dual cycle process for enzyme immobilization. Reprinted from [37] with permission from the Royal Society of Chemistry under the Creative Commons Attribution 3.0 Unported Licence.

Memon et al. synthesized hNFs using alcalase and Ca ions. When examining the reusability of the hNFs, the results showed that hNFs maintained 85.4% of their activity during seven cycles [40]. A research group investigated the reusability of hNFs produced from papain enzymes and Zn ions. The findings thus showed that the enzyme could maintain 88.8% of its original activity for ten cycles, but there was a steady decrease in its activity as the cycle number increased [41].

4. Types and Synthesis of Organic–Inorganic Hybrid Nanoflowers

As discussed earlier, hybrid nanoflowers are formed between a protein (organic component) and a metal ion (inorganic component). When proteins have a metal-binding site they can form complexes with ions through coordination interactions. For example, enzymes with nitrogen atoms in their amine and amide groups can form complexes with different metals through coordination interactions, i.e., the beginning of hybrid nanoflower synthesis [20,33,42]. Hybrid nanoflowers can be categorized based on the assembly of the particles, the type of protein/enzyme (organic component), and the metal ion (inorganic component) used. Table 1 shows the different metals and proteins used for hybrid nanoflower synthesis. The preparation of different hNFs based on the use of different metals is discussed next.

Table 1. Different hNFs based on the different metal ions and enzymes used.

Metal Ion	Enzyme	Class of Enzyme	Application	Reference
Copper (II) ions	Turkish black radish	Peroxidase	Dye decolorization	[43]
	Horseradish peroxidase	Peroxidase	Detection of hydrogen peroxide	[44]
	Horseradish peroxidase	Peroxidase	Detection of *E. coli*	[45]
	Horseradish peroxidase	Peroxidase	Detection of hydrogen peroxide and phenol	[36]
	Horseradish peroxidase	Peroxidase	Detection of amyloid	[46]
	Horseradish peroxidase	Peroxidase	-	[47]
	Horseradish peroxidase	Peroxidase	-	[42]
	Horseradish peroxidase	Peroxidase	-	[48]
	Horseradish peroxidase	Peroxidase	-	[49]
	Horseradish peroxidase	Peroxidase	-	[50]
	Chloroperoxidase (CPO)	Peroxidase	Dye decolorization	[29]
	Soybean peroxidase (SBP)	Peroxidase	-	[24]
	Lactoperoxidase (LPO)	Peroxidase	-	[51]
	Catalase	Peroxidase	Glucose biofuel cell	[52]
	Catalase	Peroxidase	Detection of hydrogen peroxide	[53]
	Catalase	Peroxidase	-	[50]
	Laccase	Laccase		[42]
	Laccase	Laccase	Degradation of the pollutant bisphenol A	[34]
	Laccase	Laccase	Decolorization of Congo Red (CR)	[32]
	Laccase	Laccase	Dye decolorization	[30]
	Laccase	Laccase	-	[54]
	Laccase	Laccase	Detection of phenol	[55]
	Laccase	Laccase	Synthesis of viniferin	[56]
	Laccase	Laccase	Synthesis of viniferin	[57]
	Laccase	Laccase	Glucose biofuel cell	[52]
	Glucose oxidase (GOx)	Carbohydrase	Detection of glucose	[58]
	Glucose oxidase (GOx)	Carbohydrase	-	[50]
	Glucose oxidase (GOx)	Carbohydrase	Glucose biofuel cell	[52]

Table 1. *Cont.*

Metal Ion	Enzyme	Class of Enzyme	Application	Reference
	Glucose oxidase (GOx)	Carbohydrase	-	[59]
	L-Xylanase	Carbohydrase	-	[60]
	Glucoamylase	Carbohydrase	-	[39]
	α-Glycosidase	Carbohydrase	Testing for α-glycosidase inhibitors	[61]
	L-Arabinose Isomerase	Carbohydrase	Preparation of two expensive rare sugar L-ribulose and D-tagatose	[62]
	Candida rugosa lipase	Lipase	-	[63]
	Candida antarctica lipase	Lipase	Epoxidation of fatty acids	[64]
	pseudomonas cepacia lipase	Lipase	-	[65]
	Bacillus subtilis lipase	Lipase	Transesterification of (R,S)-2-pentanol	[66]
	Lipase	Lipase	p-nitrophenol butyrate hydrolysis	[67]
	Lipase	Lipase	-	[27]
	Lipase	Lipase	-	[68]
	Lipase	Lipase	-	[69]
	Lipase	Lipase	Uses as green media solvent	[70]
	Lipase	Lipase	Biodiesel synthesis	[71]
	proteinase K	Protease	Detergent additive	[72]
	Alkaline protease	Protease	-	[73]
	Trypsin	Protease	Protein digestion	[23]
	Papain	Protease	-	[42]
	Papain	Protease	-	[74]
	Papain	Protease	-	[69]
	Papain	Protease	-	[75]
	Cholesterol oxidase (ChOx) and horseradish peroxidase	Dual enzyme	Detection of cholesterol	[76]
	Glucose oxidase and lipase	Dual enzyme	Epoxidation of alkenes	[77]
	Acetylcholinesterase and choline oxidase	Dual enzyme	On-site detection of the pesticide organophosphorus	[78]
	Glucose oxidase and horseradish peroxidase	Dual enzyme	Monitoring urinary tract infection (UTI) in clinical practice	[79]
	Glucose oxidase and horseradish peroxidase	Dual enzyme	Glucose sensor	[80]
	Glucose oxidase and horseradish peroxidase	Dual enzyme	Detection of glucose	[81]
	Glucose oxidase and horseradish peroxidase	Dual enzyme	Detection of glucose	[82]
	Cytochrome P450	Others	Oxidation of sulfides	[83]
	L-Arabinitol 4-dehydrogenase	Others	L-xylulose production	[84]
	Urease	Others	-	[38]
	Brevibacterium cholesterol oxidase (COD)	Others	-	[85]
	Carbonic anhydrase	Others	-	[26]
	2,4-dichlorophenol hydroxylase	Others	-	[86]

Table 1. *Cont.*

Metal Ion	Enzyme	Class of Enzyme	Application	Reference
Calcium (II) ions	chloroperoxidase (CPO)	Peroxidase	-	[87]
	Chitosan and Catalase	Peroxidase	-	[88]
	α-amylase	Carbohydrase	-	[89]
	β-Galactosidase	Carbohydrase	Protein biomarker	[90]
	Candida antarctica lipase	Lipase	-	[37]
	Porcine pancreas lipase	Lipase	-	[37]
	Thermomyces lanuginosus lipase	Lipase	-	[37]
	Burkholderia cepacia lipase (BCL)	Lipase	-	[21]
	Alcalase	Protease	-	[40]
	Bromelain	Protease	-	[37]
	Trypsin	Protease	-	[37]
	Papain	Protease	-	[37]
	α-chymotrypsin	Protease	Digestion of bovine serum albumin (BSA) and human serum albumin (HSA)	[22]
	Dual enzyme: Aldehyde ketone reductase and alcohol dehydrogenase	Dual enzyme	Production of (S)-1-(2, 6-dichloro-3-fluorophenyl) ethyl alcohol, a key chiral alcohol that is an intermediate of Crizotinib, an anti-cancer drug	[91]
	α-Acetolactate decarboxylase (ALDC)	Others	Inhibition of diacetyl formation in beer	[92]
	Elastin-like polypeptide (ELPs)	Others	Detection of H_2O_2	[93]
	Invertase	Others	*E. coli* detection from milk	[94]
	Carbonic Anhydrase	Others	-	[26]
Manganese (II) ions	L-Arabinose Isomerase	Carbohydrase	Transformation of D-Galactose to D-Tagatose	[95]
	Collagen	Others	Water oxidation	[96]
	Bovine serum albumin (BSA)	Others	Catalysis in fuel cells	[97]
	Carbonic Anhydrase	Others	-	[26]
	Ractopamine antibody	Others	Electrochemical biosensors ractopamine detection	[98]
Zinc (II) ions	Lipase	Lipase	Regioselective acylation of arbutin	[99]
	Lipase	Lipase	-	[35]
	Papain	Protease	-	[41]
	Bovine serum albumin (BSA)	Others	Adsorption of heavy metal ions	[100]
Cobalt (II) ions	Chloroperoxidase (CPO)	Peroxidase	Dye decolorization	[29]
	D-Psicose 3-Epimerase (DPEase)	Carbohydrase	-	[101]
	Lipase	Lipase	-	[102]
	ω-Transaminase	Others	-	[103]
	Bovine serum albumin (BSA)	Others	-	[104]
	Bovine serum albumin (BSA)	Others	-	[105]
	His-tagged enzyme	Others	Redox reaction cycles	[106]
Iron (II) ions	Horseradish peroxidase	Peroxidase	-	[107]
	Glucose oxidase (GOx)	Carbohydrase	-	[108]
Multi-metal (Copper+ Zinc)	Laccase	Laccase	Degradation of the pollutant bisphenol A	[109]
Non-metal (Selenium)	Pullulan (polysaccharide polymer)	Others	-	[110]

4.1. Copper-Based Hybrid Nanoflowers

For the first hNFs produced [20], an aqueous $CuSO_4$ solution was added to phosphate-buffered saline (PBS) that contained bovine serum albumin (BSA). The reaction succeeded at room temperature and was incubated for 3 days. Scientists then confirmed the synthesis process of hNFs by replacing BSA with other proteins, including α-lactalbumin, carbonic anhydrase, laccase, and lipase (Figure 8). One of the experiments performed on laccase-copper phosphate nanoflowers showed that laccase hNFs had an activity that was 4.5–6.5 higher than that of free laccase for oxidizing catecholamine syringaldazine. Additionally, they showed a good stability and reusability. Another experiment showed that the activity of carbonic anhydrase-embedded hNFs was 2.6-fold higher than the free form activity in the hydration of CO_2 [20].

Figure 8. SEM images of hybrid nanoflowers: (**a–l**) column 1, a-lactalbumin; column 2, laccase; column 3, carbonic anhydrase; column 4, lipase; at protein. Reprinted from [20] with permission from Springer Nature. Copyright © 2021, Nature Publishing Group. License Number: 5031780508604.

As the first discovered hybrid nanoflowers used protein-Cu ions, the majority of work was done on them. Another study showed the synthesis of hNFs from Turkish black radish peroxidase and copper ions. This study showed that hNFs had a better activity and stability in a wide range of pH values, as well as the ability to degrade 90% of Victoria blue dye [43]. As shown in Figure 9, hNFs are formed from glucose oxidase and copper ions embedded in amine-functionalized magnetic nanoparticle-labeled MNP-GOx NFs as antibacterial agents. The results showed that MNP-GOx NFs demonstrated an antibacterial activity with Gram-positive *S. aureus* and Gram-negative *E. coli* in a broad spectrum. This was done by disturbing the bacterial cells with the H_2O_2 produced by GOx [111]. Yang et al. produced hNFs from copper phosphate and horseradish peroxidase. These hNFs exhibited a linear detection from 100 nM to 100 μM H_2O_2. Additionally, they showed a good reusability and excellent storage stability [44]. Sun et al. [112] synthesized copper polyphosphate kinase 2 hNFs and formed an ADP regeneration approach from AMP using hNFs. The resulting hNFs had a better storage stability, in addition to a broader pH and temperature ranges. Additionally, it showed a better ADP production and retained 71.7% of its original activity after ten cycles, which showed good reusability.

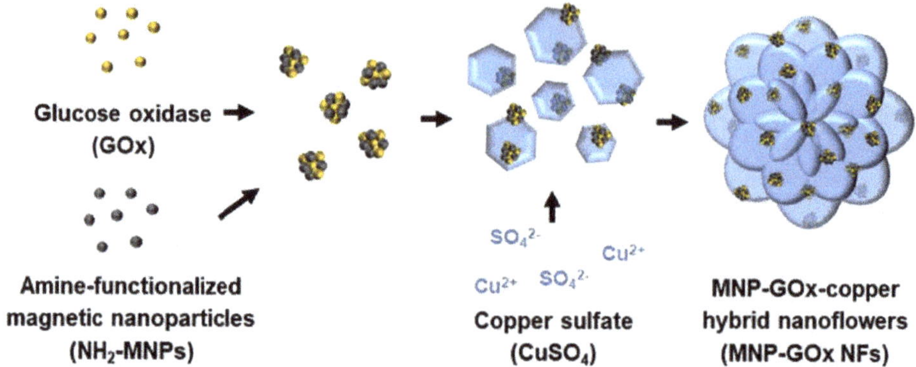

Figure 9. Preparation of MNP-GOx NFs. Reprinted from [111] with permission from Elsevier. Copyright © 2021 Elsevier B.V. License Number: 5031790647430.

Li et al. [42] produced hybrid nanoflowers on a nanofiber membrane surface from copper ions and different proteins. This led to the production of a very biocompatible and multilevel surface. The produced copper hybrid nanoflowers displayed an improved stability compared with the free protein that could stem from the protein's protection from the inorganic crystals. Figures 10 and 11 show the mechanism of how this happens. The gained stability in the hybrid nanoflowers can lead to their introduction to the application, each as biodevices and biocatalysts. The researchers found that by changing the concentration of the protein (Figure 12), the incubation time, the composition of the nanofiber membrane, and the preparation of the mineralizing solutions, the composition and structure of the copper hNFs could be controlled on the nanofiber membrane. The results showed that the different proteins tested (papain, bovine serum albumin (BSA) laccase, and horseradish peroxidase) gave different hNF morphologies, which is supported by previous studies.

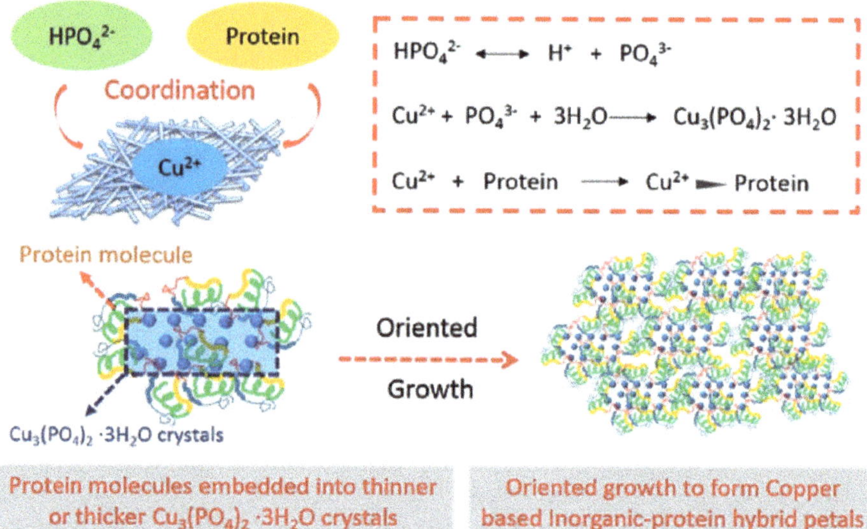

Figure 10. Formation of copper hNFs. Reprinted from [42] with permission from the American Chemical Society. Copyright © 2021 American Chemical Society.

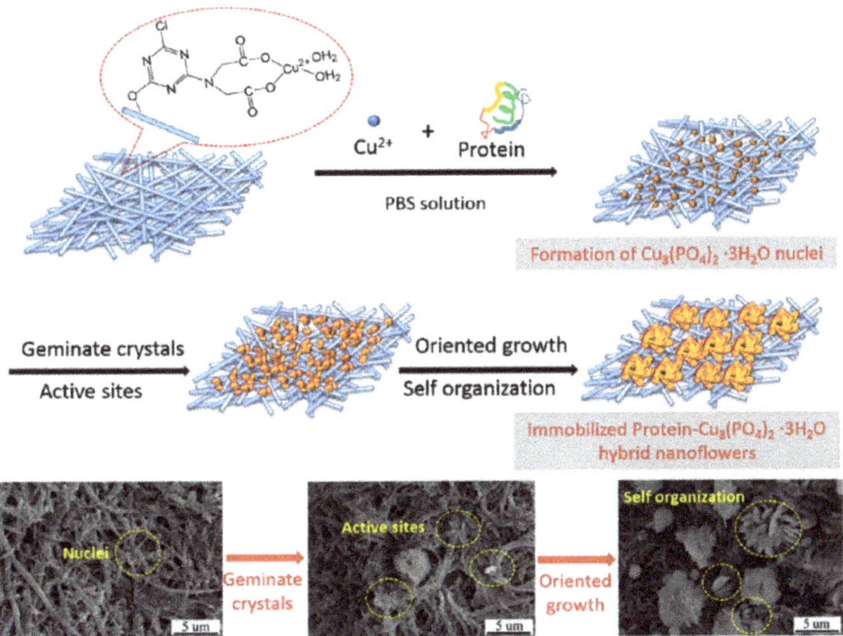

Figure 11. The growing process copper−protein hNFs. Reprinted from [42] with permission from the American Chemical Society. Copyright © 2021 American Chemical Society.

Figure 12. SEM of Cu-BSA hNFs with different protein concentrations, with an incubation time of 6 h. Reprinted from [42] with permission from the American Chemical Society. Copyright © 2021 American Chemical Society.

4.2. Calcium-Based Hybrid Nanoflowers

Even though the most commonly used metal for the synthesis of hybrid nanoflowers is copper, another metal that is heavily used to produce hNFs is calcium. Wang et al. reported the synthesis of α-amylase-CaHPO$_4$ hybrid nanomaterials, inspired by the allosteric effect. The work showed three different nanomaterial morphologies: nanoflowers, nanoplates, and parallel hexahedrons. While studying the enzymatic activity of α-amylase in the three different nanomaterial systems developed, and free α-amylase with and without calcium ions, the researchers credited two main factors that increased the enzymatic activity of α-amylase, namely: the allosteric effect of calcium ions with the amine group of the enzyme and the morphology of the nanomaterials [89]. Self-repairing hNFs were built from Ca$_3$(PO$_4$)$_2$ and chloroperoxidase (CPO) with a sodium alginate (SA) coating. The results showed that the immobilized enzyme had similar K$_m$ and K$_{cat}$ values, compared with the free enzyme. Additionally, it demonstrated the ability of the immobilized chloroperoxidase to work in acidic conditions, where it was able to maintain more than 85% of its activity after 12 cycles. Figure 13 shows the process of self-repairing SA-coated CPO-Ca$_3$(PO$_4$)$_2$ hybrid nanoflowers [87]. Zhao et al. [92] synthesized calcium hNFs by combining Ca$_3$(PO$_4$)$_2$ and α-acetolactate decarboxylase (ALDC) enzymes. These hNFs had a better activity than the free ALDC.

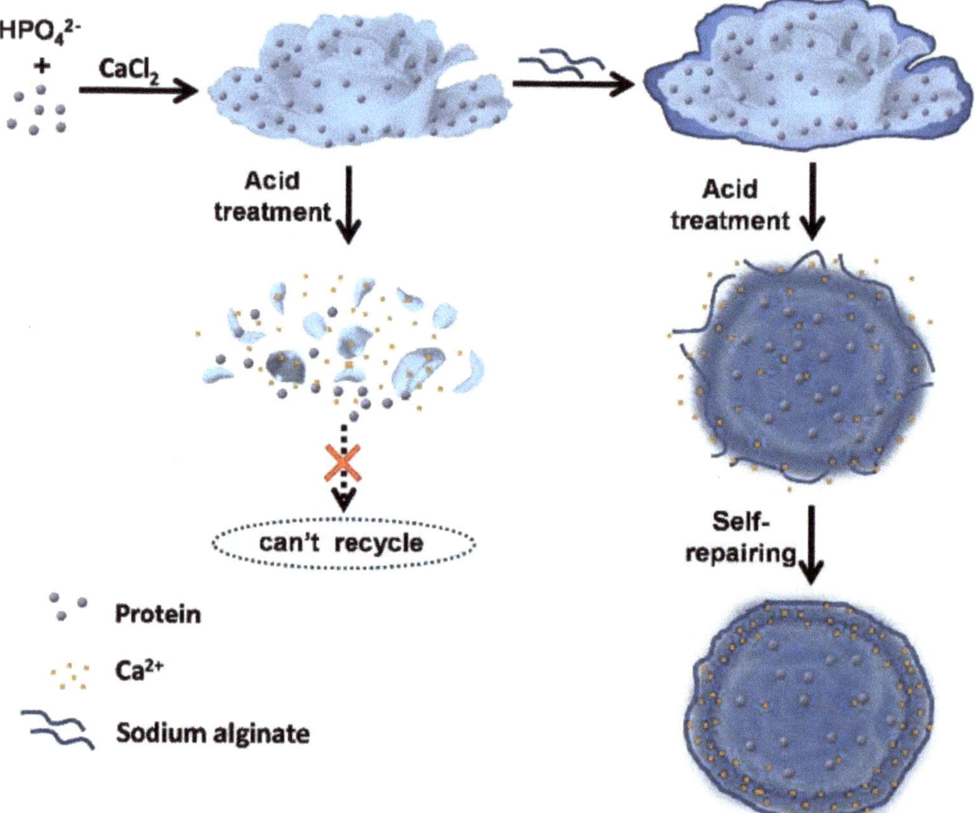

Figure 13. The process of self-repairing sodium alginate (SA)-coated CPO-Ca$_3$(PO$_4$)$_2$ hybrid nanoflowers. Reprinted from [87] with permission from the Royal Society of Chemistry. Copyright © The Royal Society of Chemistry. License Number: 1105129-1.

4.3. Manganese-Based Hybrid Nanoflowers

Nearly all of the hybrid nanoflowers produced in the literature use copper or calcium ions. Nevertheless, several studies use different metal ions. One of these metals is manganese; specifically, manganese(II) phosphate is used because of its unique electrochemical properties [113–115]. Rai et al. [95] synthesized hNFs using manganese metal as the inorganic component and L-arabinose isomerase as the organic component. Recombinant L-arabinose isomerase with 474 amino acids was synthesized into E. coli from *Lactobacillus sakai*. Hybrid nanoflowers with a spherical hierarchical morphology were produced using purified recombinant isomerase.

Several studies were performed on the L-arabinose isomerase embedded in the hybrid nanoflowers, and a circular dichroism (CD) analysis recorded no change in the isomerase structure. Compared with free L-arabinose isomerase, hNF-embedded L-arabinose isomerase demonstrated better kinetic parameters. Interestingly, L-arabinose isomerase converted approximately 50% of D-galactose to D-tagatose, a rare type of sugar, without adding more manganese to the reaction, showing the possibility of commercial production of this sugar using manganese/L-arabinose isomerase hNFs. In addition, these hNFs show good reusability and reproducibility in multiple reaction cycles. Munyeman et al. [96] reported the synthesis of collagen/manganese phosphate hNFs in an environmentally friendly biomineralization method. In this study, collagen was used as the biotemplate agent to produce hNFs. Additionally, it was used to bind the petals of the nanoflowers together. These hNFs had a great catalytic activity in relation to water oxidation. Zhang et al. [97] prepared hNFs from manganese(II) phosphate and bovine serum albumin (BSA). The results showed a good catalytic activity in the fuel cells.

4.4. Zinc-Based Hybrid Nanoflowers

Although the synthesis of copper hybrid nanoflowers is simple and easy, its three-day production process is one of its disadvantages. Thus, it is important to reduce the synthesis time to a more appropriate period. This can be done by choosing the right metal ions or by changing the synthetic method. When choosing metal ions, zinc displayed a faster reaction rate towards phosphate radicals than copper ions [35,41,100]. Zhang et al. showed that hNFs were prepared from zinc phosphate $Zn_3(PO_4)_2$ and lipase. The formation time of $Zn(PO_4)_2$/lipase hNFs took less than three hours, and the formation time of hNFs using $Cu(PO_4)_2$ took three days. $Zn(PO_4)_2$/lipase hNFs showed a great operational stability compared with free enzymes [35]. In another study, hNFs were made using $Zn(PO_4)_2$ and papain, and the resulting hNFs showed a higher activity than free papain, in addition to a better thermal stability and storage life [41]. Zhang et al. [100] prepared hNFs using Zn and bovine serum albumin (BSA) at 25 °C. These hNFs had an average size of 2.3 μm with a surface area of 146.64 cm^2/g. The hNFs were used for Cu (II) ion adsorption. The adsorption efficiency of the Zn hNFs towards copper ions was 86.33% at 5 min and 98.9% at 30 min. The highest adsorption capacity obtained with these hNFs was 6.85 mg/g. This study showed the ability of $Zn(PO_4)_2$/BSA hNFs to be used as a fast and efficient method for Cu^{2+} removal.

4.5. Cobalt-Based Hybrid Nanoflowers

Despite the various studies that investigated the synthesis of hybrid nanoflowers, only a few studies have shown the use of cobalt ions in the formation of hNFs. Kim et al. synthesized protein/cobalt hNFs using BSA and cobalt ions. The work illustrated that the BSA protein could be a template to interact with cobalt phosphate to produce protein-metal hNFs [104]. Kumar et al. [102] produced hNFs using cobalt ions and lipase enzymes. The immobilized lipase showed a 181% higher activity than the free enzyme. Additionally, it offered a better catalytic performance in harsh reaction conditions and higher temperatures.

4.6. Iron-Based Hybrid Nanoflowers

Studies that use iron ions to produce hNFs are limited. Ocsoy et al. [107] used Fe^{+2} and horseradish peroxidase (HRP) to produce hNFs. The results showed an approximately 512% increase in the activity of HRP when stored at 4 °C and an approximately 710% increase when stored at room temperature compared with the free enzyme. Additionally, the immobilized HRP lost 2.9% and 10% of its initial activity after 30 days when stored at 4 °C and room temperature, respectively. However, the free HRP lost 68% of its activity when stored at 4 °C and 91% when stored at room temperature.

4.7. Multi-Metal-Based Hybrid Nanoflowers

As previously mentioned, there is an extensive range of research and studies in the literature on using different metals to produce hierarchical hybrid nanoflower structures, such as copper, calcium, manganese, zinc, and cobalt. However, few studies have attempted the production of multi-metal-based hybrid nanoflowers; here are a few of them. In 2019, Patel et al. produced hybrid nanoflowers based on copper and zinc ions. The novel multi-metal nanoflowers were synthesized using a laccase enzyme. The Cu/Zn-laccase showed a higher encapsulation yield percentage than the copper-laccase and zinc-laccase hNFs, which were 96.5%, 87.0%, and 90.2%, respectively. The multi-metal nanoflowers (Cu-/Zn-laccase) were 1.2-, 1.5-, and 2.6-fold higher than zinc-laccase, copper-laccase, and free laccase, respectively. Interestingly, the multi-metal nanoflowers showed a charge transfer resistance that was 2.1-fold lower than zinc-laccase hNFs, and when compared with copper-laccase hNFs it was 2.7-fold lower. For the degradation of bisphenol A, the remaining multi-metal nanoflower activity was 1.9-fold higher than that of zinc-laccase hNFs and 5.1-fold higher than that of copper-laccase hNFs [109].

4.8. Non-Metal-Based Hybrid Nanoflowers

As mentioned above, there is extensive work in the literature on certain metals, such as copper, calcium, manganese, zinc, cobalt, and iron. Nevertheless, few studies have tried to produce non-metal hybrid nanoflows. In 2018 [110], selenium (non-metal) hybrid nanoflows were synthesized. In recent years, selenium nanoparticles have been studied as drug carriers because they are nontoxic and have a good biological activity and bioavailability [116–121]. There have been attempts to produce selenium nanoparticles (SeNPs) with different structures, shapes, and morphologies, including nanoplates, nanotubes, and nanospheres [61–63]. The addition of a biopolymer that is functionalized with these nanoparticles has led to an increased stability and control over the shape and size [122,123]. hNFs were synthesized using pullulan/SeNPs, and pullulan was used as a substitute for the use of proteins [110]. SeNPs were then stabilized using folic acid-decorated cationic pullulan (FA-CP), and presented a flower-like structure. The produced nanoflowers showed an excellent drug adsorption for doxorubicin and had a 142.2 mg/g loading capacity. The study showed that doxorubicin's loading capacity is three times greater in pullulan/SeNP nanoflowers than in spherical SeNPs. Additionally, these hNFs showed a better activity towards cancer cells, and they were less toxic towards normal cells.

4.9. Enhancing Hybrid Nanoflowers Synthesis

One of the most significant drawbacks of hNFs is their size, which is usually in the nano- or micro-scale range. Thus, it is difficult to separate them from their reaction mixture. Some studies have been performed to improve the synthetic processes of hybrid nanoflowers. Recently, scientists developed supported hybrid nanoflower methods to overcome some of the drawbacks of the use of hNFs. Alginate gel beads were used to entrap α-acetolactate decarboxylase/calcium hNFs. These entrapped hNFs showed a better stability and recyclability than the free enzyme [92]. Another study by Zhu et al. used a cellulose acetate membrane for laccase/copper ion hNFs. The captured hNFs showed a high reproducibility and reusability for phenol detection [55]. Cao et al. [124] produced a glassy carbon electrode (GCE) surface on which bovine serum albumin (BSA)/Ag nanoflowers were immobilized, coupled with a targeting lectin molecule for detecting human colon cancer cells. This sensor showed specificity for a cell expressing sialic acid. Therefore, it has a possible application for monitoring tumor cells.

4.10. Morphology of Hybrid Nanoflowers

hNFs commonly have a hierarchical structure and nanoplate petals that look like flower petals. These nanoplates/nanopetals are made of enzyme and metal phosphate. From the first hybrid nanoflower produced by J. Ge et al. in 2012, there have been numerous synthesis techniques to produce hNFs, which have resulted in different hNFs with different microstructures [125–127]. Table 2 shows the most common hNF microstructure morphologies, which are spherical, rosette, and rhombic, with their equivalent flowers in nature. Typically, hNFs have a diameter range between 1–30 µm. There are many factors and reaction conditions that control the shape and size of hNFs; these factors include (I) the type of enzyme used; (II) the type of metal ion; and (III) the reaction condition, which includes the pH value, the reaction temperature, and time [69]. Table 3 shows how morphology differs with changing the factors and conditions.

Table 2. Characteristic of hybrid nanoflowers.

Nature Flower	Shape	Example		SEM Image	Size	Reference
	Spherical	The enzyme: *Brevibacterium* cholesterol oxidase (COD)	The metal: copper		5 µm	[85]
		The enzyme: laccase	The metal: copper	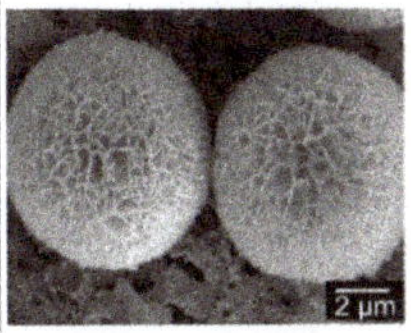	1 min sonication: ~2 µm; 5 min sonication: ~8 µm; 7 min sonication: No additional increase in the size	[54]
		The enzyme: glucose oxidase (GOx) + horseradish peroxidase (HRP)	The metal: copper	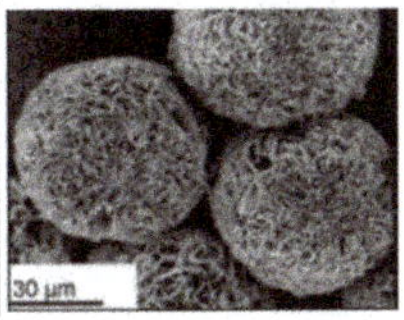	30 µm	[80]
	Rosette	The enzyme: α-glycosidase	The metal: copper	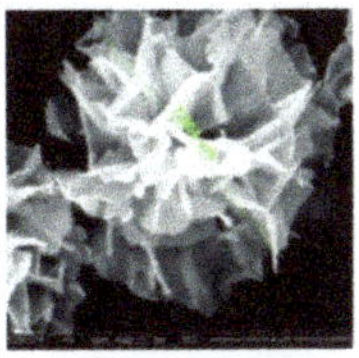	7.564 µm	[61]
		The enzyme: catalase	The metal: copper	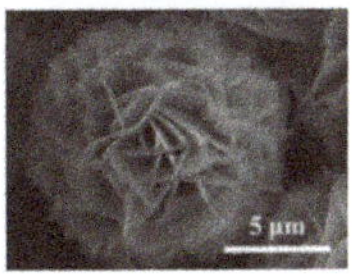	10 to 20 µm	[53]

Table 2. *Cont.*

Nature Flower	Shape	Example		SEM Image	Size	Reference
		The enzyme: streptavidin + horseradish peroxidase (HRP)	The metal: copper		5 μm	[49]
	Rhombic	The enzyme: D-psicose 3-epimerase (DPEase)	The metal: cobalt		7 μm	[101]
		The enzyme: papain	The metal: zinc		-	[41]
		The enzyme: laccase	The metal: copper		-	[20]

Table 3. Different factors and conditions that affect the morphology of hNFs.

Enzyme	Morphology				Reference
Type of enzyme used	1. Enzyme: Glucose oxidase (GOx)	2. Enzyme: Laccase	3. Enzyme: Catalase	4. Enzyme: Lipase	[35,52]
Different amount of the enzyme # Example 1	5. 0.01 g Lipase	6. 0.025 g Lipase	7. 0.05 g Lipase	8. 0.10 g Lipase	[35]
Different amount of the enzyme # Example 2	9. 0.025 g papain	10. 0.05 g papain	11. 0.1 g papain	12. 0.25 g papain	[41]

Table 3. *Cont.*

Enzyme	Morphology			Reference
Metal ion	Morphology			
Type of metal ion used	13. Copper 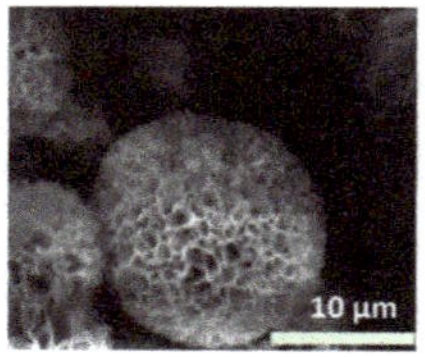	14. Cadmium 	15. Cobalt 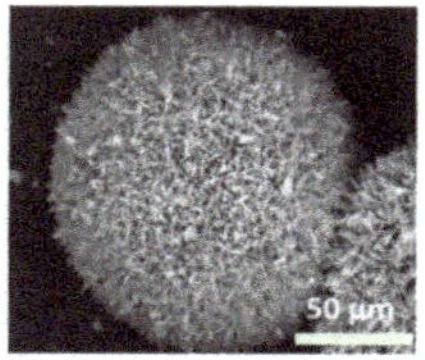	[29]
Reaction pH	Morphology			
Different pH values	16. pH 6 	17. pH 7.4 	18. pH 8  — 19. pH 9 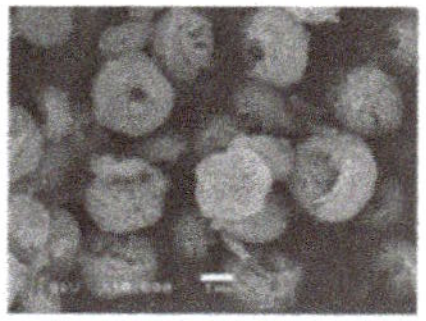	[69]
Reaction temperature	Morphology			
Different reaction temperature # Example 1	20. Temperature: Below T_t, at 4 °C ELP/**Ca** hNFs 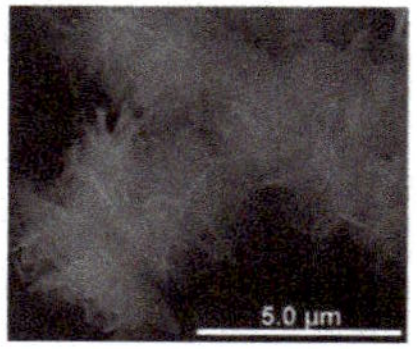	21. Temperature: Above T_t, at 37 °C ELP/**Ca** hNFs 	22. Temperature: Below T_t, at 4 °C ELP/**Cu** hNFs 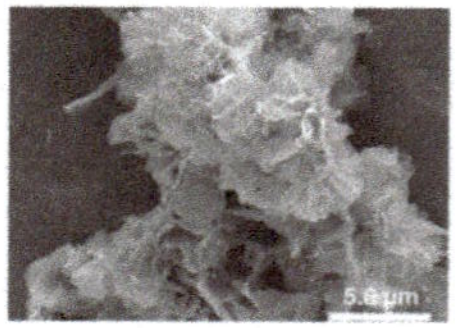 — 23. Temperature: Above T_t, at 37 °C ELP/**Cu** hNFs 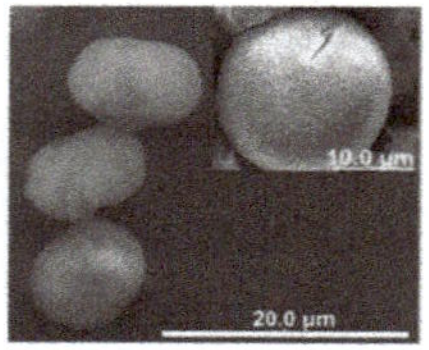	[93]

Table 3. *Cont.*

Enzyme	Morphology				Reference
Different reaction temperature # Example 2	24. Temperature: 20 °C	25. Temperature: 30 °C	26. Temperature: 40 °C	27. Temperature: 60 °C	[35]
Different reaction temperature # Example 3	28. Temperature: 4 °C	29. Temperature: 20 °C	-	-	[51]
Reaction time	Morphology				
Different reaction time	30. Time: 2 h	31. Time: 8 h	32. Time: 24 h	-	[39]

4.11. The Type of Enzyme Used

As mentioned above, for hNFs to form, the presence of an enzyme is essential. Enzymes attach to the petals of various metal phosphate nanoplates. Numerous studies have shown that different enzymes lead to the formation of different hNF morphologies [21–24]. Chung et al. [52] developed hNFs using copper ions with various enzymes, including glucose oxidase (GOx), laccase, and catalase. Each enzyme gave a unique microstructure morphology of the hNFs, as illustrated in Table 3 (1, 2, and 3).

Zhang et al. [35] showed the use of lipase enzyme to form hNFs (Table 3 (4)), which has a completely different morphology from the three previously mentioned enzymes of glucose oxidase (GOx), laccase, and catalase. Another study used enzymes, laccase, papain, and horseradish peroxidase (HRP) to synthesize different shapes and sizes of hNFs [42]. Research explains that the diversity of hNF morphology arises from the various amide groups on the surface of the enzyme used. Thus, it presents different nucleation sites with varied geometries and densities for phosphate metal nanoplates to bind [20,33]. In addition, Lin et al. revealed that the employed enzyme's molecular weight has an impact on the hNF structure [36].

Additionally, it has been determined that the concentration of the enzyme plays a role in the final morphology of hNFs, where it can affect the size and density of the nanoplates [56]. Several studies have indicated that the lower the enzyme concentration used, the larger the size of the hNFs formed with a lightly constructed structure. In addition, an increase in the employed enzyme content leads to a denser packed structure, although the size will decrease if the enzyme concentration reaches a particular value [72,86,99]. The reason behind this is the subsequent increase in the number of nucleation sites [68,103]. Table 3 (5–12) shows the synthesis of two different hybrid nanoflowers using two different enzymes (lipase and papain) with the same metal ion (Zn). As seen from the table, the morphology of both hNFs dramatically changes with an increased amount of enzyme, from a rhombus shape to a square, oval shape. With the continuous addition of the enzyme (an excess amount), the shape changes into a dense cluster structure with cracks instead of a well-formed flower shape [35,41].

4.12. The Type of Metal Ion Used

In the formation of hNFs, metal ions play an essential role in the primary crystal nucleation step and metal-enzyme coordination to produce hNFs. Table 3 (13–15) shows a study where hNFs were synthesized using a chloroperoxidase enzyme with three different metals (copper, cadmium, and cobalt). It was observed that copper and cobalt hNFs had similar morphologies, while cadmium hNFs were completely different. Copper and cobalt hNFs had a spherical flower-like structure, whereas cadmium hNFs had a butterfly-like structure [29]. The effect of the metal ion concentration on the morphology of hNFs was also studied. The results show that with an increase in the metal ion concentration, the hNF morphology becomes much denser [38,85].

Rai et al. [95] produced hNFs using $_L$-arabinose isomerase as the organic component, and 13 metal ions, including manganese, cobalt, magnesium calcium, potassium, sodium, nickel, and lithium, as the inorganic component. It was determined that manganese and cobalt ions can substantially improve the catalytic activity of the enzyme. In contrast, the magnesium calcium, potassium, sodium, nickel, and lithium ions had no impact on the enzyme activity.

4.13. Reaction Condition (pH, Temperature, and Time)

It is well known that the charge of the enzyme varies at different pH values. Thus, leading to the enzyme's different interaction capabilities in the formation of hNFs will influence the morphology of the hNFs [74]. Nadar et al. [39] examined the effect of changing the pH on hNFs synthesized using copper ions and glucoamylase enzymes. The study tested the pH range from 3.5 to 9.5. The net change in free glucoamylase enzymes is neutral, and it has an isoelectric point (pI) of approximately 6. The enzyme is expected to have a positive charge below the isoelectric point, and a negative charge above the isoelectric threshold. No nanoflower formation was observed at pH values of 3.5, 4.5, and 5.5 (below the isoelectric point). This can be explained by the fact that the positively charged protonated glucoamylase enzyme had an extremely strong repulsion with the positive copper metal ions at a low pH. Hence, no nanoflowers were formed. At pH values higher than the pI point, the charge on the glucoamylase enzyme surface was negative because of deprotonation. Glucoamylase/Cu^{2+} ions hNFs were formed in this pH range. It was observed that at pH 7.5 the nanoflowers were less packed, which is attributed to the small rise in the negative charge on the enzyme. At pH 9.5, nanoflowers did not form. This is because of the increased repulsion between the negative charges that are highly dense on the surface of the glucoamylase enzyme. Therefore, $Cu_3(PO_4)_2 \cdot 3H_2O$ petals repel one another rather than attaching. Table 3 (16–19) shows the morphology of the synthesized lipase/copper hNFs at different pH values, where the enzyme concentration is 1.0 mg/mL. The pH values of phosphate-buffered saline (PBS) were adjusted to 6.0, 7.4, 8.0, and 9.0. It was noted that the density of the petals decreased as the pH value increased. However, the diameters and the size of the formed nanoflowers remained the same [69].

Another important reaction condition that affects the morphology of hNFs is temperature. Temperature can play a vital role by initiating the various diffusion activities of the enzyme at different applied temperatures. Thus, nanoflowers have different degrees of density in petals and alter their size and diameter [43]. Table 3 (20–23) shows the effect of temperature on the production of calcium ions/elastin-like polypeptide (ELP) hNf and copper ions/ELP hNFs. ELPs are a class of polypeptides derived from an amino acid sequence of naturally accusing elastin in humans. Their pentapeptide sequence is Val-Pro-Gly-X-Gly, where X is any amino acid, other than from Pro. ELPs have a key transition temperature (T_t), where ELPs are soluble in the solution below it. Nevertheless, above this temperature, ELPs will suffer a phase transition that is destructive to the polypeptide. As seen from the table, the morphology of the hNFs synthesized below T_t (4 °C) had larger, more expanded petals. However, the hNFs synthesized above T_t (37 °C) had a more closed structure [93]. Another study, shown in Table 3 (24–27), demonstrated that hNFs formed at 20 °C had a more oval spherical shape with cracks. However, when the temperature increased above 40 °C, the shape of the hNFs changed to a more sheet-like structure, and as the temperature further increased, the hNF sheet morphology increased [35]. Altinkaynak et al. [51] showed the production of hNFs from copper ions and lactoperoxidase (LPO), and it was concluded that at a lower temperature, the petals of the hNFs became more compacted. This result is shown in Table 3 (28–29). For this, the ideal temperature for the synthesis of the best hNF structure must be investigated.

An additional factor that can affect the hNF morphology is the reaction time. The reaction time of the hNFs depends on the method used to produce them. Since the discovery of hNFs, there have been different time intervals used when producing them. Some of these time intervals include 3 days [20,30,38,44,68,111], 24 h [34,39,109], 3 h [35], and 5 min [52,58]. Table 3 (30–32) shows the different stages of hNF formation. The first step, nucleation, which is the formation of primary copper phosphate crystals, occurs between 0 and 2 h. The second step, growth, is when the metal ion and the enzyme form large agglomerates, which are the primary petals. This occurs between 2 to 8 h. In the last step, complete hNFs are formed, which occurs between 8 to 24 h [39].

5. Applications for Hybrid Nanoflowers

The excellent catalytic properties of hNFs have provided a wide selection of applications. Some of these applications are in the fields of biosensors, biomedical, bioremediation, and industrial biocatalysts [45,128–130]. Different examples of these applications are illustrated in Figure 14.

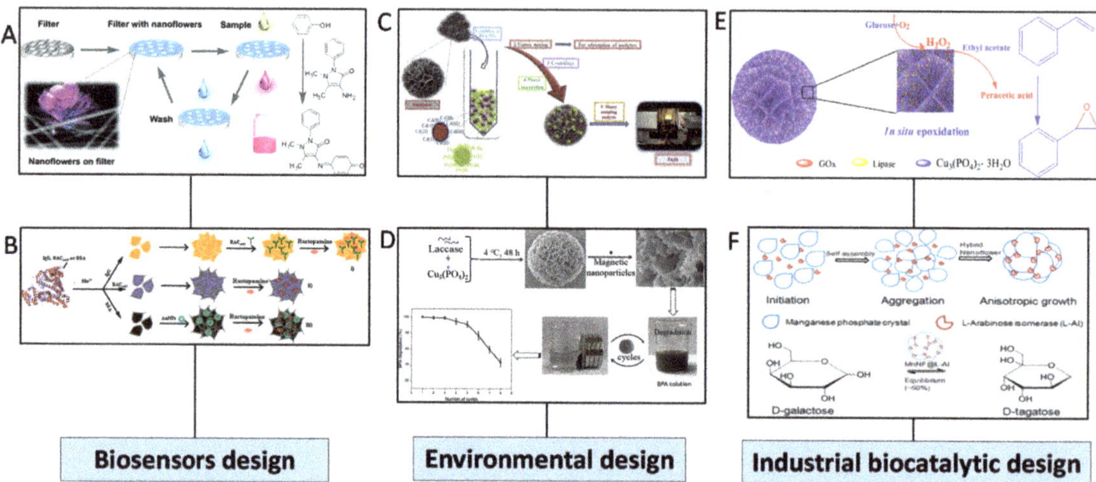

Figure 14. Summary of different applications used for hNFs. (**A**) Fast detection of phenol using laccase/cooper phosphate hNFs integrated into a membrane. Reprinted from [55] with permission from John Wiley and Sons. Copyright © 2021 WILEY-VCH Verlag GmbH and Co. KGaA, Weinheim. License Number: 5031800087304. (**B**) Detection of ractopamine using a protein/manganese ion hNF electrochemical biosensor. Reprinted from [98] with permission from Elsevier. Copyright © 2021 Elsevier B.V. License Number: 5031800271904. (**C**) Selective separation of cadmium and lead in water, cigarette, and hair samples using BSA/copper ions hNFs. Reprinted from [131] with permission from Elsevier. Copyright © 2021 Elsevier B.V. License Number: 5031800434687. (**D**) Degradation of bisphenol A using laccase/cooper phosphate hNFs. Reprinted from [34] with permission from Elsevier. Copyright © 2021 Elsevier B.V. License Number: 5031800600650. (**E**) Epoxidation of alkenes using dual enzyme system GOx and lipase/copper ions hNFs. Reprinted from [77] with permission from Elsevier. Copyright © 2021 Elsevier Ltd. License Number: 5031800753597. (**F**) Transformation of D-Galactose to D-Tagatose using L-Arabinose Isomerase/manganese ions hNFs. Reprinted from [95] with permission from the American Chemical Society. Copyright © 2021 American Chemical Society.

5.1. Biosensors

A biosensor is an analytical device employed to detect chemical substances and has recognition elements (biological components), such as enzymes, antibodies, microorganisms, and DNA [132,133]. There is a higher demand for cheap, easy, fast, and sensitive analytical biosensors [134]. Synthesized organic–inorganic hybrid nanoflowers have recently been used as novel biosensors with different classes of enzymes and metal ions because of their high surface area and sensitivity [135].

Zhu et al. produced hNFs for the fast detection of aqueous phenol using laccase enzyme and copper phosphate integrated into a membrane. The results showed the ability for the rapid on-site detection of phenol in water [55]. Zhang et al. reported the production of an electrochemical biosensor using Mn ions/protein hNFs. These hNFs were able to detect ractopamine and showed a high activity and an excellent electrochemical performance [98]. Additionally, copper/HRP hNFs could be used as a sensor for dopamine detection [31]. Sun et al. [80] used dual enzyme (HRP + GOx) hNFs as colorimetric sensors for glucose detection.

5.2. Environmental Applications

Environmental pollutants are a rising concern worldwide. Multiple methods have been used to eliminate these pollutants in the environment. One of these methods is bioremediation, which utilizes microorganisms and enzymes to reduce the concentrations of these harmful pollutants to an acceptable range [136–139]. Although enzymatic treatment has been shown to be effective, it faces some challenges, including a low activity, stability, and sustainability. Recently efforts have been made to immobilize enzymes on different support materials that can eliminate some of the challenges of using free enzymes [140,141]. However, traditional immobilization techniques can lead to a reduced activity of the enzyme compared with the free enzyme. Thus, enzyme immobilization of hNFs has been shown to increase the activity, stability, and reusability of the enzyme [24,37]. Yilmaz et al. produced hNFs from BSA/copper ions for the selective separation of cadmium (Cd) and lead (Pb) in water, cigarette, and hair samples. The method reported a low detection limit for Cd and Pb compared with the other existing methods [131]. The Rong group synthesized hNFs using laccase/copper ions and loaded them onto a treated Cu foil surface. These hNFs showed a high decolorization efficiency and rate on Congo red (CR) dye compared with the free enzyme. In addition, the hNFs showed a good stability and reusability [32]. Fu et al. [34] produced enzyme hNFs that could degrade 100% of bisphenol A within 5 min.

5.3. Industrial Biocatalytic Applications

As a result of the great activity, stability, and reusability of enzymes immobilized on hNFs, there has been increasing interest in their use in industrial biocatalytic applications, making them a prominent research area [142]. Zhang et al. [77] used hNFs made from a dual enzyme system GOx + lipase/copper ions for the epoxidation of alkenes. The hNFs showed an excellent reusability, retaining 82% of their activity after ten reaction cycles. First, hydrogen peroxide is produced by GOx from glucose and is then directly used by lipase. As a consequence of lipase, carboxylic ester is converted into peracid. After that, peracid is used in the epoxidation of alkenes.

Hybrid nanoflowers are also used in the food manufacturing industry [62,95]. Bai et al. and Xu et al. were able to synthesize two expensive rare sugars: L-ribulose and D-tagatose. These sugars are monosaccharides, but are very difficult to find in nature. Multiple chemical reactions have been employed to manufacture L-ribulose and D-tagatose from disaccharides and polysaccharides. They used L-arabinose isomerase as the organic component in hNFs and manganese ions [95] and copper ions [62] to produce sugars. The conviction rates were 50% and 61.88% for the manganese hNFs and copper hNFs, respectively. In addition, hNFs can be used in the brewing industry. hNFs made from α-acetolactate decarboxylase (ALDC)/calcium ions can be used to inhibit diacetyl formation in beer, which results in a buttery off-flavor [92].

6. Cost Assessment and Industrial Integration Aspects

As discussed above in the representative sections/subsections, the development of immobilized enzyme-based biocatalysts using different nanoscale materials, including hybrid nanoflowers (hNFs), has gained remarkable interest. The rationale behind this rising research trend in immobilized enzyme-based biocatalysts is the unique structural, physicochemical, and functional promises that enzyme-loaded hNFs offer with industrial integration potentialities. However, early-stage cost assessment and market analysis of enzyme-loaded biocatalytic nano-constructs, both pristine and hybrid, are equally essential to govern their industrial appropriateness [143,144]. The overall cost of the enzyme-based product and market analysis can be made via life-cycle assessments (LCAs) and/or techno-economic scrutinizes as powerful tools. By taking the added value of these tools, an array of immobilized enzyme-based biocatalysts have been or are being assessed for commercial-scale biocatalytic processes [145,146]. More specifically, from the cost and sustainability considerations, LCAs provide deep insight into the material, energy consumption, wasteful protection, and deprotection points in order to ensure the sustainability of the entire industrial process. The cost-effective ratio related to the energy consumption, chemical inputs, wasteful protection, and deprotection steps involved in the traditional industrial processes can be controlled effectively by implementing immobilized enzyme-based biocatalysts, e.g., enzyme-loaded hNFs, which are recoverable and reusable. For example, in several industrial processes, a massive amount of heat energy is required in the form of steam to preheat the feedstock for treatment purposes. In contrast, this can be significantly reduced by alternating the process with immobilized enzyme-driven treatment [144]. According to the Global Industrial Enzymes Market Size and Regional Forecasts 2020–2027, in 2019, the global enzymes market was worth $8636.8 Million USD, which is projected to reach up to $14,507.6 Million USD with a compound annual growth rate of 6.5% over the duration of 2020–2027 [147]. Conclusively, the large-scale industrial integration and/or scale-up deployment of immobilized enzyme-based biocatalysts technology requires a proper understanding of both technological and economic aspects, as well as a good perception of the larger market forces.

7. Conclusions and Future Perspectives

Since the discovery of organic–inorganic hybrid nanoflowers in 2012, the topic has become a prominent research area and numerous types of biomolecules/inorganic hybrid nanoflowers have been explored. These hybrid nanoflowers have recently shown their ability to become stable, easy, fast, efficient, and recyclable for different biomolecules, specifically enzymes immobilizing host platforms. Additionally, hybrid nanoflower applications have been extended to biosensor designs, environmental treatment, bioassays, and different industrial biocatalysis. Studies have shown that hybrid nanoflowers lead to an enhancement in the immobilized enzyme's catalytic activity, resulting from the higher surface area of hNFs, a reduced mass transfer limitation, and favorable enzyme conformation in hNFs. Another potential advantage is that in some cases, enzymes immobilized on hybrid nanoflowers behaved better than other immobilization platforms, especially in terms of their reusability. For example, lipase enzyme immobilized on hybrid nanoflowers showed an excellent operation stability as they retained up to 94.5% of their activity, even after eight cycles of reaction [35]. Similar results were obtained by Li et al., where they could efficiently use the lipase/calcium nanoflowers up to six times [148]. On the other hand, when lipase was immobilized on the Fe_3O_4@MIL-100(Fe) composite it lost around 15% of its activity after only five cycles [149]. Similarly, when the alcalase enzyme was immobilized on hybrid nanoflowers [40] and glyoxyl supports [150], the nanoflowers could be recycled efficiently up to seven times, as opposed to the agarose beads (only five times). In addition, HRP immobilized on NH_2-modified magnetic Fe_3O_4/SiO_2 particles was only able to do four cycles before losing most of its activity [151], whereas HRP immobilized on hybrid nanoflowers only lost 25% of its activity after six reaction cycles [49].

Although there has been a rapid increase in the use of hybrid nanoflowers in recent years, there are still some key challenges that need to be addressed. First, the interaction between the organic component (enzymes) and the inorganic component (metal ions) needs to be examined more intensively, providing more information about the hybrid nanoflower design with a well-retained biological activity. This will help with controlling the morphology as well as adjusting the properties of the synthesized hybrid nanoflowers. In addition, this would facilitate the production of new kinds of hybrid nanoflowers for a particular application purpose. In addition, the nucleation step in the inorganic phase is still not described quantitatively in the presence of different amounts of enzymes. Second, hybrid nanoflowers with multiple enzymes or dual enzyme systems have not been thoroughly investigated. Third, more research needs to be done for examining the production of hNFs in organic media, as most of the synthesized hNFs are in aqueous media. This would be beneficial for industrial biocatalysis applications. Finally, the industrial application of hybrid nanoflowers can expand to include energy application like fuel cell fabrication and biodiesel. In summary, we believe that organic–inorganic hybrid nanoflower research will expand the application arena of hNFs and lead to smart solutions for modern problems.

Author Contributions: Conceptualization, S.S.A.; data curation, K.A.A.-M., A.A. and I.S.; writing—original draft preparation, K.A.A.-M. and A.A.; writing—review and editing, M.B. and H.M.N.I.; supervision, S.S.A.; project administration, S.S.A.; funding acquisition, S.S.A. All authors have read and agreed to the published version of the manuscript.

Funding: This research received no external funding.

Data Availability Statement: Not applicable.

Acknowledgments: The authors are thankful to their representative universities for supplying funds for this work. Partial funding for K.A.A. was allocated by the PhD fund (no. 31S389 to I.S.) from the College of Graduate Studies, UAE University. Generous support from Khalifa University to S.S.A. (CIRA-2020-046) is also graciously acknowledged. Consejo Nacional de Ciencia y Tecnología (CONACYT) is thankfully acknowledged for partially supporting this work under the Sistema Nacional de Investigadores (SNI) program awarded to Hafiz M. N. Iqbal (CVU: 735340).

Conflicts of Interest: The authors declare no conflict of interest.

References

1. Zdarta, J.; Meyer, A.S.; Jesionowski, T.; Pinelo, M. A General Overview of Support Materials for Enzyme Immobilization: Characteristics, Properties, Practical Utility. *Catalysts* **2018**, *8*, 92. [CrossRef]
2. Mohamad, N.R.; Marzuki, N.H.C.; Buang, N.A.; Huyop, F.; Wahab, R.A. An Overview of Technologies for Immobilization of Enzymes and Surface Analysis Techniques for Immobilized Enzymes. *Biotechnol. Biotechnol. Equip.* **2015**, *29*, 205–220. [CrossRef] [PubMed]
3. Basso, A.; Serban, S. Industrial Applications of Immobilized Enzymes—A Review. *Mol. Catal.* **2019**, *479*, 110607. [CrossRef]
4. DiCosimo, R.; McAuliffe, J.; Poulose, A.J.; Bohlmann, G. Industrial Use of Immobilized Enzymes. *Chem. Soc. Rev.* **2013**, *42*, 6437–6474. [CrossRef] [PubMed]
5. Truppo, M.D.; Hughes, G. Development of an Improved Immobilized CAL-B for the Enzymatic Resolution of a Key Intermediate to Odanacatib. *Org. Process Res. Dev.* **2011**, *15*, 1033–1035. [CrossRef]
6. Al-Maqdi, K.A.; Hisaindee, S.; Rauf, M.A.; Ashraf, S.S. Detoxification and Degradation of Sulfamethoxazole by Soybean Peroxidase and UV + H_2O_2 Remediation Approaches. *Chem. Eng. J.* **2018**, *352*, 450–458. [CrossRef]
7. Almaqdi, K.A.; Morsi, R.; Alhayuti, B.; Alharthi, F.; Ashraf, S.S. LC-MSMS Based Screening of Emerging Pollutant Degradation by Different Peroxidases. *BMC Biotechnol.* **2019**, *19*, 83. [CrossRef]
8. Jun, L.Y.; Yon, L.S.; Mubarak, N.M.; Bing, C.H.; Pan, S.; Danquah, M.K.; Abdullah, E.C.; Khalid, M. An Overview of Immobilized Enzyme Technologies for Dye and Phenolic Removal from Wastewater. *J. Environ. Chem. Eng.* **2019**, *7*, 102961. [CrossRef]
9. Liu, D.-M.; Chen, J.; Shi, Y.-P. Advances on Methods and Easy Separated Support Materials for Enzymes Immobilization. *TrAC Trends Anal. Chem.* **2018**, *102*, 332–342. [CrossRef]
10. Girelli, A.M.; Astolfi, M.L.; Scuto, F.R. Agro-Industrial Wastes as Potential Carriers for Enzyme Immobilization: A Review. *Chemosphere* **2020**, *244*, 125368. [CrossRef] [PubMed]
11. Hartmann, M.; Kostrov, X. Immobilization of Enzymes on Porous Silicas—Benefits and Challenges. *Chem. Soc. Rev.* **2013**, *42*, 6277–6289. [CrossRef] [PubMed]
12. Datta, S.; Christena, L.R.; Rajaram, Y.R.S. Enzyme Immobilization: An Overview on Techniques and Support Materials. *3 Biotech* **2013**, *3*, 1–9. [CrossRef] [PubMed]
13. Homaei, A.A.; Sariri, R.; Vianello, F.; Stevanato, R. Enzyme Immobilization: An Update. *J. Chem. Biol.* **2013**, *6*, 185–205. [CrossRef] [PubMed]
14. Zucca, P.; Sanjust, E. Inorganic Materials as Supports for Covalent Enzyme Immobilization: Methods and Mechanisms. *Molecules* **2014**, *19*, 14139–14194. [CrossRef] [PubMed]
15. Klein, M.P.; Nunes, M.R.; Rodrigues, R.C.; Benvenutti, E.V.; Costa, T.M.H.; Hertz, P.F.; Ninow, J.L. Effect of the Support Size on the Properties of β-Galactosidase Immobilized on Chitosan: Advantages and Disadvantages of Macro and Nanoparticles. *Biomacromolecules* **2012**, *13*, 2456–2464. [CrossRef] [PubMed]
16. Wong, J.K.H.; Tan, H.K.; Lau, S.Y.; Yap, P.-S.; Danquah, M.K. Potential and Challenges of Enzyme Incorporated Nanotechnology in Dye Wastewater Treatment: A Review. *J. Environ. Chem. Eng.* **2019**, *7*, 103261. [CrossRef]
17. Mukhopadhyay, A.; Dasgupta, A.K.; Chakrabarti, K. Enhanced Functionality and Stabilization of a Cold Active Laccase Using Nanotechnology Based Activation-Immobilization. *Bioresour. Technol.* **2015**, *179*, 573–584. [CrossRef]
18. Gupta, M.N.; Kaloti, M.; Kapoor, M.; Solanki, K. Nanomaterials as Matrices for Enzyme Immobilization. *Artif. Cells Blood Substit. Biotechnol.* **2011**, *39*, 98–109. [CrossRef]
19. Liu, W.; Wang, L.; Jiang, R. Specific Enzyme Immobilization Approaches and Their Application with Nanomaterials. *Top. Catal.* **2012**, *55*, 1146–1156. [CrossRef]
20. Ge, J.; Lei, J.; Zare, R.N. Protein–Inorganic Hybrid Nanoflowers. *Nat. Nanotechnol.* **2012**, *7*, 428–432. [CrossRef] [PubMed]
21. Ke, C.; Fan, Y.; Chen, Y.; Xu, L.; Yan, Y. A New Lipase–Inorganic Hybrid Nanoflower with Enhanced Enzyme Activity. *RSC Adv.* **2016**, *6*, 19413–19416. [CrossRef]
22. Yin, Y.; Xiao, Y.; Lin, G.; Xiao, Q.; Lin, Z.; Cai, Z. An Enzyme–Inorganic Hybrid Nanoflower Based Immobilized Enzyme Reactor with Enhanced Enzymatic Activity. *J. Mater. Chem. B* **2015**, *3*, 2295–2300. [CrossRef] [PubMed]

23. Lin, Z.; Xiao, Y.; Wang, L.; Yin, Y.; Zheng, J.; Yang, H.; Chen, G. Facile Synthesis of Enzyme–Inorganic Hybrid Nanoflowers and Their Application as an Immobilized Trypsin Reactor for Highly Efficient Protein Digestion. *RSC Adv.* **2014**, *4*, 13888–13891. [CrossRef]

24. Yu, Y.; Fei, X.; Tian, J.; Xu, L.; Wang, X.; Wang, Y. Self-Assembled Enzyme–Inorganic Hybrid Nanoflowers and Their Application to Enzyme Purification. *Colloids Surf. B Biointerfaces* **2015**, *130*, 299–304. [CrossRef]

25. Wu, X.; Hou, M.; Ge, J. Metal–Organic Frameworks and Inorganic Nanoflowers: A Type of Emerging Inorganic Crystal Nanocarrier for Enzyme Immobilization. *Catal. Sci. Technol.* **2015**, *5*, 5077–5085. [CrossRef]

26. Duan, L.; Li, H.; Zhang, Y. Synthesis of Hybrid Nanoflower-Based Carbonic Anhydrase for Enhanced Biocatalytic Activity and Stability. *ACS Omega* **2018**. [CrossRef]

27. Cui, J.; Zhao, Y.; Liu, R.; Zhong, C.; Jia, S. Surfactant-Activated Lipase Hybrid Nanoflowers with Enhanced Enzymatic Performance. *Sci. Rep.* **2016**, *6*, 27928. [CrossRef]

28. Li, K.; Wang, J.; He, Y.; Abdulrazaq, M.A.; Yan, Y. Carbon Nanotube-Lipase Hybrid Nanoflowers with Enhanced Enzyme Activity and Enantioselectivity. *J. Biotechnol.* **2018**, *281*, 87–98. [CrossRef]

29. Wang, S.; Ding, Y.; Chen, R.; Hu, M.; Li, S.; Zhai, Q.; Jiang, Y. Multilayer Petal-like Enzymatic-Inorganic Hybrid Micro-Spheres [CPO-(Cu/Co/Cd)3(PO4)2] with High Bio-Catalytic Activity. *Chem. Eng. Res. Des.* **2018**, *134*, 52–61. [CrossRef]

30. Patel, S.K.S.; Otari, S.V.; Li, J.; Kim, D.R.; Kim, S.C.; Cho, B.-K.; Kalia, V.C.; Kang, Y.C.; Lee, J.-K. Synthesis of Cross-Linked Protein-Metal Hybrid Nanoflowers and Its Application in Repeated Batch Decolorization of Synthetic Dyes. *J. Hazard. Mater.* **2018**, *347*, 442–450. [CrossRef] [PubMed]

31. Altinkaynak, C.; Tavlasoglu, S.; Ÿzdemir, N.; Ocsoy, I. A New Generation Approach in Enzyme Immobilization: Organic-Inorganic Hybrid Nanoflowers with Enhanced Catalytic Activity and Stability. *Enzyme Microb. Technol.* **2016**, *93–94*, 105–112. [CrossRef]

32. Rong, J.; Zhang, T.; Qiu, F.; Zhu, Y. Preparation of Efficient, Stable, and Reusable Laccase–Cu3(PO4)2 Hybrid Microspheres Based on Copper Foil for Decoloration of Congo Red. *ACS Sustain. Chem. Eng.* **2017**, *5*, 4468–4477. [CrossRef]

33. Cui, J.; Jia, S. Organic–Inorganic Hybrid Nanoflowers: A Novel Host Platform for Immobilizing Biomolecules. *Coord. Chem. Rev.* **2017**, *352*, 249–263. [CrossRef]

34. Fu, M.; Xing, J.; Ge, Z. Preparation of Laccase-Loaded Magnetic Nanoflowers and Their Recycling for Efficient Degradation of Bisphenol A. *Sci. Total Environ.* **2019**, *651*, 2857–2865. [CrossRef] [PubMed]

35. Zhang, B.; Li, P.; Zhang, H.; Wang, H.; Li, X.; Tian, L.; Ali, N.; Ali, Z.; Zhang, Q. Preparation of Lipase/Zn3(PO4)2 Hybrid Nanoflower and Its Catalytic Performance as an Immobilized Enzyme. *Chem. Eng. J.* **2016**, *291*, 287–297. [CrossRef]

36. Lin, Z.; Xiao, Y.; Yin, Y.; Hu, W.; Liu, W.; Yang, H. Facile Synthesis of Enzyme-Inorganic Hybrid Nanoflowers and Its Application as a Colorimetric Platform for Visual Detection of Hydrogen Peroxide and Phenol. *ACS Appl. Mater. Interfaces* **2014**, *6*, 10775–10782. [CrossRef]

37. Yu, J.; Wang, C.; Wang, A.; Li, N.; Chen, X.; Pei, X.; Zhang, P.; Gang Wu, S. Dual-Cycle Immobilization to Reuse Both Enzyme and Support by Reblossoming Enzyme–Inorganic Hybrid Nanoflowers. *RSC Adv.* **2018**, *8*, 16088–16094. [CrossRef]

38. Somturk, B.; Yilmaz, I.; Altinkaynak, C.; Karatepe, A.; Özdemir, N.; Ocsoy, I. Synthesis of Urease Hybrid Nanoflowers and Their Enhanced Catalytic Properties. *Enzyme Microb. Technol.* **2016**, *86*, 134–142. [CrossRef]

39. Nadar, S.S.; Gawas, S.D.; Rathod, V.K. Self-Assembled Organic-inorganic Hybrid Glucoamylase Nanoflowers with Enhanced Activity and Stability. *Int. J. Biol. Macromol.* **2016**, *92*, 660–669. [CrossRef] [PubMed]

40. Memon, A.H.; Ding, R.; Yuan, Q.; Wei, Y.; Liang, H. Facile Synthesis of Alcalase-Inorganic Hybrid Nanoflowers Used for Soy Protein Isolate Hydrolysis to Improve Its Functional Properties. *Food Chem.* **2019**, *289*, 568–574. [CrossRef] [PubMed]

41. Zhang, B.; Li, P.; Zhang, H.; Fan, L.; Wang, H.; Li, X.; Tian, L.; Ali, N.; Ali, Z.; Zhang, Q. Papain/Zn3(PO4)2 Hybrid Nanoflower: Preparation, Characterization and Its Enhanced Catalytic Activity as an Immobilized Enzyme. *RSC Adv.* **2016**, *6*, 46702–46710. [CrossRef]

42. Li, M.; Luo, M.; Li, F.; Wang, W.; Liu, K.; Liu, Q.; Wang, Y.; Lu, Z.; Wang, D. Biomimetic Copper-Based Inorganic–Protein Nanoflower Assembly Constructed on the Nanoscale Fibrous Membrane with Enhanced Stability and Durability. *J. Phys. Chem. C* **2016**, *120*, 17348–17356. [CrossRef]

43. Altinkaynak, C.; Tavlasoglu, S.; Kalin, R.; Sadeghian, N.; Ozdemir, H.; Ocsoy, I.; Özdemir, N. A Hierarchical Assembly of Flower-like Hybrid Turkish Black Radish Peroxidase-Cu2+ Nanobiocatalyst and Its Effective Use in Dye Decolorization. *Chemosphere* **2017**, *182*, 122–128. [CrossRef] [PubMed]

44. Yang, C.; Zhang, M.; Wang, W.; Wang, Y.; Tang, J. UV-Vis Detection of Hydrogen Peroxide Using Horseradish Peroxidase/Copper Phosphate Hybrid Nanoflowers. *Enzyme Microb. Technol.* **2020**, *140*, 109620. [CrossRef] [PubMed]

45. Wei, T.; Du, D.; Zhu, M.-J.; Lin, Y.; Dai, Z. An Improved Ultrasensitive Enzyme-Linked Immunosorbent Assay Using Hydrangea-Like Antibody–Enzyme–Inorganic Three-in-One Nanocomposites. *ACS Appl. Mater. Interfaces* **2016**, *8*, 6329–6335. [CrossRef] [PubMed]

46. Wang, C.; Tan, R.; Wang, Q. One-Step Synthesized Flower-like Materials Used for Sensitively Detecting Amyloid Precursor Protein. *Anal. Bioanal. Chem.* **2018**, *410*, 6901–6909. [CrossRef] [PubMed]

47. He, G.; Hu, W.; Li, C.M. Spontaneous Interfacial Reaction between Metallic Copper and PBS to Form Cupric Phosphate Nanoflower and Its Enzyme Hybrid with Enhanced Activity. *Colloids Surf. B Biointerfaces* **2015**, *135*, 613–618. [CrossRef] [PubMed]

48. Somturk, B.; Hancer, M.; Ocsoy, I.; Özdemir, N. Synthesis of Copper Ion Incorporated Horseradish Peroxidase-Based Hybrid Nanoflowers for Enhanced Catalytic Activity and Stability. *Dalton Trans.* **2015**, *44*, 13845–13852. [CrossRef]

49. Liu, Y.; Chen, J.; Du, M.; Wang, X.; Ji, X.; He, Z. The Preparation of Dual-Functional Hybrid Nanoflower and Its Application in the Ultrasensitive Detection of Disease-Related Biomarker. *Biosens. Bioelectron.* **2017**, *92*, 68–73. [CrossRef] [PubMed]

50. Lang, X.; Zhu, L.; Gao, Y.; Wheeldon, I. Enhancing Enzyme Activity and Immobilization in Nanostructured Inorganic–Enzyme Complexes. *Langmuir* **2017**, *33*, 9073–9080. [CrossRef]

51. Altinkaynak, C.; Yilmaz, I.; Koksal, Z.; Özdemir, H.; Ocsoy, I.; Özdemir, N. Preparation of Lactoperoxidase Incorporated Hybrid Nanoflower and Its Excellent Activity and Stability. *Int. J. Biol. Macromol.* **2016**, *84*, 402–409. [CrossRef]

52. Chung, M.; Nguyen, T.L.; Tran, T.Q.N.; Yoon, H.H.; Kim, I.T.; Kim, M.I. Ultrarapid Sonochemical Synthesis of Enzyme-Incorporated Copper Nanoflowers and Their Application to Mediatorless Glucose Biofuel Cell. *Appl. Surf. Sci.* **2018**, *429*, 203–209. [CrossRef]

53. Zhang, M.; Yang, N.; Liu, Y.; Tang, J. Synthesis of Catalase-Inorganic Hybrid Nanoflowers via Sonication for Colorimetric Detection of Hydrogen Peroxide. *Enzyme Microb. Technol.* **2019**, *128*, 22–25. [CrossRef]

54. Batule, B.S.; Park, K.S.; Kim, M.I.; Park, H.G. Ultrafast Sonochemical Synthesis of Protein-Inorganic Nanoflowers. *Int. J. Nanomed.* **2015**, *10*, 137–142. [CrossRef]

55. Zhu, L.; Gong, L.; Zhang, Y.; Wang, R.; Ge, J.; Liu, Z.; Zare, R.N. Rapid Detection of Phenol Using a Membrane Containing Laccase Nanoflowers. *Chem. Asian J.* **2013**, *8*, 2358–2360. [CrossRef] [PubMed]

56. Zhu, P.; Wang, Y.; Li, G.; Liu, K.; Liu, Y.; He, J.; Lei, J. Preparation and Application of a Chemically Modified Laccase and Copper Phosphate Hybrid Flower-like Biocatalyst. *Biochem. Eng. J.* **2019**, *144*, 235–243. [CrossRef]

57. Wu, Z.; Li, H.; Zhu, X.; Li, S.; Wang, Z.; Wang, L.; Li, Z.; Chen, G. Using Laccases in the Nanoflower to Synthesize Viniferin. *Catalysts* **2017**, *7*, 188. [CrossRef]

58. Batule, B.S.; Park, K.S.; Gautam, S.; Cheon, H.J.; Kim, M.I.; Park, H.G. Intrinsic Peroxidase-like Activity of Sonochemically Synthesized Protein Copper Nanoflowers and Its Application for the Sensitive Detection of Glucose. *Sens. Actuators B Chem.* **2019**, *283*, 749–754. [CrossRef]

59. Huang, Y.; Ran, X.; Lin, Y.; Ren, J.; Qu, X. Self-Assembly of an Organic-Inorganic Hybrid Nanoflower as an Efficient Biomimetic Catalyst for Self-Activated Tandem Reactions. *Chem. Commun.* **2015**, *51*, 4386–4389. [CrossRef]

60. Kumar, A.; Patel, S.K.S.; Mardan, B.; Pagolu, R.; Lestari, R.; Jeong, S.-H.; Kim, T.; Haw, J.R.; Kim, S.-Y.; Kim, I.-W.; et al. Immobilization of Xylanase Using a Protein-Inorganic Hybrid System. *J. Microbiol. Biotechnol.* **2018**, *28*, 638–644. [CrossRef]

61. Qian, K.; Wang, H.; Liu, J.; Gao, S.; Liu, W.; Wan, X.; Zhang, Y.; Liu, Q.-S.; Yin, X.-Y. Synthesis of α-Glycosidase Hybrid Nano-Flowers and Their Application for Enriching and Screening α-Glycosidase Inhibitors. *New J. Chem.* **2018**, *42*, 429–436. [CrossRef]

62. Xu, Z.; Wang, R.; Liu, C.; Chi, B.; Gao, J.; Chen, B.; Xu, H. A New l-Arabinose Isomerase with Copper Ion Tolerance Is Suitable for Creating Protein–Inorganic Hybrid Nanoflowers with Enhanced Enzyme Activity and Stability. *RSC Adv.* **2016**, *6*, 30791–30794. [CrossRef]

63. Lee, H.R.; Chung, M.; Kim, M.I.; Ha, S.H. Preparation of Glutaraldehyde-Treated Lipase-Inorganic Hybrid Nanoflowers and Their Catalytic Performance as Immobilized Enzymes. *Enzyme Microb. Technol.* **2017**, *105*, 24–29. [CrossRef]

64. Hua, X.; Xing, Y.; Zhang, X. Enhanced Promiscuity of Lipase-Inorganic Nanocrystal Composites in the Epoxidation of Fatty Acids in Organic Media. *ACS Appl. Mater. Interfaces* **2016**, *8*, 16257–16261. [CrossRef]

65. Ren, W.; Fei, X.; Tian, J.; Li, Y.; Jing, M.; Fang, H.; Xu, L.; Wang, Y. Multiscale Immobilized Lipase for Rapid Separation and Continuous Catalysis. *New J. Chem.* **2018**, *42*, 13471–13478. [CrossRef]

66. Wu, Z.; Li, X.; Li, F.; Yue, H.; He, C.; Xie, F.; Wang, Z. Enantioselective Transesterification of (R,S)-2-Pentanol Catalyzed by a New Flower-like Nanobioreactor. *RSC Adv.* **2014**, *4*, 33998–34002. [CrossRef]

67. Escobar, S.; Velasco-Lozano, S.; Lu, C.-H.; Lin, Y.-F.; Mesa, M.; Bernal, C.; López-Gallego, F. Understanding the Functional Properties of Bio-Inorganic Nanoflowers as Biocatalysts by Deciphering the Metal-Binding Sites of Enzymes. *J. Mater. Chem. B* **2017**, *5*, 4478–4486. [CrossRef]

68. Ren, W.; Li, Y.; Wang, J.; Li, L.; Xu, L.; Wu, Y.; Wang, Y.; Fei, X.; Tian, J. Synthesis of Magnetic Nanoflower Immobilized Lipase and Its Continuous Catalytic Application. *New J. Chem.* **2019**, *43*, 11082–11090. [CrossRef]

69. Li, Y.; Fei, X.; Liang, L.; Tian, J.; Xu, L.; Wang, X.; Wang, Y. The Influence of Synthesis Conditions on Enzymatic Activity of Enzyme-Inorganic Hybrid Nanoflowers. *J. Mol. Catal. B Enzym.* **2016**, *133*, 92–97. [CrossRef]

70. Papadopoulou, A.A.; Tzani, A.; Polydera, A.C.; Katapodis, P.; Voutsas, E.; Detsi, A.; Stamatis, H. Green Biotransformations Catalysed by Enzyme-Inorganic Hybrid Nanoflowers in Environmentally Friendly Ionic Solvents. *Environ. Sci. Pollut. Res.* **2018**, *25*, 26707–26714. [CrossRef] [PubMed]

71. Jiang, W.; Wang, X.; Yang, J.; Han, H.; Li, Q.; Tang, J. Lipase-Inorganic Hybrid Nanoflower Constructed through Biomimetic Mineralization: A New Support for Biodiesel Synthesis. *J. Colloid Interface Sci.* **2018**, *514*, 102–107. [CrossRef]

72. Gulmez, C.; Altinkaynak, C.; Özdemir, N.; Atakisi, O. Proteinase K Hybrid Nanoflowers (P-HNFs) as a Novel Nanobiocatalytic Detergent Additive. *Int. J. Biol. Macromol.* **2018**, *119*, 803–810. [CrossRef]

73. Zhang, H.; Fei, X.; Tian, J.; Li, Y.; Zhi, H.; Wang, K.; Xu, L.; Wang, Y. Synthesis and Continuous Catalytic Application of Alkaline Protease Nanoflowers–PVA Composite Hydrogel. *Catal. Commun.* **2018**, *116*, 5–9. [CrossRef]

74. Yu, J.; Chen, X.; Jiang, M.; Wang, A.; Yang, L.; Pei, X.; Zhang, P.; Wu, S.G. Efficient Promiscuous Knoevenagel Condensation Catalyzed by Papain Confined in Cu3(PO4)2 Nanoflowers. *RSC Adv.* **2018**, *8*, 2357–2364. [CrossRef]

75. Liang, L.; Fei, X.; Li, Y.; Tian, J.; Xu, L.; Wang, X.; Wang, Y. Hierarchical Assembly of Enzyme-Inorganic Composite Materials with Extremely High Enzyme Activity. *RSC Adv.* **2015**, *5*, 96997–97002. [CrossRef]
76. Chung, M.; Jang, Y.J.; Kim, M.I. Convenient Colorimetric Detection of Cholesterol Using Multi-Enzyme Co-Incorporated Organic–Inorganic Hybrid Nanoflowers. *J. Nanosci. Nanotechnol.* **2018**, *18*, 6555–6561. [CrossRef] [PubMed]
77. Zhang, L.; Ma, Y.; Wang, C.; Wang, Z.; Chen, X.; Li, M.; Zhao, R.; Wang, L. Application of Dual-Enzyme Nanoflower in the Epoxidation of Alkenes. *Process Biochem.* **2018**, *74*, 103–107. [CrossRef]
78. Jin, R.; Kong, D.; Zhao, X.; Li, H.; Yan, X.; Liu, F.; Sun, P.; Du, D.; Lin, Y.; Lu, G. Tandem Catalysis Driven by Enzymes Directed Hybrid Nanoflowers for On-Site Ultrasensitive Detection of Organophosphorus Pesticide. *Biosens. Bioelectron.* **2019**, *141*, 111473. [CrossRef] [PubMed]
79. Li, Y.; Xie, G.; Qiu, J.; Zhou, D.; Gou, D.; Tao, Y.; Li, Y.; Chen, H. A New Biosensor Based on the Recognition of Phages and the Signal Amplification of Organic-Inorganic Hybrid Nanoflowers for Discriminating and Quantitating Live Pathogenic Bacteria in Urine. *Sens. Actuators B Chem.* **2018**, *258*, 803–812. [CrossRef]
80. Sun, J.; Ge, J.; Liu, W.; Lan, M.; Zhang, H.; Wang, P.; Wang, Y.; Niu, Z. Multi-Enzyme Co-Embedded Organic–Inorganic Hybrid Nanoflowers: Synthesis and Application as a Colorimetric Sensor. *Nanoscale* **2014**, *6*, 255–262. [CrossRef]
81. Zhu, X.; Huang, J.; Liu, J.; Zhang, H.; Jiang, J.; Yu, R. A Dual Enzyme–Inorganic Hybrid Nanoflower Incorporated Microfluidic Paper-Based Analytic Device (MPAD) Biosensor for Sensitive Visualized Detection of Glucose. *Nanoscale* **2017**, *9*, 5658–5663. [CrossRef]
82. Ariza-Avidad, M.; Salinas-Castillo, A.; Capitán-Vallvey, L.F. A 3D MPAD Based on a Multi-Enzyme Organic–Inorganic Hybrid Nanoflower Reactor. *Biosens. Bioelectron.* **2016**, *77*, 51–55. [CrossRef]
83. He, X.; Chen, L.; He, Q.; Xiao, H.; Zhou, X.; Ji, H. Cytochrome P450 Enzyme-Copper Phosphate Hybrid Nano-Flowers with Superior Catalytic Performances for Selective Oxidation of Sulfides. *Chin. J. Chem.* **2017**, *35*, 693–698. [CrossRef]
84. Patel, S.S.K.; Otari, S.V.; Kang, Y.C.; Lee, J.-K. Protein–Inorganic Hybrid System for Efficient His-Tagged Enzymes Immobilization and Its Application in l-Xylulose Production. *RSC Adv.* **2017**, *7*, 3488–3494. [CrossRef]
85. Hao, M.; Fan, G.; Zhang, Y.; Xin, Y.; Zhang, L. Preparation and Characterization of Copper-Brevibacterium Cholesterol Oxidase Hybrid Nanoflowers. *Int. J. Biol. Macromol.* **2019**, *126*, 539–548. [CrossRef]
86. Fang, X.; Zhang, C.; Qian, X.; Yu, D. Self-Assembled 2,4-Dichlorophenol Hydroxylase-Inorganic Hybrid Nanoflowers with Enhanced Activity and Stability. *RSC Adv.* **2018**, *8*, 20976–20981. [CrossRef]
87. Liu, Y.; Zhang, Y.; Li, X.; Yuan, Q.; Liang, H. Self-Repairing Metal–Organic Hybrid Complexes for Reinforcing Immobilized Chloroperoxidase Reusability. *Chem. Commun.* **2017**, *53*, 3216–3219. [CrossRef] [PubMed]
88. Wang, X.; Shi, J.; Li, Z.; Zhang, S.; Wu, H.; Jiang, Z.; Yang, C.; Tian, C. Facile One-Pot Preparation of Chitosan/Calcium Pyrophosphate Hybrid Microflowers. *ACS Appl. Mater. Interfaces* **2014**, *6*, 14522–14532. [CrossRef] [PubMed]
89. Wang, L.-B.; Wang, Y.-C.; He, R.; Zhuang, A.; Wang, X.; Zeng, J.; Hou, J.G. A New Nanobiocatalytic System Based on Allosteric Effect with Dramatically Enhanced Enzymatic Performance. *J. Am. Chem. Soc.* **2013**, *135*, 1272–1275. [CrossRef]
90. Liu, Y.; Wang, B.; Ji, X.; He, Z. Self-Assembled Protein-Enzyme Nanoflower-Based Fluorescent Sensing for Protein Biomarker. *Anal. Bioanal. Chem.* **2018**, *410*, 7591–7598. [CrossRef] [PubMed]
91. Chen, X.; Xu, L.; Wang, A.; Li, H.; Wang, C.; Pei, X.; Zhang, P.; Wu, S.G. Efficient Synthesis of the Key Chiral Alcohol Intermediate of Crizotinib Using Dual-Enzyme@CaHPO4 Hybrid Nanoflowers Assembled by Mimetic Biomineralization. *J. Chem. Technol. Biotechnol.* **2019**, *94*, 236–243. [CrossRef]
92. Zhao, F.; Wang, Q.; Dong, J.; Xian, M.; Yu, J.; Yin, H.; Chang, Z.; Mu, X.; Hou, T.; Wang, J. Enzyme-Inorganic Nanoflowers/Alginate Microbeads: An Enzyme Immobilization System and Its Potential Application. *Process Biochem.* **2017**, *57*, 87–94. [CrossRef]
93. Ghosh, K.; Balog, E.R.M.; Sista, P.; Williams, D.J.; Kelly, D.; Martinez, J.S.; Rocha, R.C. Temperature-Dependent Morphology of Hybrid Nanoflowers from Elastin-like Polypeptides. *APL Mater.* **2014**, *2*, 021101. [CrossRef]
94. Ye, R.; Zhu, C.; Song, Y.; Song, J.; Fu, S.; Lu, Q.; Yang, X.; Zhu, M.-J.; Du, D.; Li, H.; et al. One-Pot Bioinspired Synthesis of All-Inclusive Protein–Protein Nanoflowers for Point-of-Care Bioassay: Detection of E. Coli O157:H7 from Milk. *Nanoscale* **2016**, *8*, 18980–18986. [CrossRef]
95. Rai, S.K.; Narnoliya, L.K.; Sangwan, R.S.; Yadav, S.K. Self-Assembled Hybrid Nanoflowers of Manganese Phosphate and l-Arabinose Isomerase: A Stable and Recyclable Nanobiocatalyst for Equilibrium Level Conversion of d-Galactose to d-Tagatose. *ACS Sustain. Chem. Eng.* **2018**, *6*, 6296–6304. [CrossRef]
96. Claude Munyemana, J.; He, H.; Ding, S.; Yin, J.; Xi, P.; Xiao, J. Synthesis of Manganese Phosphate Hybrid Nanoflowers by Collagen-Templated Biomineralization. *RSC Adv.* **2018**, *8*, 2708–2713. [CrossRef]
97. Zhang, Z.; Zhang, Y.; He, L.; Yang, Y.; Liu, S.; Wang, M.; Fang, S.; Fu, G. A Feasible Synthesis of Mn3(PO4)2@BSA Nanoflowers and Its Application as the Support Nanomaterial for Pt Catalyst. *J. Power Sources* **2015**, *284*, 170–177. [CrossRef]
98. Zhang, Z.; Zhang, Y.; Song, R.; Wang, M.; Yan, F.; He, L.; Feng, X.; Fang, S.; Zhao, J.; Zhang, H. Manganese(II) Phosphate Nanoflowers as Electrochemical Biosensors for the High-Sensitivity Detection of Ractopamine. *Sens. Actuators B Chem.* **2015**, *211*, 310–317. [CrossRef]
99. Zheng, L.; Xie, X.; Wang, Z.; Zhang, Y.; Wang, L.; Cui, X.; Huang, H.; Zhuang, H. Fabrication of a Nano-Biocatalyst for Regioselective Acylation of Arbutin. *Green Chem. Lett. Rev.* **2018**, *11*, 55–61. [CrossRef]
100. Zhang, B.; Li, P.; Zhang, H.; Li, X.; Tian, L.; Wang, H.; Chen, X.; Ali, N.; Ali, Z.; Zhang, Q. Red-Blood-Cell-like BSA/Zn3(PO4)2 Hybrid Particles: Preparation and Application to Adsorption of Heavy Metal Ions. *Appl. Surf. Sci.* **2016**, *366*, 328–338. [CrossRef]

101. Zheng, L.; Sun, Y.; Wang, J.; Huang, H.; Geng, X.; Tong, Y.; Wang, Z. Preparation of a Flower-Like Immobilized D-Psicose 3-Epimerase with Enhanced Catalytic Performance. *Catalysts* **2018**, *8*, 468. [CrossRef]

102. Kumar, A.; Kim, I.-W.; Patel, S.K.S.; Lee, J.-K. Synthesis of Protein-Inorganic Nanohybrids with Improved Catalytic Properties Using Co3(PO4)2. *Indian J. Microbiol.* **2018**, *58*, 100–104. [CrossRef]

103. Cao, G.; Gao, J.; Zhou, L.; He, Y.; Li, J.; Jiang, Y. Enrichment and Coimmobilization of Cofactors and His-Tagged ω-Transaminase into Nanoflowers: A Facile Approach to Constructing Self-Sufficient Biocatalysts. *ACS Appl. Nano Mater.* **2018**, *1*, 3417–3425. [CrossRef]

104. Kim, K.H.; Jeong, J.-M.; Lee, S.J.; Choi, B.G.; Lee, K.G. Protein-Directed Assembly of Cobalt Phosphate Hybrid Nanoflowers. *J. Colloid Interface Sci.* **2016**, *484*, 44–50. [CrossRef]

105. He, L.; Zhang, S.; Ji, H.; Wang, M.; Peng, D.; Yan, F.; Fang, S.; Zhang, H.; Jia, C.; Zhang, Z. Protein-Templated Cobaltous Phosphate Nanocomposites for the Highly Sensitive and Selective Detection of Platelet-Derived Growth Factor-BB. *Biosens. Bioelectron.* **2016**, *79*, 553–560. [CrossRef] [PubMed]

106. López-Gallego, F.; Yate, L. Selective Biomineralization of Co3(PO4)2-Sponges Triggered by His-Tagged Proteins: Efficient Heterogeneous Biocatalysts for Redox Processes. *Chem. Commun.* **2015**, *51*, 8753–8756. [CrossRef] [PubMed]

107. Ocsoy, I.; Dogru, E.; Usta, S. A New Generation of Flowerlike Horseradish Peroxides as a Nanobiocatalyst for Superior Enzymatic Activity. *Enzyme Microb. Technol.* **2015**, *75–76*, 25–29. [CrossRef] [PubMed]

108. Guo, J.; Wang, Y.; Zhao, M. A Self-Activated Nanobiocatalytic Cascade System Based on an Enzyme-Inorganic Hybrid Nanoflower for Colorimetric and Visual Detection of Glucose in Human Serum. *Sens. Actuators B Chem.* **2019**, *284*, 45–54. [CrossRef]

109. Patel, S.K.S.; Choi, H.; Lee, J.-K. Multimetal-Based Inorganic–Protein Hybrid System for Enzyme Immobilization. *ACS Sustain. Chem. Eng.* **2019**, *7*, 13633–13638. [CrossRef]

110. Nonsuwan, P.; Puthong, S.; Palaga, T.; Muangsin, N. Novel Organic/Inorganic Hybrid Flower-like Structure of Selenium Nanoparticles Stabilized by Pullulan Derivatives. *Carbohydr. Polym.* **2018**, *184*, 9–19. [CrossRef]

111. Lee, I.; Cheon, H.J.; Adhikari, M.D.; Tran, T.D.; Yeon, K.-M.; Kim, M.I.; Kim, J. Glucose Oxidase-Copper Hybrid Nanoflowers Embedded with Magnetic Nanoparticles as an Effective Antibacterial Agent. *Int. J. Biol. Macromol.* **2020**, *155*, 1520–1531. [CrossRef]

112. Sun, X.; Niu, H.; Song, J.; Jiang, D.; Leng, J.; Zhuang, W.; Chen, Y.; Liu, D.; Ying, H. Preparation of a Copper Polyphosphate Kinase Hybrid Nanoflower and Its Application in ADP Regeneration from AMP. *ACS Omega* **2020**, *5*, 9991–9998. [CrossRef]

113. Ben Brahim, F.; Boughzala, H. Crystal Structure, Vibrational Spectra and Thermal Analysis of a New Centrosymmetric Transition Metal Phosphate Compound, Mn(H2PO4)2·4H2O. *J. Mol. Struct.* **2013**, *1034*, 336–345. [CrossRef]

114. Ferdov, S.; Lopes, A.M.L.; Lin, Z.; Ferreira, R.A.S. New Template-Free Layered Manganese(III) Phosphate: Hydrothermal Synthesis, Ab Initio Structural Determination, and Magnetic Properties. *Chem. Mater.* **2007**, *19*, 6025–6029. [CrossRef]

115. Zhang, Y.; Ge, J.; Liu, Z. Enhanced Activity of Immobilized or Chemically Modified Enzymes. *ACS Catal.* **2015**, *5*, 4503–4513. [CrossRef]

116. Yu, B.; Zhou, Y.; Song, M.; Xue, Y.; Cai, N.; Luo, X.; Long, S.; Zhang, H.; Yu, F. Synthesis of Selenium Nanoparticles with Mesoporous Silica Drug-Carrier Shell for Programmed Responsive Tumor Targeted Synergistic Therapy. *RSC Adv.* **2015**, *6*, 2171–2175. [CrossRef]

117. Wu, H.; Li, X.; Liu, W.; Chen, T.; Li, Y.; Zheng, W.; Man, C.W.-Y.; Wong, M.-K.; Wong, K.-H. Surface Decoration of Selenium Nanoparticles by Mushroom Polysaccharides–Protein Complexes to Achieve Enhanced Cellular Uptake and Antiproliferative Activity. *J. Mater. Chem.* **2012**, *22*, 9602–9610. [CrossRef]

118. Yang, F.; Tang, Q.; Zhong, X.; Bai, Y.; Chen, T.; Zhang, Y.; Li, Y.; Zheng, W. Surface Decoration by Spirulina Polysaccharide Enhances the Cellular Uptake and Anticancer Efficacy of Selenium Nanoparticles. *Int. J. Nanomed.* **2012**, *7*, 835–844. [CrossRef]

119. Kumar, A.; Sevonkaev, I.; Goia, D.V. Synthesis of Selenium Particles with Various Morphologies. *J. Colloid Interface Sci.* **2014**, *416*, 119–123. [CrossRef]

120. Luesakul, U.; Komenek, S.; Puthong, S.; Muangsin, N. Shape-Controlled Synthesis of Cubic-like Selenium Nanoparticles via the Self-Assembly Method. *Carbohydr. Polym.* **2016**, *153*, 435–444. [CrossRef]

121. Yin, H.; Xu, Z.; Bao, H.; Bai, J.; Zheng, Y. Single Crystal Trigonal Selenium Nanoplates Converted from Selenium Nanoparticles. *Chem. Lett.* **2004**, *34*, 122–123. [CrossRef]

122. Song, C.; Li, X.; Wang, S.; Meng, Q. Enhanced Conversion and Stability of Biosynthetic Selenium Nanoparticles Using Fetal Bovine Serum. *RSC Adv.* **2016**, *6*, 103948–103954. [CrossRef]

123. Xia, Y.; You, P.; Xu, F.; Liu, J.; Xing, F. Novel Functionalized Selenium Nanoparticles for Enhanced Anti-Hepatocarcinoma Activity In Vitro. *Nanoscale Res. Lett.* **2015**, *10*, 349. [CrossRef]

124. Cao, H.; Yang, D.-P.; Ye, D.; Zhang, X.; Fang, X.; Zhang, S.; Liu, B.; Kong, J. Protein-Inorganic Hybrid Nanoflowers as Ultrasensitive Electrochemical Cytosensing Interfaces for Evaluation of Cell Surface Sialic Acid. *Biosens. Bioelectron.* **2015**, *68*, 329–335. [CrossRef]

125. Li, Y.; Wu, H.; Su, Z. Enzyme-Based Hybrid Nanoflowers with High Performances for Biocatalytic, Biomedical, and Environmental Applications. *Coord. Chem. Rev.* **2020**, *416*, 213342. [CrossRef]

126. Lee, S.W.; Cheon, S.A.; Kim, M.I.; Park, T.J. Organic–Inorganic Hybrid Nanoflowers: Types, Characteristics, and Future Prospects. *J. Nanobiotechnol.* **2015**, *13*, 54. [CrossRef]

127. Bilal, M.; Asgher, M.; Shah, S.Z.H.; Iqbal, H.M.N. Engineering Enzyme-Coupled Hybrid Nanoflowers: The Quest for Optimum Performance to Meet Biocatalytic Challenges and Opportunities. *Int. J. Biol. Macromol.* **2019**, *135*, 677–690. [CrossRef] [PubMed]

128. Zhang, W.; Yu, X.; Li, Y.; Su, Z.; Jandt, K.D.; Wei, G. Protein-Mimetic Peptide Nanofibers: Motif Design, Self-Assembly Synthesis, and Sequence-Specific Biomedical Applications. *Prog. Polym. Sci.* **2018**, *80*, 94–124. [CrossRef]

129. Gong, C.; Sun, S.; Zhang, Y.; Sun, L.; Su, Z.; Wu, A.; Wei, G. Hierarchical Nanomaterials: Via Biomolecular Self-Assembly and Bioinspiration for Energy and Environmental Applications. *Nanoscale* **2019**, *11*, 4147–4182. [CrossRef]

130. Celik, C.; Tasdemir, D.; Demirbas, A.; Katı, A.; Tolga Gul, O.; Cimen, B.; Ocsoy, I. Formation of Functional Nanobiocatalysts with a Novel and Encouraging Immobilization Approach and Their Versatile Bioanalytical Applications. *RSC Adv.* **2018**, *8*, 25298–25303. [CrossRef]

131. Yilmaz, E.; Ocsoy, I.; Ozdemir, N.; Soylak, M. Bovine Serum Albumin-Cu(II) Hybrid Nanoflowers: An Effective Adsorbent for Solid Phase Extraction and Slurry Sampling Flame Atomic Absorption Spectrometric Analysis of Cadmium and Lead in Water, Hair, Food and Cigarette Samples. *Anal. Chim. Acta* **2016**, *906*, 110–117. [CrossRef]

132. Songa, E.A.; Okonkwo, J.O. Recent Approaches to Improving Selectivity and Sensitivity of Enzyme-Based Biosensors for Organophosphorus Pesticides: A Review. *Talanta* **2016**, *155*, 289–304. [CrossRef]

133. Mehrotra, P. Biosensors and Their Applications–A Review. *J. Oral Biol. Craniofacial Res.* **2016**, *6*, 153–159. [CrossRef]

134. Zhao, W.-W.; Xu, J.-J.; Chen, H.-Y. Photoelectrochemical Enzymatic Biosensors. *Biosens. Bioelectron.* **2017**, *92*, 294–304. [CrossRef] [PubMed]

135. Zhang, M.; Zhao, X.; Zhang, G.; Wei, G.; Su, Z. Electrospinning Design of Functional Nanostructures for Biosensor Applications. *J. Mater. Chem. B* **2017**, *5*, 1699–1711. [CrossRef] [PubMed]

136. Thatoi, H.; Das, S.; Mishra, J.; Rath, B.P.; Das, N. Bacterial Chromate Reductase, a Potential Enzyme for Bioremediation of Hexavalent Chromium: A Review. *J. Environ. Manag.* **2014**, *146*, 383–399. [CrossRef] [PubMed]

137. Corsaro, D.; Pages, G.S.; Catalan, V.; Loret, J.-F.; Greub, G. Biodiversity of Amoebae and Amoeba-Associated Bacteria in Water Treatment Plants. *Int. J. Hyg. Environ. Health* **2010**, *213*, 158–166. [CrossRef] [PubMed]

138. Al-Maqdi, K.A.; Hisaindee, S.M.; Rauf, M.A.; Ashraf, S.S. Comparative Degradation of a Thiazole Pollutant by an Advanced Oxidation Process and an Enzymatic Approach. *Biomolecules* **2017**, *7*, 64. [CrossRef] [PubMed]

139. Wang, M.; Mohanty, S.K.; Mahendra, S. Nanomaterial-Supported Enzymes for Water Purification and Monitoring in Point-of-Use Water Supply Systems. *Acc. Chem. Res.* **2019**, *52*, 876–885. [CrossRef]

140. Bilal, M.; Asgher, M.; Iqbal, H.M.N.; Hu, H.; Zhang, X. Bio-Based Degradation of Emerging Endocrine-Disrupting and Dye-Based Pollutants Using Cross-Linked Enzyme Aggregates. *Environ. Sci. Pollut. Res.* **2017**, *24*, 7035–7041. [CrossRef]

141. Sun, H.; Yang, H.; Huang, W.; Zhang, S. Immobilization of Laccase in a Sponge-like Hydrogel for Enhanced Durability in Enzymatic Degradation of Dye Pollutants. *J. Colloid Interface Sci.* **2015**, *450*, 353–360. [CrossRef]

142. Ildiz, N.; Baldemir, A.; Altinkaynak, C.; Özdemir, N.; Yilmaz, V.; Ocsoy, I. Self Assembled Snowball-like Hybrid Nanostructures Comprising Viburnum Opulus L. Extract and Metal Ions for Antimicrobial and Catalytic Applications. *Enzyme Microb. Technol.* **2017**, *102*, 60–66. [CrossRef]

143. Olofsson, J.; Barta, Z.; Börjesson, P.; Wallberg, O. Integrating enzyme fermentation in lignocellulosic ethanol production: Life-cycle assessment and techno-economic analysis. *Biotechnol. Biofuels* **2017**, *10*, 1–14. [CrossRef]

144. Bilal, M.; Iqbal, H.M. Sustainable bioconversion of food waste into high-value products by immobilized enzymes to meet bio-economy challenges and opportunities—A review. *Food Res. Int.* **2019**, *123*, 226–240. [CrossRef] [PubMed]

145. Raman, J.K.; Ting, V.F.W.; Pogaku, R. Life cycle assessment of biodiesel production using alkali, soluble and immobilized enzyme catalyst processes. *Biomass Bioenerg.* **2011**, *35*, 4221–4229. [CrossRef]

146. Chapman, J.; Ismail, A.E.; Dinu, C.Z. Industrial applications of enzymes: Recent advances, techniques, and outlooks. *Catalysts* **2018**, *8*, 238. [CrossRef]

147. Manjrekar, S.; Wadekar, T.; Sumant, O. Enzymes Market Type (Protease, Carbohydrase, Lipase, Polymerase and Nuclease, and Other Types), Source (Microorganisms, Plants, and Animals), Reaction Type (Hydrolase, Oxidoreductase, Transferase, Lyase, and Other Reaction Types), and Application (Food and Beverages, Household Care, Bioenergy, Pharmaceutical and Biotechnology, Feed, and Other Applications)—Global Opportunity Analysis and Industry Forecast, 2020–2027. 2021. Available online: https://www.alliedmarketresearch.com/enzymes-market (accessed on 26 May 2021).

148. Li, C.; Zhao, J.; Zhang, Z.; Jiang, Y.; Bilal, M.; Jiang, Y.; Jia, S.; Cui, J. Self-Assembly of Activated Lipase Hybrid Nanoflowers with Superior Activity and Enhanced Stability. *Biochem. Eng. J.* **2020**, *158*, 107582. [CrossRef]

149. Xie, W.; Huang, M. Enzymatic Production of Biodiesel Using Immobilized Lipase on Core-Shell Structured Fe3O4@MIL-100(Fe) Composites. *Catalysts* **2019**, *9*, 850. [CrossRef]

150. Hussain, F.; Arana-Peña, S.; Morellon-Sterling, R.; Barbosa, O.; Ait Braham, S.; Kamal, S.; Fernandez-Lafuente, R. Further Stabilization of Alcalase Immobilized on Glyoxyl Supports: Amination Plus Modification with Glutaraldehyde. *Molecules* **2018**, *23*, 3188. [CrossRef] [PubMed]

151. Chang, Q.; Tang, H. Immobilization of Horseradish Peroxidase on NH2-Modified Magnetic Fe3O4/SiO2 Particles and Its Application in Removal of 2,4-Dichlorophenol. *Molecules* **2014**, *19*, 15768–15782. [CrossRef] [PubMed]

nanomaterials

MDPI

Review

Transition Metal Phosphides (TMP) as a Versatile Class of Catalysts for the Hydrodeoxygenation Reaction (HDO) of Oil-Derived Compounds

Latifa Ibrahim Al-Ali [1,2,†], Omer Elmutasim [1,2,†], Khalid Al Ali [2,3], Nirpendra Singh [2,4,*] and Kyriaki Polychronopoulou [1,2,*]

[1] Mechanical Engineering Department, Khalifa University, Abu Dhabi 127788, United Arab Emirates; 920020079@ku.ac.ae (L.I.A.-A.); omer.elfaki@ku.ac.ae (O.E.)
[2] Center for Catalysis and Separations, Khalifa University, Abu Dhabi 127788, United Arab Emirates; khalid.alali@ku.ac.ae
[3] Chemical Engineering Department, Khalifa University, Abu Dhabi 127788, United Arab Emirates
[4] Physics Department, Khalifa University, Abu Dhabi 127788, United Arab Emirates
* Correspondence: nirpendra.singh@ku.ac.ae (N.S.); kyriaki.polychrono@ku.ac.ae (K.P.)
† These authors contributed equally to this work.

Abstract: Hydrodeoxygenation (HDO) reaction is a route with much to offer in the conversion and upgrading of bio-oils into fuels; the latter can potentially replace fossil fuels. The catalyst's design and the feedstock play a critical role in the process metrics (activity, selectivity). Among the different classes of catalysts for the HDO reaction, the transition metal phosphides (TMP), e.g., binary (Ni2P, CoP, WP, MoP) and ternary Fe-Co-P, Fe-Ru-P, are chosen to be discussed in the present review article due to their chameleon type of structural and electronic features giving them superiority compared to the pure metals, apart from their cost advantage. Their active catalytic sites for the HDO reaction are discussed, while particular aspects of their structural, morphological, electronic, and bonding features are presented along with the corresponding characterization technique/tool. The HDO reaction is critically discussed for representative compounds on the TMP surfaces; model compounds from the lignin-derivatives, cellulose derivatives, and fatty acids, such as phenols and furans, are presented, and their reaction mechanisms are explained in terms of TMPs structure, stoichiometry, and reaction conditions. The deactivation of the TMP's catalysts under HDO conditions is discussed. Insights of the HDO reaction from computational aspects over the TMPs are also presented. Future challenges and directions are proposed to understand the TMP-probe molecule interaction under HDO process conditions and advance the process to a mature level.

Keywords: HDO reaction; transition metal phosphides; structure; acidity; characterization

Citation: Al-Ali, L.I.; Elmutasim, O.; Al Ali, K.; Singh, N.; Polychronopoulou, K. Transition Metal Phosphides (TMP) as a Versatile Class of Catalysts for the Hydrodeoxygenation Reaction (HDO) of Oil-Derived Compounds. *Nanomaterials* **2022**, *12*, 1435. https://doi.org/10.3390/nano12091435

Academic Editor: Ioannis V. Yentekakis

Received: 8 February 2022
Accepted: 23 March 2022
Published: 22 April 2022

Publisher's Note: MDPI stays neutral with regard to jurisdictional claims in published maps and institutional affiliations.

1. Introduction

With the world-wide energy demands increasing in an almost uncontrollable manner, the need to produce fuels in an eco-friendly way is imperative for a population that keeps growing. Renewable sources, e.g., wind, biomass, solar, are used to tackle the energy challenges. Using biomass waste to [1–3] produce fuels in the absence of carbon dioxide is a promising route to the future energy quota for replacing fossil fuels. By 2030, 32% of energy is expected to be covered by renewable sources [1].

Bio-ethanol and biodiesel are the so-called *first-generation bio-diesel*, and they suffer from low efficiency. The *second-generation biofuels* utilize the inedible biomass; in this case, the destruction of lignocellulosic biomass remains a challenge for the process. The bio-fuels derived from cyanobacteria and microalgae are categorized as *third-generation*.

Liquid and gaseous fuels can be derived from biomass applying different pathways: hydrolysis, pyrolysis and Fischer-Tropsch. Bio-oils derived from the pyrolysis can be subjected to a versatile upgrading procedure, such as zeolite upgrading [4–6], steam reforming [7–12] and hydrodeoxygenation [13,14] (Figure 1).

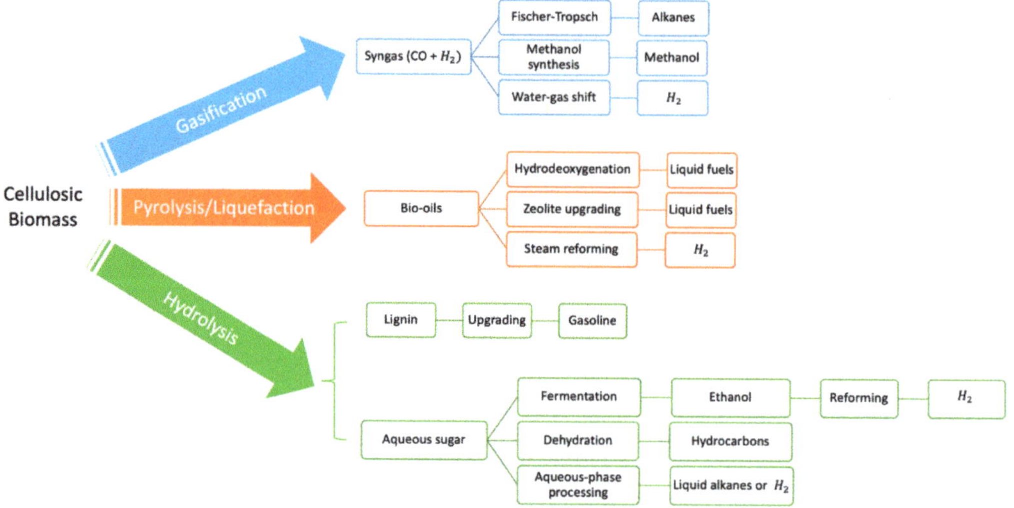

Figure 1. Process for the fuel production starting from cellulosic biomass.

Pyrolysis oil is an attractive renewable fuel used as an alternative to petroleum (fossil) fuel to minimize the environmental impact of the latter. However, the pyrolysis oil contains a high amount of oxygen, between 20% to 40%, which leads to low energy density, high acidity, and thermal and low chemical stability [15–18]. Therefore, the oxygen should be extracted from the oil; the reaction of hydrodeoxygenation (HDO), specifically the direct deoxygenation method, as described by the reactions below (for the case of fatty acids), can be used to remove the oxygen content in the presence of H_2 (Figure 2).

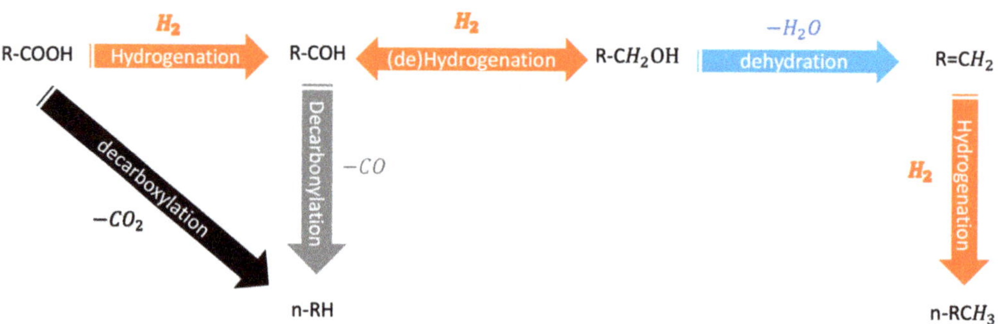

Figure 2. Hydrodeoxygenation reaction network.

The HDO reaction can happen by two possible approaches: *direct HDO and two-step HDO*. In the first process, unsaturated bonds are hydrogenated using the high pressure of H_2 (100 bar, ~573 K); separation of hydrogen from green diesel follows. Downstream, green diesel is separated from molecules such as excess hydrogen, water, CO_2, as well as an amount of C2-C3 alkanes. Dehydration of glycerol (co-product) follows [19]. Higher capital costs are involved in the latter case and lesser hydrogen consumption. The direct HDO has higher H_2 consumption due to oxygen removal from glycerol backbone towards propane formation. The co-product amount in the second process is much higher than in the first one. In the direct HDO, the electricity consumption has been reported to be higher than the two-step process due to the presence of a compressor for pressurizing the H_2. Overall, the cost for green diesel manufacturing has been reported to be lower in the two-step process (0.798 USD/kg) than the one step HDO (0.84 USD/kg) for plant operating 0.12 MMT Karanja oil/annum. Assuming a rate of return equal to 8.5%, the minimum selling price of green diesel (1.21 USD/kg, 2-step process) compared to direct HDO (1.186 USD/kg, direct) [14]. The feedstock used brings up to 75% of the production cost [14,19]. Traditionally, metal sulfides (e.g., NiMoS, CoMoS) have been used for the HDO reaction due to their exceptional hydrogenation activity [14–16,19–21], although they suffer from deactivation in the absence of sulfur (structural deterioration) as is the case for bio-oils [22,23]. Another family of hydrogenation catalysts that are good candidates for HDO reaction are the noble metals (e.g., Pt, Pd, Rh, Ru), although they are costly and exhibit low tolerance towards poisoning in the presence of sulfur (S) and nitrogen (N) compounds [24–26]. Recently, transition metal carbides, nitrides, borides and especially phosphide catalysts have become essential in hydrotreatment reactions [27–30].

In recent years, transition metal phosphides (TMP) have been reported in the open literature as active hydrotreatment catalysts and are associated with increasing demand in the processing of biomass-derived zero-carbon renewable raw materials into fuels and added-value chemicals. TMP are already known catalysts, as they have been extensively used for many catalytic reactions in the energy conversion and catalysis fields, such as HDS [31,32], HDN [33,34], WGS [35,36], photocatalysis [37,38], and HER [39,40]. The superiority of TMPs for the reactions above can be found in particular structural features of the TMPs, such as metal site density at the surface, metal electron density, and BrØnsted acidity. For example, among the S-containing compounds in the HDS reaction are thiol, thioether, thiophene, benzothiophene (BT), dibenzothiophene (DBT), etc. The conventional catalysts for HDS are MoS_2 and WS_2 (layered structures) supported on alumina. The metal sites are only located at the edge of the catalysts, introducing an intrinsic limitation in the activity over those catalysts. On the other hand, TMPs with non-layered structure, presented high performances in HDS with Ni_2P being almost the best HDS catalyst. The structure of TMPs is strongly associated with the methods of preparation [41–43]; the latter gives rise to unique electronic properties and coordination/bonding environment. TMPs combine the physical properties of the phosphorus (P) and the transition metal (TM), such as ceramic hardness and strength with electronic properties, such as metal conductivity. At the same time, they have an obvious advantage in terms of cost-effectiveness and environmental-friendliness over the other active metallic phases.

In the present review article, we focus on the unique structural and electronic properties of the transition metal phosphides (TMPs) and their reactivity towards the HDO reaction. In Section 2, the structural versatility of TMPs is presented with all the different classifications (e.g., compositional, structural, active sites). In Section 3, necessary characterization tools and analytical techniques are critically discussed in terms of enriching our knowledge on structural, compositional, and morphological features of TMPs. Techniques, such as X-ray Photoelectron Spectroscopy (XPS), X-ray Absorption Near Edge Spectroscopy (XANES), Nuclear Magnetic Resonance (NMR), Chemisorption, Transmission Electron Microscopy (TEM) are discussed. Moving to Section 4, the catalytic activity of a plethora of TMPs is discussed for three representative families of bio-oil derivatives with different functionality, namely phenolics, vegetable oils, and furans. As a consequence of the nature of the HDO reaction, deactivation happens; the latter critical issue is discussed in Section 5. To get a deep insight into the TMP surface–probe molecule interactions, computational aspects of the HDO reaction are discussed in Section 6. The last part of this review, Section 7, discusses the outlook and future perspectives of the HDO field using TMPs and the way forward towards the mature knowledge-based design of functional TMPs.

2. A Brief on the Transition Metal Phosphide (TMP)

2.1. Structural Concepts

The compounds formed between phosphorus and any d- (e.g., Ni, Mo, W, Co, Fe) or f-metal are called phosphides. Phosphorous can adopt any oxidation state between 0 and 3, resulting in a plethora of structural configurations. The TMPs are refractory metallic compounds exhibiting both metallic and acidic sites [44,45] due to the alloying of TM with P atoms (Figure 3); their classification can be found in Table 1, whereas their physical properties are shown in Table 2. Thorough review articles on the structure of TMPs can be found in the literature [46–52]. The most frequently used elements in TMPs are iron (Fe), cobalt (Co), nickel (Ni), and molybdenum (Mo). Each TMPs have a variety of morphology and particle size, leading to different properties depending on many crucial parameters, such as the method of preparation followed, P source used, capping agents, temperature, etc. Due to the versatility of their structures (Table 3), TMPs have been used for many catalytic reactions [53]. The catalytic performance of TMPs catalysts, together with the reaction conditions for hydrodeoxygenation reaction of various feedstock are tabulated in Table 4.

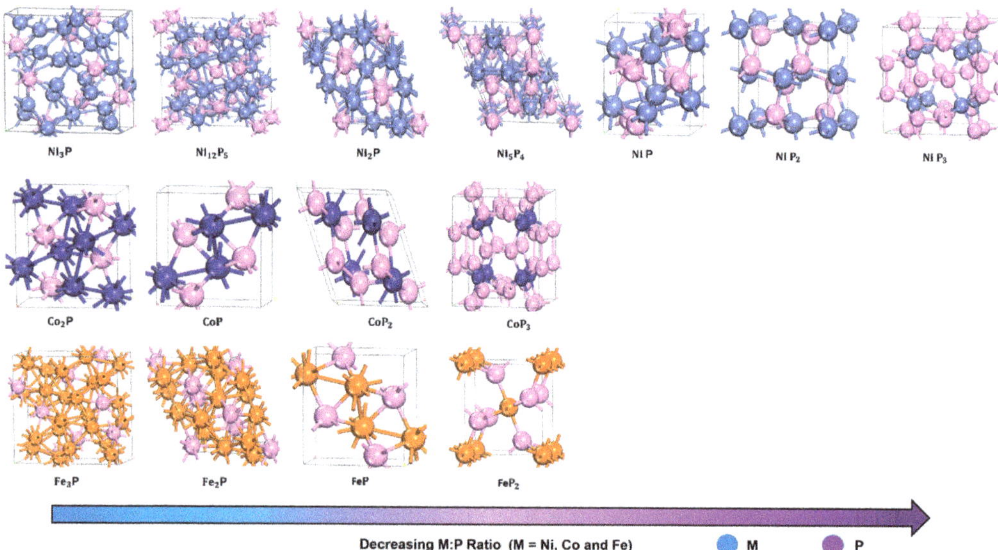

Figure 3. Versatility of Crystal Structures of TMPs.

Table 1. Categorization of metal phosphides.

Metal Phosphide Classes	Examples of Metal Phosphides
Metal-rich phosphides	$M_xP_y (x > y)$; Ni_2P, Co_2P, Ni_3P, etc.
Monophosphides	$M_xP_y (x = y)$; MoP, WP, CoP, etc.
Phosphorous-rich phosphides	$M_xP_y (x < y)$; NiP_2, FeP_2, etc.
Ionic phosphides	$M_x^{n+}P_y^{n-}$; Th_3P_4, etc.

Table 2. Physical properties of metal-rich phosphides.

Ceramic Properties			Metallic Properties		
Melting point	°C	830–1530	Electrical resistivity	$\mu\Omega\text{cm}$	900–25,000
Microhardness	kg·mm^{-2}	600–1100	Magnetic susceptibility	$10^6\text{cm}^3\text{·mol}^{-1}$	110–620
Heat of formation	kJ·mol^{-2}	30–180	Heat capacity	J·(mol·°C)^{-1}	20–50

Table 3. Different TMPs and their crystalline structure.

Phosphide Phase	Crystal System, Space Group	Number of Formula Units	References
Ni_2P	Hexagonal $P\bar{6}2m$ $Ni/P = 2$	3	[1–40]
$Ni_{12}P_5$	Tetragonal $\bar{I}4m$ $Ni/P = 2.4$	2	[1–40]
Ni_3P	Tetragonal $\bar{I}4$ $Ni/P = 3$	8	[1–40]
Ni_5P_4	Hexagonal $P6_3mc$ $Ni/P = 1.25$	4	[1–40]
CoP	Orthorhombic Pnma $Co/P = 1$	4	JCPDS No.29-0497
Co_2P	Orthorhombic Pnma $Co/P = 2$	4	JCPDS No.89-3030
MoP	Hexagonal $P\bar{6}2m$ $Mo/P = 1$	1	[1–40]
Mo_3P	Tetragonal $\bar{I}42m$ $Mo/P = 3$	8	[1–40]
WP	Orthorhombic Pnma $W/P = 1$	4	JCPDS No.29-1364

Transition metal phosphides (TMPs) have a formula of M_xP_y which eliminates the electron delocalization around the metal atom due to the slightly higher electronegativity of P than the metal; this leads to electron transfer with a direction from metal to P. The characteristics of the metal phosphide rely on the difference in the M-P electronegativity and the M:P ratio. In terms of bonding, TMPs exhibit a combination of covalent and ionic nature bonds due to the small electronegativity differences that give the metal a small positive charge (δ^+) and the phosphorus a small negative charge (δ^-). Therefore, this strong bond leads to high thermal and chemical stability and hardness [54].

There are two types of phosphides: (a) phosphorus-rich phosphides ($MxPy, x < y$) and (b) metal-rich phosphides ($MxPy, x \geq y$) (see Table 1). In particular, the metal-rich catalysts are of great interest as they resemble the metals' properties (e.g., good thermal conductivity, high thermal and chemical stability) [49]. In the metal-rich phosphides ($x : y \geq 1$) the electrons do not fully surround the phosphorus atom, so intensive M-M interactions can develop. Of particular interest are the metal-rich TMPs as they can present a plethora of crystal structures and compositions (Figure 3). For example, nickel phosphide can be crystallized in Ni_2P, Ni_5P_4, $Ni_{12}P_5$, Ni_3P, where Ni_2P adopts Fe_2P-type structure [42], whereas Ni_3P adopts Fe_3P-type structure [55].

Phosphorus-rich ($x : y < 1$) do not have any M-M bonding, particularly when $x : y < 1/2$ because of the poor electrical conductivity. Most of the P atoms can exist in the form of oligomeric chains and clusters and any increase in the content of phosphorus leads to higher reactivity, along with poor thermal stability. As a sequence, at high a temperature, it can decompose to elemental phosphorus and metal-rich phosphides [54]. Especially in the case of alkali metals, alkaline earth metals and some of the transition metals, e.g., zinc phosphides, a large difference in the M-P electronegativity can be noticed and thus high ionicity of M-P bonds ready to hydrolyze.

Other TMP classifications: Another way to categorize the TMPs is to split them into binary, ternary, and supported ones [56]. Different nanostructures of binary TMPs have been reported in the open literature, such as CoP in the form of nanotubes, nanoparticles, nanosheets, nanorods [57–59], Co_2P in the form of nanoflowers, nanoparticles, nanosheets [60,61], Cu_3P in the form of nanoarrays, nanowires [62,63], and MoP in the form of nanoflakes, nanoparticles [64,65].

Ternary phosphides are exceptional catalysts for various reactions, presenting interesting morphologies. Some examples include: $NiCo_2Px$ (nanowires) [66,67], Ni–Fe–P (nanocubes) [68], CoMoP (core-shell) [69], NiCu-P (porous) [70]. To prepare the above phosphides, a plethora of P sources have been used including NaH_2PO_2, TOP, Red P, whereas different methods of synthesis have been followed [71,72].

Supported TMPs are famous for increasing the dispersion and stabilizing the active phase (e.g., TMPs); alumina, silica, and carbon are among the most used supports. Additionally, the supports can offer active reaction sites (e.g., Bronsted acidity) and thus participate in the reaction (bi-functional catalysis) [73–75]. There is a wide amount of literature covering the use of some supports and the challenges imposed on the process due to their chemistry. For instance, the traditionally industrial support, Alumina, facilitates the Mo-O-Al linkage formation which seems to deactivate part of the active MoP phase, ultimately deteriorating the catalytic activity. Additionally, alumina is subjected to phase transformation (boehmite) when exposed to water (part of HDO reaction conditions). Al_2O_3 contains a high water level (30 wt.%), leading to intrinsic oxidation of the metal and the P in the TMPs. Other supports are the mesoporous silica, such as MCM-41 and SBA-15, where both possess high surface area and acid site density. Activated carbons have already been used [73–75], but they suffer from the presence of micropores (unsuitable in the HDO reaction) and poor mechanical stability.

It is well-established that support plays key roles in supported catalysts. On the one hand, the oxygen vacancy sites and surface acidity/basicity might contribute to the activation of reactants and consequently affect products distribution. On the other hand, the support affects the structure of the active phase, such as electronic properties and dispersion. The literature reported that the type of support influences the nature of the metal phosphide phase formed since the oxide support can react with the P precursor salt resulting in the generation of phosphate species [76–79]. For instance, Alumina interacts favorably with phosphorus-producing $AlPO_4$, which prevents the synthesis of Ni_2P phase and yields various Ni phases [76–79].

HDO activity and product selectivity are primarily contingent on the nature of the metal phosphide phase and as well as the type of support. For example, Shi and co-workers [76–79] synthesized various nickel phosphide catalysts supported on TiO_2, CeO_2, SiO_2, Al_2O_3, HY and SAPO-11. The results indicated that the Ni_2P phase was observed on TiO_2, CeO_2, SiO_2 and SAPO-11. The Ni_2P and $Ni_{12}P_5$ were formed on HY, whereas $Ni_{12}P_5$ and Ni_3P were obtained on Al_2O_3. Clearly, the supported ones exhibited different interactions with P and Ni atoms giving rise to the generation of various Ni_xP_y phases. The activity for deoxygenation of methyl laurate followed the order: Ni_2P/SiO_2 > $Ni_{12}P_5$-Ni_3P /Al_2O_3 > Ni_2P/TiO_2 > $Ni_2P/SAPO$-11 > $Ni_{12}P_5$-Ni_2P /HY > Ni_2P/CeO_2, and it was revealed that the influence of supports mainly originates from their reducibility and acidity.

Moon et al. [79] examined the HDO of guaiacol on Ni_2P supported on active carbon (AC), SiO_2 and ZrO_2 under a temperature of 573 K and 30 atm. Unlike active carbon (AC) and ZrO_2 support, the in situ XAFS analysis revealed that the Ni_2P/SiO_2 catalyst experienced oxidation to generate nickel phosphate during the course of reaction. Ni_2P/SiO_2 has demonstrated higher conversion (87%) than Ni_2P/ZrO_2 (72%) and Ni_2P/AC (46%) catalysts.

2.2. Functionality (Acidity) of TMPs

TMP surface has a mild acidity, as proved by NH_3 temperature programmed desorption (TPD) studies [80]. For example, one of the most popular TMP, Ni_2P, contains not only the metal sites where the hydrogenation takes place but also two types of acid sites, namely: $Ni^{\delta+}$ sites and P-OH sites, which are categorized as Lewis and Bronsted acid sites, respectively. Later, how these acid sites participate in the HDO reaction mechanistic steps will be discussed. Unsupported TMPs (e.g., MoP) and supported ones (e.g., Ni_xP, Co_xP, W_xP, Mo_xP, Fe_xP and Ru_xP) are efficient catalysts for HDO since they contain BrØnsted and Lewis acid sites. TM develops a small positive charge $M^{\delta+}$ which act as Lewis acid sites and thus favors reactions such as hydrogenation, hydrogenolysis, and demethylation. BrØnsted acid sites are generated on the P-OH sites that produce active hydrogen species; the hydroxyl groups of P-OH promote hydrogen diffusion (spillover) and contribute to the stability of these species.

2.3. A Close Look on the TMP Active Sites for the HDO Reaction

Two examples of widely used TMP in catalysis are the metal-rich phosphides Ni_2P and MoP. The crystal structure of MoP is composed of hexagonal layers of P having Mo in the trigonal prismatic positions. All Mo atoms are equivalent; the same applies to the P atoms [81,82]. Two types of metal sites can be found in the Ni_2P: 4-coordinated distorted tetrahedron and 5-coordinated square pyramidal. The P atoms are located in a face-capped trigonal prismatic environment, though two types of P sites can be found based on the coordination with 4- and 5-coordinated Ni atoms [81,82].

Depending on the position of Metal and P in the phosphide lattice, the Ni and P charges can vary. In the case of MoP, Mo and P present only slightly positive and negative charges, whereas, in the case of Ni_2P, Ni can exhibit either positive or negative charge with P carrying a slightly positive charge [83]. Based on the charge distribution, both phosphides exhibit covalent bonding and metal-like character.

The so-called *ligand effects* describe the Ni $\rightarrow$ P charge transfer [84,85]. In addition, P can exert structural effects on the metals leading to an increase in the M-M site distance compared to the pure metal lattice [86]. These effects lower the reactivity compared to the pure transition metals and suppress the phase transitions (e.g., retarding metal sulfides formation in S-containing environments). In turn, P atoms may act as a reservoir of H atoms; the latter can be provided for hydrogenation and hydrogenolysis reactions for the species adsorbed at the metal atoms.

The bifunctional character of the phosphide catalysts is being constituted by the metal sites (metal atoms or metal-P clusters), and the P-OH groups (acidic sites) at the surface of phosphides due to the presence of strong P-O bonds environment [87]. On the other hand, the selectivity of the hydrogenolysis and hydrogenation reactions have been linked to the population of the OH groups [88].

2.4. TMPs Preparation Strategies

The TMP can be synthesized following different methods and a plethora of their modifications. Methods, such as solution–phase reactions [89,90], gas–solid reactions [91], and solvothermal reactions [92], have been applied for the preparation of unsupported and supported TMPs.

Gas-Solid synthesis: A typical procedure of applying temperature-programmed reduction (TPR) method in TMPs synthesis has been presented by Cecilia et al. (2009) [93] for the preparation of CoP nanoparticles; $Co(OH)_2$ and H_2PO_3H were used as the cobalt and phosphorus precursors, respectively. By mixing the two precursors, Cobalt(II) dihydrogenophosphite ($Co(HPO_3H)_2$) was the resultant product. Different supports were used in this study, such as MCM-41 (CoP-10 (Si)), MCM-41 zirconium doped (CoP-10(Zr)), Cabosil silica (CoP-10 (Cab)), and alumina (CoP-10 (Al)). The TPR step aimed to decompose the above precursor to phosphide in a reducing atmosphere in the 100 to 800 °C range [93].

Liquid phase synthesis: In another report by Zhang et.al. (2017), CoP nanoparticles were prepared using liquid phase synthesis instead. In this procedure, $Co(acac)_2$, 1-octadecene, and TOP were mixed and subjected to heating at 300 °C for 5 h. Following the cooling of the particles, washing was performed using hexane and drying was completed under vacuum at 80 °C [94]. Following the protocol of liquid synthesis, Jiang et al. (2015) [95], synthesized CoP nanoparticles using $Co(acac)_2$ as Co sources, whereas Triphenylphosphine (PPh_3) was used as a P-source in this case. The heating of the precursors in a furnace at 400 °C for 1 h was performed. However, phosphine is a toxic gas so using an alternative is recommended, such as hypophosphites, phosphites, or phosphates. At medium temperatures, hypophosphites decompose (approximately 250 °C), leading to the formation of PH_3, which subsequently reacts with metal precursor such as metals, metal oxides, hydroxides, salts, and other compounds. In contrast, phosphites or phosphates need hydrogen to create phosphine. Additionally, CoP and Co_2P nanoparticles were produced by Ha et. al. (2011) [96] using the solution phase arrested precipitation.

Lu et al. have reported on the controlled synthesis of Ni_xP_y nanoparticles encapsulated by silica following solution-based synthesis in oleylamine (OAm) with trioctylphine (TOP); the method is anticipated to effectively control the morphology and phase of MP nanoparticles due to the high boiling point and ease in controlling the metals' complexity property. For example, OAm can have a triple role as a surfactant agent, a solvent medium and a reducing agent and stabilizer for the synthesis of MP nanoparticles. Additionally, TOP acts as the P source, surfactant and solvent. There are two stages for a typical synthesis of Ni_2P nanoparticles in OAm where the first stage produces Ni nanoparticles, and in the second stage, Ni_2P is formed. In the first stage, the Ni nanoparticles are formed by decomposition of Ni precursor such as $Ni(acac)_2$ above 200 °C. During the second stage, P insertion is taking place to produce Ni_2P above 300 °C. The ultrafine Ni_2P nanoparticles were confined in mesoporous silica so to design a sintering-resistant catalyst for the catalytic reduction of SO_2 to sulfur in the H_2 presence [89,90].

D'Accriscio et al. reported on the synthesis of Fe_xP nanoparticles using tetrakis(acyl) cyclotetraphosphane, $P_4(MesCO)_4$; the latter P source is a very stable one and not air-sensitive like TOP or white phosphorous [97] (Figure 4a). At a temperature as low as 250 °C, they reported the synthesis of FeP and Fe_2P nanoparticles. The local order in the materials was studied by using XPS and atomic pair distribution function (PDF). The authors reported that crystalline FeP is formed via an intermediate amorphous one. The FeP nanoparticles exhibited a spherical hollow shape of 8.9 nm size. TEM studies on the Fe_2P showed that they had spherical hollow shapes with two shells with the inner shell being about 1 nm thickness (Figure 4a). Huiming Li et al. [98] reported on a systematic effort to synthesize nanocrystals of Ni-P with different stoichiometries (Ni_5P_4, Ni_2P, $Ni_{12}P_5$) by adjusting the P/Ni precursor ratio, the temperature and the reaction duration (Figure 4b) and starting from OAm and TOP. Ni_5P_4 monodisperse with a size of 5.6 nm was achieved; such nanocrystals, due to the high surface areas, are anticipated to exhibit promising catalytic activity. Shanfu Sun et al. [99] (Figure 4c) synthesize $(Ni_xFe_{1-x})_2P$ using a two-step process. Starting from a NiFe foam which was subjected to corrosion, they succeeded in growing NiFe-LDH nanosheets; the latter were subsequently phosphorized using NaH_2PO_2 under an inert atmosphere (N_2).

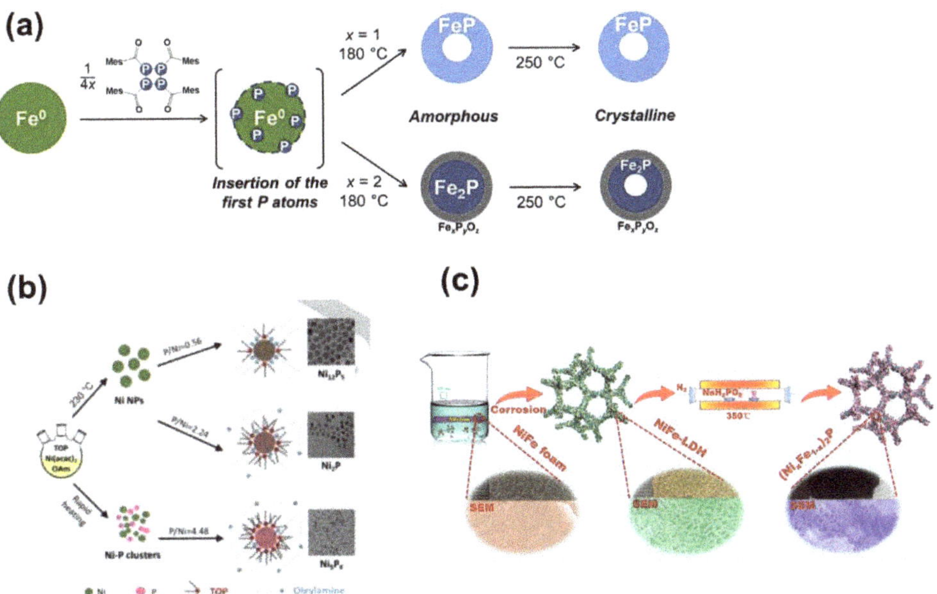

Figure 4. (**a**) Fe-P synthesis. Reprinted with permission from Ref. [97], Copyright 2020 Wily Online; (**b**) NiP nanocrystals formation. Reprinted with permission from Ref. [98], Copyright 2018 Wily Online; (**c**) Synthesis of $(Ni_xFe_{1-x})_2P$ nanosheets via phosphorization of the NiFe-LDH precursors. Reprinted with permission from Ref. [99], Copyright 2020 ACS Publication.

An example of ternary phosphide synthesis has been reported by Ye et al. [100], who synthesized $Co_{2-x}Fe_xP$ using TOP and OAm. It was reported that the synthesis temperature had a crucial role to adjust the morphology of the resultant phosphide; $Co_{1.5}Fe_{0.5}P$ (rice-shaped) or $Co_{1.7}Fe_{0.3}P$ (split). In another study by Lawes et al. [101], the synthesis of the $Co_xFe_{2-x}P$ in the compositional range $0 < x < 2$ was investigated. In the first step of the process, starting from carbonyl precursors of both metals, Co-Fe alloy nanoparticles were prepared, which were subjected to reaction with the TOP in the 330–350 °C temperature range. The authors noticed that the Co-rich compositions prefer to crystallize at low temperatures, whereas the Fe-rich follow an opposite temperature profile. At intermediate compositions (values of x), the predominant growth mechanism is aggregation. Another procedure to synthesize $(Co_xFe_{1-x})_2P$ is through the synthesis of the relevant ferrite ($CoFe_2O_4$), which is subjected into reaction with TOP [102]. Many different compositions of ternary TMPs have been reported, such as Ni-Co-P, Fe-Ni-P, Fe-Mn-P, Co-Mn-P, Co-Rh-P, and Ni-Ru-P [103]. It is known that their synthesis is rather demanding due to the combination of the reactivity from two metal compounds, whereas manipulating the stoichiometry is important as the latter can change their catalytic chemistry of them. The content of the metals in the synthesis solution frequently happens does not coincide with the stoichiometry of the final phosphide. That is the reason behind the strategies followed to synthesis of the ternary phosphides. Usually, the bimetallic or the oxide are prepared and their reaction with the P source follows (Figure 5).

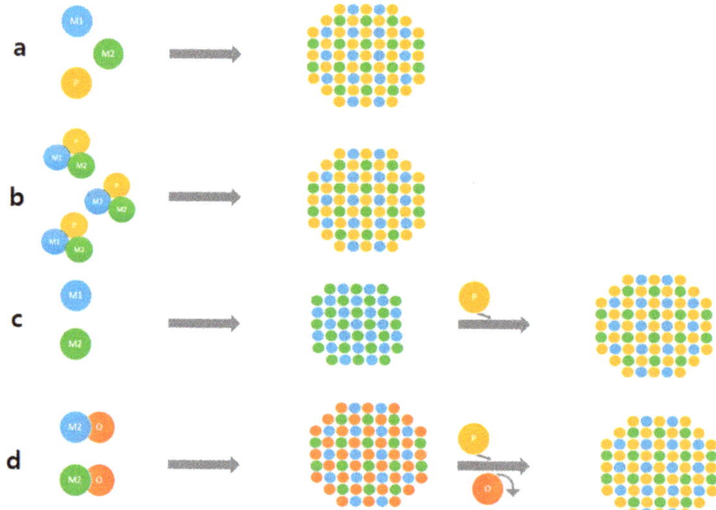

Figure 5. Colloidal synthesis of ternary TMPs. Adapted from Ref [41], (**a**) reaction of a phosphorus precursor (P) with both transition metal sources (M1 and M2), (**b**) Thermal decomposition of a single precursor consisting of phosphorus and both metals, (**c**) generation of bi-metallic nanoparticles their subsequent phosphorisation reaction, (**d**) generation of oxidic nanoparticles and their subsequent phosphorisation.

Table 4. The catalytic performance of different TMPs catalysts used in deoxygenation reactions of various reactants.

Catalysts	Reactants	Reaction Conditions		Coversion (%)	Selectivtiy (%)	Reference
		T (°C)	P (bar)			
$Pd - Ni_2P$ /SiO$_2$	Phenol	220	20	~100	~92 Cyclohexane	[104]
MoP/SiO$_2$				15	14 Pentane 70 Pentenes 7 Pentadienes	[105]
CoP/SiO$_2$	2-methyltetrahydrofuran	275	1	11	40 Pentane 15 Pentenes 25 C14	
Ni$_2$P /SiO$_2$				12	67 pentane 2 pentanone 15 C14	[106]
CoP/SiO$_2$	dibenzofuran	275	30	~90	~72 Bicyclohexane	[107]
Ni$_2$P				92.9	14.1 Phenol 58.7 Benzene	
Ni$_{12}$P$_5$	guaiacol	300	1	93.9	31.5 Phenol 34.9 Benzene	[108]
Ni$_3$P				99.8	23.9 Phenol 22.8 Benzene	
Ni$_2$P/ pillared -ZSM-5		260	40	78	~82 cyclohexane	[109]

Table 4. *Cont.*

Catalysts	Reactants	Reaction Conditions		Coversion (%)	Selectivtiy (%)	Reference
		T (°C)	**P (bar)**			
FeMoP	Anisole	400	21	>99	90 Benzene 10 Cyclohexane	[110]
$Ni_3P- Ni_{12}P_5/\gamma - Al_2O_3$	methyl laurate	340	20	~98	~96 C11 + C12~12 C11/C12	[111]
Ni_2P supported on SiO_2, MCM-41		340	20-30	97-99	~100 C11 + C12	[112]
$Ni_2P/$ SBA-15	methyl oleate	270	30	~84	~45 C18/(C17+C18)	[113]
Ni_2P/SiO_2	methyl laurate	340	30	~97	~87 C11 + ~12 C12	[114]
MoP/SiO_2		340	30	~90	~4 C11 + ~80 C12	
$NiMoP/SiO_2$		340	30	~98	~51 C11 + ~49 C12	
$Ni_{1.5}P/$activated carbon	palmitic acid	350	1	100	57 C15 7 C11-C14 21 alkenes	[115]
Ni_xP /HZSM-22		350	1	99.6	-	[116]
10 wt% $CoP/\gamma - Al_2O_3$	2-furyl methyl ketone	400	1	100	~100 Methyl Cyclopentane	[117]

3. Characterization of TMP Features towards Unveiling Outstanding Performance

3.1. Structural Features (XRD, Raman, TEM, SAED)

To understand the structure–property correlation for the TMP, X-ray diffraction (XRD) is one of the primary techniques to be utilized to be able to assess the crystallinity of the TMP nanostructures. Along the same lines, Selected Area Electron Diffraction (SAED) collected in the TEM analysis provides complementary information with XRD. XRD has been extensively used to validate the success of synthesis of TMPs, e.g., CoP/SiO_2 prepared using the TPR method [105]. Using XRD analysis after the TPR synthesis can advocate the reduction process efficiency and conditions, the phenomena involved, and the suitability of the preparation method for the scale up of the process. In some cases, fluoresce can happen, deteriorating the quality of the XRD pattern, e.g., the case of FeP, though the presence of FeP could still be identified. Following a TPR synthesis protocol on different P precursors (phosphite $[HPO_3]^{2-}$ vs. phosphate $[PO_4]^{3-}$) can be challenging for the TMPs' quality (absence of impurity phases) and XRD can offer an assessment tool. Crystallinity of the TMP can also be different based on the nature of the metal (e.g., better quality crystals are formed for CoP vs. WP, as the latter starts at a higher oxidation state) [105]. The presence of pure metallic phases as impurities have also been reported [105]. The crystallite size of TMPs can be calculated using the Scherrer formula. An example is presented by H. Y. Zhao et al. [108], who studied supported TMPs, such as MoP/SiO_2, Ni_2P/SiO_2, WP/SiO_2 and Co_2P/SiO_2. Among the catalysts studied, Ni and Mo catalysts had the smallest particle size. In the case of MoP, no XRD peaks were detected, indicating the high dispersion of the MoP phase on silica and small particle size (less than 5nm). In the case of Ni_2P, a broad XRD pattern was received, showing the small crystallites size. Additionally, nickel phosphide showed three peaks that line up with the iron pattern but with more narrow peaks, which means Fe has a larger particle than Ni.

Advanced diffraction techniques, such as in situ XRD, have been employed to monitor the TMP phase evolution; Inocencio et al. [118] showed that this technique could shed light on the preparation mechanism and phase transformations that are taking place. The method of preparation followed was for TPR of phosphates ($Ni_xP_yO_z$; $Mo_xP_yO_z$; $Co_xP_yO_z$ and $Fe_xP_yO_z$, two-step synthesis). The presence of Ni_2P, MoP, CoP/CoP_2, FeP and WP was identified using in situ XRD. Under the reduction atmosphere, the phosphate precursor was decomposed towards the formation of metal oxide (Mo case) or pyrophosphate phases (Ni, Fe) with no formation of metallic phase. In an attempt to understand the TPR over phosphates, in the case of $Mo_xP_yO_z$, two peaks at 437 and 554 °C are exhibited and this finding is in agreement with a previous study by Bui et al. [105]. Based on the in situ studies, the $Mo(OH)_3PO_4$ phase is initially formed and then likely decomposed to MoO_3 and $MoOPO_4$ and some amorphous portion. The CoP_xO_y TPR presented a peak in the 500–530 °C range and a broad peak at 580 °C reflecting the reduction of Co-phosphate to metal and PH_3, which reacts with the metal towards CoP formation. Peroni et al. [119] studied the catalyst's particles sizes either using X-ray or statistical analysis of TEM micrographs (see table below). From TEM results, MoP catalysts tend to have smaller particles, i.e., 29.2 nm compared to the Ni_2P particles, i.e., 132.4 nm. However, the low temperature methods tend to have a smaller particle size than the TPR approach. In addition, the citric acid that is used during the synthesis reduced the Ni_2P crystal size from 132.2 nm to 49 nm because of the suppression of the sintering of active phases. In addition, Raman spectroscopy was used to study ternary solid solutions phosphides of $Fe_{1-x}Ni_xP$ ($0 < X < 0.3$) composition. Raman spectroscopy showed that variation of Ni content in the solid solution caused an almost linear shift of several peaks. Experimental peak positions were found to match with the phonon frequencies, giving a better understanding of the frequency-composition dependence [120].

3.2. Dispersion of the TMPs over a Support

There are many reports where the TMPs are dispersed on a high surface area support, preferably with acid sites so as to modify the ultimate acidity of the final structure and increase the active phase (TMP) exposure and bifunctionality of the catalyst. Such supports can be MCM-41, SBA-15, other types of silica, alumina, and carbon. In the cases of supported TMPs, the dispersion of TMP needs to be evaluated for appropriate assessment of the catalytic activity. For evaluating the dispersion, CO chemisorption [94] along with CO infrared studies are reported. CO is a typical molecule used for such studies due to its small size and its C-O stretching frequency, which is well-defined for the case where CO adsorption takes place on metal surfaces [121,122]. Based on the CO uptake values ($\mu mols/g$), blockage of active metal sites can be predicted and correlated with the synthesis conditions followed. It has been observed in the case of the phosphate method where the P^{5+} is not easily reducible and higher temperatures are needed to convert into phosphide that unreduced phosphate stays bound on the metal, reducing the active sites [123]. In addition, smaller crystallite size corroborates with a higher dispersion, giving rise to higher CO uptakes ($\mu mols/g$).

Infrared spectra (IR) of CO adsorbed on the surface of reduced Ni_2P/SiO_2 catalysts gives rise to four species, as follows: (1) CO adsorption on atop Ni sites, (2) CO adsorption on bridge Ni sites, (3) formation of adsorbed species $Ni(CO)_4$, and (4) formation of adsorbed $P=C=O$ species on the surface, terminal attachment of CO species to P [121]. The relative CO site densities and the $\Delta H_{ads}(CO)$ are reported. In the case of MoP [124], CO chemisorption on different Mo sites is shown in Figure 6.

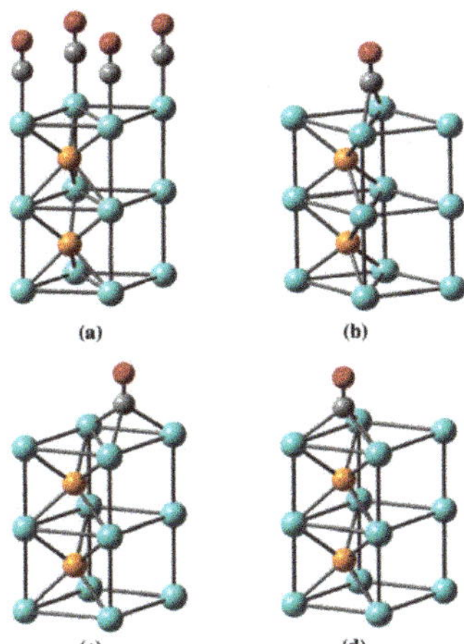

Figure 6. CO chemisorption on MoP-(001)-(1 × 1) surface: (**a**) on-top Mo, (**b**) bridge, (**c**) fcc site, and (**d**) hcp hollow site. Reprinted with permission from Ref. [107]. Copyright 2017 Springer Nature.

3.3. Surface Studies and Coordination Environment (XPS, XANES, NMR)

The surface chemistry of TMPs can be studied using X-ray photoelectron spectroscopy (XPS). A closer look at the binding energies (BE) can shed light on the charge transfer as well as the hybridization of phosphorous. In general, in the case of P 2p, the peak at 129.5 and 130.5 eV assigned to P $2p_{1/2}$ and P $2p_{3/2}$ are the fingerprint of TMP. Other peaks originating from P have been reported at 132.5, 133.6, 134.2, and 135.5 and are linked to P-C, P-O, P-O-M, and metaphosphates [125,126]. In an attempt to fully utilize the XPS data, someone has to have a careful look at the main and satellite peaks of the metal as well. Usually, positive shifts of BE on the metal can imply a partial positive charge on the metal site, e.g., $M^{\delta+}$; the latter is usually accompanied by a negative shift of BE for P, attributing a partial negative charge to the latter ($P^{\delta-}$) and dictating the direction of charge transfer.

A compile of XPS as a tool for TMPs studies is discussed in what follows. Burns et.al. (2008), [127] used XPS spectra to study the electronic environment of the Co_2P/SiO_2 catalyst. The peak at BE 781.6 eV is linked to Co^{2+} and the one at BE 134 eV to P^{5+}. Those two peaks are originated due to the passivation step that was completed at the end of the synthesis. Based on the BE 778 and 129.2 eV, which are associated with elemental cobalt and phosphorus, respectively, it can be stated that Co is positively partially charged, and the P is negatively partially charged, and this corroborates the Co → P charge transfer. In the same study, CoP/SiO_2 and Co_2P/SiO_2 are also presented. The relative intensities of Co and P in the Co_2P are opposite compared to the ones of CoP, which indicates that CoP has a P-rich passivation layer. In addition, the peak shift by 0.8 eV at for CoP to higher energies indicates that there is higher charge transfer from cobalt to phosphorus in CoP than in Co_2P [127].

In their study of optimizing the synthesis of TMPs starting with phosphates and phosphites and following a controlled reduction environment, Bui et al. [94] reported a detailed XPS study of Ni_2P and WP. In the case of Ni_2P, a P peak at 128.6 eV (P $2p_{3/2}$) is the main proof for the Ni_2P phase formation [128], while the P $2p_{3/2}$ at higher BE (135.0 eV) was assigned to PO_4^{3-} (as a result of passivation of the phosphide layer). In the case of the phosphite method of synthesis, unreduced phosphites can also give rise to a P $2p_{3/2}$ at 133.6 eV. So, this reveals another benefit of using XPS: assessing the purity of the TMP. Similar peaks were observed for the WP.

Near-Edge X-ray absorption fine structure (NEXAFS), also known as XANES, is a great tool to probe parameters, such as chemical bonding, electronic structure, and interactions in the TMPs. The collected XANES spectra are usually fitted with standard compounds.

Yun et al. [14] used EXAFS due to the XRD limitations imposed by the significantly small Ni_2P size (3 nm), making it undetectable in XRD. The FT spectrum shows two peaks at 0.180 and 0.228 nm, which are assigned to the Ni-P and Ni-Ni chemical interaction [129]. A three-shell curve fitting was followed, leading to Ni-P(I), Ni-P(II), and Ni-Ni bonds lengths of 0.2207, 0.2344 and 0.2551 nm, which are associated with 1.8, 2.8, and 2.0 coordination numbers. The data allow us to assess that Ni-Ni coordination is smaller, whereas Ni-P is higher in small Ni_2P crystallites.

Transition metal phosphides possess versatile electronic structures and properties that are anticipated to give rise to unique signals in the ^{31}P NMR, including difficulties in tracing the signal or the impossibility to carry out such characterization. NMR spectra features can unveil the characteristics of the TMPs electronic structures. This is showcased in the rigorous research conducted by Bekaert et al. [130] on a series of V, Fe, Co, and Ni phosphides, combining ^{31}P MAS NMR and first-principles calculations. The study shows that the electronic structure of the TMPs (diamagnetic, paramagnetic, metallic) and the ^{31}P NMR signal are correlated. More attention needs to be paid in the case of paramagnetic TMPs (e.g., Co_2P), where shorter relaxation delays, a decrease in the signal, and line broadening, can lead to a perplex spectrum. In the case of Co_2P, short Co-Co distances lead to a strong Co-Co bond interaction giving rise to metallic conductivity and high values for NMR shifts; this amount of high shift makes it possible to differentiate between phosphides and phosphates [131]. In addition, ssNMR nanocrystallography has been used for facets analysis of TMP [132]. In particular, ^{31}P DFT-assisted solid-state NMR coupled with HRTEM and XRD were used to evaluate the exposed crystal facets of the terminal surfaces in the case of Ni_2P nanoparticles. Results showed that two facets (0001-A) and $(10\bar{1}0)$ are the predominant; the former is the major one. These exposed facets cause a shift in the Ni d-band towards the Fermi level, enhancing the catalytic activity.

Wanmolee et al. [133] monitored the phase transformation of $CuHPO_4 \cdot H_2O$ precursor into Cu_3P phase using in situ XAS and XRD characterization techniques. The atomic coordination and electronic structure were investigated via EXAFS and XANES. The H_2-TPR was performed in situ for the calcined catalyst by collecting the in situ XAS spectrum at the temperature range of 25–650 °C. According to the XANES results, at temperatures below 110 °C, the divalent copper ions (Cu^{2+}), in $CuHPO_4 \cdot H_2O$ precursor, were predominantly observed. Notably, successive dehydration of water from the precursor is observed with a temperature rise up to 200 °C; thereby, the intermediate precursor $Cu_2P_2O_7$ is formed. Above 250 °C, the Cu^{2+} species completely disappeared, while the Cu^0 and Cu^{1+} species became prevalent, signifying the formation of metal Cu and $Cu(PO_3)_2$. Above 320 °C, the Cu_3P phase is formed which was borne out by the predominance of Cu^{1+} species. Interestingly, the Cu_3P phase experienced various transformations in the space group, as conformed by XRD analysis; however, the crystal structure was finally relaxed to $P6_3cm$ space group following the cooling down. The FT-EXAFS revealed that copper atoms are thoroughly bonded with the phosphorus atoms with Cu-Cu radial length of 2.18 Å.

4. TMPs as Catalysts for the Upgrading of Bio-Oil through HDO Reaction

Hydrodeoxygenation (HDO) reaction is used to remove the oxygen in bio-oil without losing carbon (C) and hydrogen (H) atoms. The commercial catalysts, such as Ni/Co-MoS_2 supported on alumina, suffer from oxygen-induced deterioration in the absence of H_2S, leading to the ultimate deactivation of the catalysts. Moreover, the acidic sites of alumina tend to favor the coke deposition and deactivation of the catalysts. TMPs are suggested as thermally and chemically stable catalysts for the HDO reaction and as good alternatives to noble metal catalysts. Each TMP may have a different reaction path leading to different products. The HDO mechanism over TMPs and noble metals are compared in Figures 7 and 8; the former is more advantageous cost-wise and selectivity-wise (less prone to secondary reactions, such as C-C bond hydrogenolysis and methanation). The benefit of using TMPs can be summarized as follows: for instance, in the case of popular Ni_2P, Ni contains a slight positive charge and phosphorus has a slight negative charge because of the partial transfer of electron density from nickel to phosphorus (ligand effect). The metal-rich phosphide catalysts contain metal sites (responsible for hydrogenation and hydrogenolysis) and two types of acid sites (Lewis acid sites, Br∅nsted acid sites). Ligand effect is responsible for the Lewis acid sites ($Ni^{+\delta}$), while the Br∅nsted acid sites (P-OH groups) exist on the surface of phosphide. In brief, the advantage of TMPs is the *ligand effect (electronic)* and the *ensemble effect (geometric)* [134].

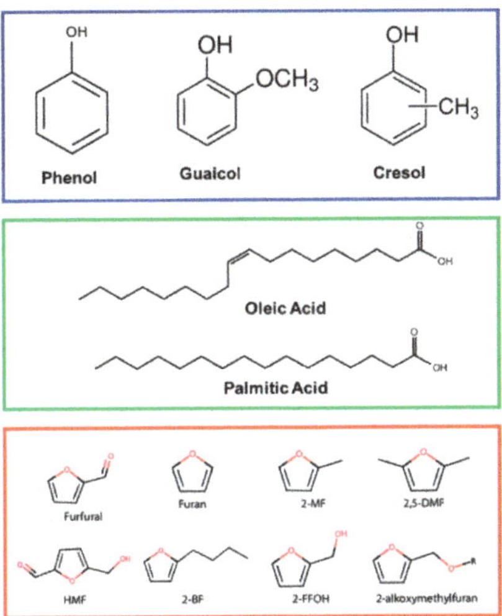

Figure 7. Representative compounds from lignin phenols, fatty acids and cellulose furans.

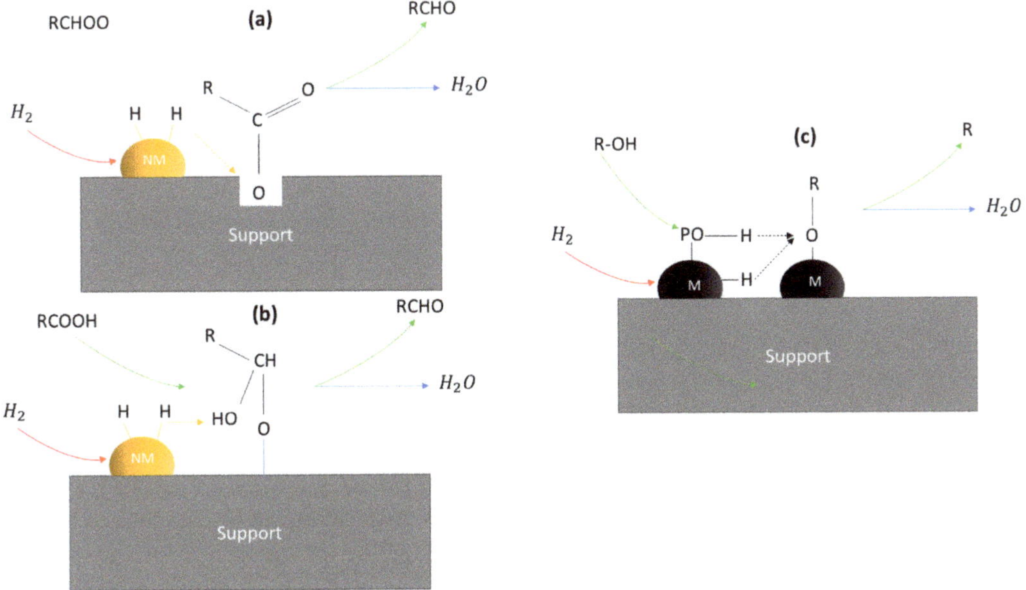

Figure 8. (**a**,**b**) HDO mechanisms over noble metal (NM) supported catalysts. Adapted from Refs [135,136]. (**c**) Proposed HDO mechanisms over TMPs catalysts. Adapted from Refs [137,138], M stands for metals.

4.1. Lignin Phenolic Derivatives

Lignin is a natural polymeric material composed of the monomeric units syringyl (S), guaiacyl (G), and p-hydroxyphenyl (H). Lignin phenolic derivatives have been studied as products of bio-oil processing. TMPs have been extensively used for the HDO reaction of model compounds of this category, such as phenol [139], guaiacol [140], catechol [141], and cresols [142].

Under realistic HDO conditions, a temperature range of 380–430 °C and high H_2 pressure is used. Many reactions, including cleavage of interunit linkages, deoxygenation reaction, ring hydrogenation reaction, as well as the removal of alkyl and methoxyl moieties, occur. The hydrogen pressure, typically 50–150 bar, strongly influences the oil yield. Catalysts design criteria can be as follows: (a) high activity for hydrogenolysis and/or cleavage of C-O-C and C-C bonds (metal sites function), (b) low activity for ring hydrogenation; (c) good selectivity for aromatic compounds (d) tolerance against coking and sulfur. Another critical parameter is the active metal particle size; the latter regulates the C-O cleavage activity and tunes the selectivity to aromatic compounds. For example, tiny metal particles suppress the co-adsorption of the aromatic ring and H_2 and thus the hydrogenation of the first. Apart from the metal particle size, the geometry of adsorption is also a crucial factor. For example, it has been reported that using Pt(111) surface favors the anisole horizontal orientation (parallel to the surface), while adding Zn to the Pt(111) can induce a tilted mode of adsorption. These two completely different geometries of adsorption favor the ring saturation and the selectivity to aromatic compounds, respectively [143].

On a comparative note, in a metal-supported catalyst, the metal sites provide the function of hydrogenation, whereas the support provides the oxophilic sites; the latter tune the C-O cleavage activity of the catalyst. Popular oxophilic metal oxides with strong Lewis acid character are the VOx, NbOx, MoOx, ReOx, WOx, and TaOx. However, studies by Jing 2019b [126] showed that the catalytic activity does not coincide with the Lewis acidity order.

Trying to get an insight and consider phenol as the model compound of this category, there are three different mechanistic pathways for removing -OH in the phenolic ring. In particular, these routes are: (a) direct hydrogenolysis of -OH in the phenol ring, (b) hydrogenation followed by dehydration and then dehydrogenation, and (c) tautomerization followed by hydrogenation and dehydration route [143].

The main focus of the research in this front is to design selective catalysts that allow the lowering of the use of hydrogen while preserving the aromatic character of the lignin compound. HDO of lignin model compounds can be efficiently performed over a copper chromite catalyst (Deutsch and Shanks, 2012). For the benzyl alcohol, the hydroxymethyl group is very reactive under HDO. In the case of anisole, demethoxylation is the primary reaction pathway compared to demethylation and transalkylation. The latter is predominantly for conventional hydrotreating catalysts. The hydroxyl group of phenol contributed to the activation of the aromatic ring toward cyclohexanol and cyclohexane production.

Guaiacol is another popular model compound of pyrolysis oil with a challenging molecule to fully deoxygenate due to the co-presence of phenolic and methoxy groups. In their study, Zhao et al. [144] reported that HDO of guaiacol leads to benzene and phenol as the major products with a small amount of methoxybenzene. Moreover, the formation of reaction intermediates, such as catechol and cresol, were traced. Guaiacol (methoxyphenol) is used due to the high lignin concentration that tend to be highly coke. In the same study, the performed comparison of the TMP catalysts with two commercial ones is noteworthy to be discussed. In particular, 5% Pd/Al_2O_3 catalyst activity was higher at lower contact time, but catechol was the sole product. A sulfide catalyst, commercial $CoMoS/Al_2O_3$, had low activity for the HDO of guaiacol under the studied conditions.

The TOF was found to follow the order $Ni_2P/SiO_2 > Co_2P/SiO_2 > FeP/SiO_2 > WP/SiO_2 > MoP/SiO_2$. The nickel and cobalt TMPs had a smooth activity which is an indication of no deactivation. However, Fe has a high variation of TOF with a small range of temperature, which means it underwent deactivation, which may occur due to coking. These results propose Ni and Co as the most reactive and stable TMP in the HDO of guaiacol.

The reaction network of HDO of guaiacol HDO involves the hydrogenolysis of the methyl–oxygen bond (O-CH$_3$ group) leading to catechol formation, followed by the hydroxyl groups removal towards phenol production and water and then hydrocarbons (benzene or cyclohexene) [145–147]. Activation energies for the overall reaction of guaiacol were found to be in the 23–63 kJ/mol range; these values were lower compared to a network involving carbon-oxygen bond breaking (>240 kJ/mol) [144]. These low activation energies suggest that another pathway is followed for the oxygen removal: that of hydrogenation of aromatic double bonds followed by water elimination [145–147]. Recently, another mechanistic scheme has been proposed for the HDO of guaiacol, eliminating the production of catechol through the direct hydrogenolysis leading to phenol production from guaiacol [148].

In another study, Inocencio et al. [118] (Figure 9) investigated the HDO performance of TMPs (Ni, Mo, Co, Fe, W) for the phenol HDO reaction. The highest activity was found over the FeP, whereas the order of activity was FeP > MoP = Ni$_2$P > WP = CoP. The TOF (min^{-1}) was found to be 1.94, 0.67, 0.28 and 0.44, respectively. Benzene was the main product for all the under-study catalysts, along with cyclohexene, cyclohexane, cyclohexanone, and C5-C6 hydrocarbons.

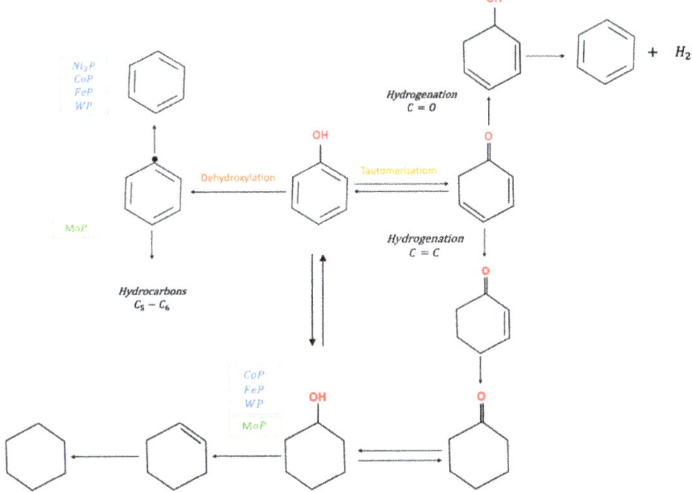

Figure 9. Reaction pathway for HDO of phenol over various supported TMP catalysts. Adapted from Ref. [118].

Inocencio attempted to correlate the measured activity results with the structure of the TMPs used. In particular, the crystal structures for the different TMPs are: MoP (WC type hexagonal); Ni$_2$P (Fe$_2$P-type Orthorhombic); WP, CoP, FeP (MnP-type Orthohombic). It was found that the distribution of products depends on the type of TMP crystal structure. The direct deoxygenation reaction is running over WC (hexagonal) and Fe$_2$P (Orthorhombic), whereas MnP (Orthorhombic) structure seems to facilitate the aromatic ring hydrogenation. The following selectivity order was found: Ni$_2$P > MoP > CoP $\approx$ WP > FeP. This result has to be seen associated with the P/Metal ratio for each crystal structures. WC (hexagonal) and Fe$_2$P (Orthorhombic) exhibit a higher phosphorus/metal ratio than MnP (Orthorhombic), which promotes stronger adsorption of the model molecule on the Ni cation through the oxygen atom. Gonçalves et al. [149] compared the activity of Ni and Ni$_2$P supported catalysts on SiO$_2$ and ZrO$_2$ for the direct deoxygenation of m-cresol to toluene. Among the catalysts investigated, the Ni$_2$P/ZrO$_2$ had the highest activity. It was also shown that the support nature (SiO_2, ZrO_2) did not impact the direct deoxygenation route. The acidic nature of zirconia was found to affect the aromatic ring or double bond hydrogenation along with the dehydration and isomerization reactions.

4.2. Vegetable Oils

Triglycerides of higher fatty acids are the main constituents of vegetable oils. Typically, vegetable oils contain triglycerides of palmitic acid (C16:0), oleic acid (C18:0), and linoleic acid (C18:2) [143]. The carbon chain length in the acids resembles the length of the carbon backbone of the diesel fuel and this makes the vegetable oils an ideal candidate for green diesel production. There is extensive literature on metal-supported catalysts for the HDO reaction of the higher fatty acids [9,10]. Shi et al. [150] studied supported Ni_2P catalyst on SiO_2, CeO_2, TiO_2, SAPO-11, γ-Al_2O_3, and HY zeolite, with a Ni/P ratio of 1.0 for the HDO reaction of methyl laurate to yield C11 and C12 hydrocarbons. It was found that Ni_2P/SiO_2 had the highest conversion rate at all temperatures, which related to the SiO_2 support effect (e.g., reduced acidity or reducibility), demonstrating the vital support role on the TMPs activity. Activity and selectivity were both found to increase with temperature.

M. Peroni et al. [151] (Figure 10) studied different TMPs (WP, MoP, and Ni2P) for the catalytic hydrodeoxygenation of palmitic acid. The exposed metal sites are directly correlated with the catalytic activity. The activation energies were found to be in the 57 kJ mol^{-1} for MoP to 142 kJ mol^{-1} for WP. On WP, the conversion of palmitic acid proceeds via R-$CH_2COOH \rightarrow R$-$CH_2CHO \rightarrow R$-$CH_2CH_2OH \rightarrow R$-$CHCH_2 \rightarrow R$-CH_2CH_3 (hydrodeoxygenation). Decarbonylation of the intermediate aldehyde (R-$CH_2COOH \rightarrow R$-$CH_2CHO \rightarrow R$-CH_3) was a crucial intermediate in the case of MoP and Ni_2P. An alternative pathway was followed on Ni_2P: that of dehydration to a ketene, followed by decarbonylation. The important role of Lewis acid sites was revealed as the alcohol dehydration (R-$CH_2CH_2OH \rightarrow R$-$CHCH_2$) was found to depend on their concentration.

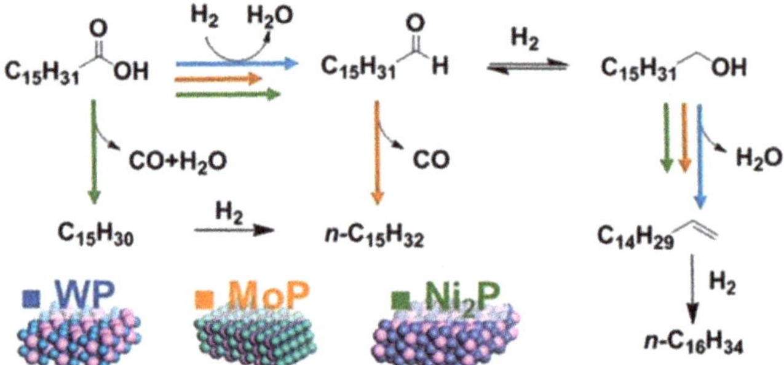

Figure 10. Different mechanistic pathways of the reaction. Reprinted with permission from Ref. [151]. Copyright 2017 ACS Publication.

4.3. Cellullose and Semi-Cellullose Furan Derivatives

Acid-catalyzed transformation of cellulose leads to the formation of furanic derivatives and by-products because of the acid process conditions. The HDO reaction of furfural [152], 2-methylfuran [153], valerolactone [19], and levulinic acid [154] has been investigated over TMPs.

Bui et al. [105] studied the TMPs catalytic activity for HDO of 2-Methyltetrahydrofuran (2-MTHF); the TMPs were prepared following phosphite and phosphate method. Catalysts prepared by phosphide method (I) have a slight difference at TOF (s^{-1}) with an order of Ni_2P/$SiO_2 >$ CoP/$SiO_2 =$ MoP/$SiO_2 \sim$ WP/SiO_2 with expectation of FeP(I) that has a very low activity. On the other hand, the phosphate method catalysts (A) have different activity following Ni_2P/$SiO_2 >$ WP/$SiO_2 >$ CoP/$SiO_2 \sim$ MoP/$SiO_2 >$ FeP/SiO_2. As observed from the results, Ni_2P/SiO_2 is the most effective catalyst while FeP/SiO_2 is the lowest one. In addition, CoP and MoP are more active using the phosphite method (I) while for Ni_2P, WP and FeP is the opposite where Ni_2P (A) is 1.4 times than Ni_2P (I). The difference between phosphate and phosphite results are not significant so using the phosphite method is more practical since it does not require calcination whereas use of lower reduction temperature preserves porosity. The same results were found at a lower temperature (275 °C) where the order of activity of the catalyst by phosphite method is WP > Ni_2P > MoP > CoP > FeP and by phosphate method is Ni_2P > WP > MoP > CoP > FeP. However, the phosphite method can affect the HDO reaction pathway caused by the deposition of excess phosphorus on the surface. In all cases, the phosphide method has higher conversions except for Ni_2P and FeP and needs further studies. The average total conversions of the catalysts are $Ni_2P(16\%)$ > WP(13.5%) > MoP(5.5%) > CoP(3%) > FeP(0.7%). In addition, the study shows that the catalyst did not have any activity at 275 °C and deactivate at 300 °C therefore the apparent activation energy E_a calculated at range of 300-325 °C. The activation energy range was between 88–137 kJ/mol which is lower than the C-O bond dissociation energy (360 kJ/mol). The average HDO conversions of catalysts in order are MoP > FeP > WP > Ni_2P > CoP. Fe, Co, Ni and Pd generate C4 products by decarbonylation mechanism while Mo and W do not. MoP (I) and (A) has the highest HDO activity but not the highest in the total conversion.

The HDO of furfural towards the production of 2-methylfuran (MF) was investigated over different TMPs (Figures 11 and 12), having as a particular focus of the study to establish the role of P and M/P stoichiometry in the TMP on the catalytic activity. Furfural is an important chemical in the resins industry. Metal-based catalysts (e.g., Co [155], Pt [156], Pd [157] are somewhat too active in their interaction with the furan ring leading to the ring opening or ring hydrogenation reactions. This issue seems to be efficiently addressed using TMPs. With a closer look at the furan ring-metal (Ni or NiP) interaction, the following remarks can be made: the interaction involved d-band of Ni and π^* of furan ring. In the case of Ni/SiO_2 catalyst, Ni holds most of the electronic density, making the interaction with the furan ring severe (opening). As P is introduced, Ni electronic density drops and the d-π interaction between Ni-P and furan ring becomes milder. The authors performed furfural infrared spectroscopy (IR) studies where the bands at 1675 and 1695 cm^{-1} were both assigned to the C=O stretching vibration of the carbonyl group, whereas also bands corresponding to the C=C stretching vibration coming from the furan ring were identified. It is interesting to note that as the P increases (P/Ni ratio) the carbonyl bands were found to shift to smaller frequency; namely from 1675 (Ni/SiO_2 case) to 1655 cm^{-1} (Ni_2P/SiO_2 case); the latter corroborates for a boost in the carbonyl-surface interaction with the P increase. Regarding the planar adsorption geometry of the furan ring, this seems to be disturbed by the P increase in the TMP. The authors also provided another explanation on the role of P as BrØnsted acid site in the catalytic activity. As it is known that BrØnstead sites close to metal sites favor the C-O hydrogenolysis reaction by protonating the -OH or -OCH$_3$ group, similarly it is inferred that P-OH sites can facilitate the C=O bond hydrogenolysis.

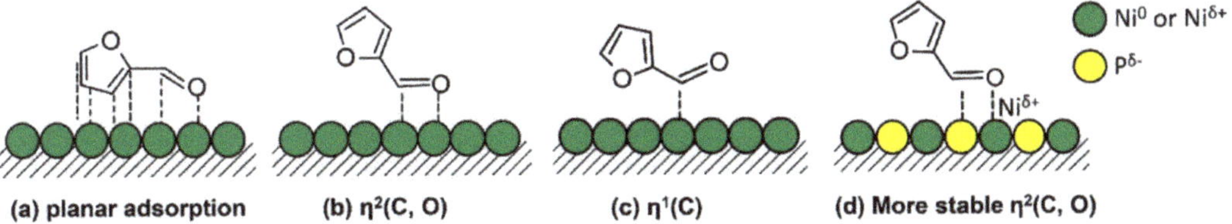

Figure 11. Adsorption configurations of furfural: (**a**) planar adsorption on Ni surface, (**b**) furfural adsorbed through its C and carbonyl O atoms on Ni surface, (**c**) furfural coordinated to Ni surface via C atom on Ni surface and (**d**) furfural adsorption via C and carbonyl O atoms on NiP(x) surface. Reprinted with permission from Ref. [98]. Copyright 2018 Wily Online.

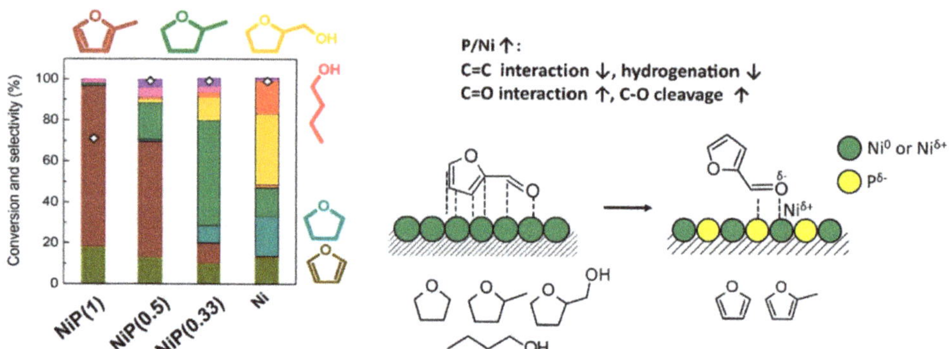

Figure 12. Effect of P on furfural HDO over Ni and nickel phosphide catalysts. Reprinted with permission from Ref. [98]. Copyright 2018 Wily Online.

Another member of the furans family, benzofuran, was investigated by Movick et al. [158] for the HDO reaction over a ternary TMP, CoPdP, with varying Co/Pd ratios and being supported on ultra-stable Y zeolite (KUSY). The Co-Co, Co-P, Pd-Pd, Pd-P interactions were found by analysis of the X-ray absorption spectroscopy spectra of the catalysts. The binary Pd containing TMP most likely is present as Pd_6P, while the ternary TMP had higher catalytic activity compared to both the binary ones (Co-P, Pd-P). According to the CO infrared spectroscopy, Co was found to be the active site, whereas TPR studies showed that Pd facilitates Co reduction.

Along the same lines, dibenzofuran was studied by Molina et al. [159] towards HDO reaction over Ni and Co TMPs. According to the main findings of the study, the Ni_2P catalyst was found to be a better catalyst than the CoP one, achieving conversion of more than 90% of DBF. Contact time, hydrogen pressure and H_2/feed ratio were studied. An increase in contact time increased the catalytic activity and the production of oxygen-free products. In particular, at low values of contact time intermediates, such as tetrahydrodibenzofuran (THDBF) and hexahydrodibenzofuran (HHDBF) were found to be formed, while higher contact time gave rise to the formation of mainly bicyclohexane (BCH), proposing the following pathway DBF → HHDBF → THDBF → 2-CHP → BCH. When it comes to H_2 pressure, both Ni_2P and CoP catalysts were found to be active in intermediate values of pressure. Finally, the H_2/DBF ratio is more impactful in the case of CoP rather than the Ni_2P catalyst.

4.4. On the HDO Activity of the Multi-Elemental TMPs

In catalysis, a single catalyst is rarely capable of carrying out different types of elementary steps needed to deoxygenate oil-derived compounds. Therefore, bi-functional TMP catalysts have been recently receiving great interest, wherein two different TMPs are employed to provide diverse catalytic active sites for various reaction steps. The coordination of two TMPs phases in a catalytic structure can provide a variety and richer active sites as well as modify the electronic structure of each TMP via interfacial interaction for adjusting their propensity to catalyze different elementary reaction steps within the overall reaction. For instance, Liu and co-workers synthesized heterostructured NI_2P-NiP_2 nanoparticles which performed better than the corresponding single TMP components [160]. Density functional theory calculations showed that a strong charge redistribution took place at the interface, since the average valence charge of P (in NiP_2), near the interface, decreased from 5.22 to 5.05 eV. Thus, the valence electron state of active catalytic sites could be optimized owing to the existence of hetero-interface.

Multi-functionality is crucial in catalytic hydrodeoxygenation reactions due to the fact that the HDO catalyst is responsible for multiple reactions such as: C—O bond scission via hydrogenolysis, dissociative adsorption of hydrogen molecule. Mixed phase of TMPs catalysts have exhibited synergistic effect for variety of HDO reactions [161–163]. Bonita et al., [161] synthesized a variety of molybdenum-based bimetallic phosphides (MMoP, M = Ru, Ni, Co and Fe), via temperature-programmed reduction (TPR) method, and tested their application in hydrodeoxygenation reaction. The results revealed that charge sharing between the phosphorus and metals controls the relative oxidation of Mo and hence the products distribution. As regards MMoP (M = Fe, Co, Ni), the more oxidized the molybdenum, the higher the selectivity towards benzene from phenol via direct deoxygenation at 400 °C and 750 psig. Cho et al. [163] prepared bimetallic NiFeP catalysts supported on potassium ion-exchanged USY zeolite through TPR method and investigated their activity in HDO reaction of 2-methyltetrahydrofuran (2-MTHF) at temperature range of 250–300 °C. It was found that the NiFeP catalysts comprised of Ni_2P and FeP phases as revealed by the XRD results. Notably, the bimetallic NiFeP catalysts presented a weak ligand effect on the reaction rate but a considerable ensemble effect on the products distribution in the HDO of 2-MTHF. Particularly, the NiFeP supported catalysts demonstrated a lower selectivity towards butane as compared to Ni_2P supported catalysts, signifying the decarboxylation route is suppressed over the former.

It should be noted that the controlled synthesis of multi-metallic phosphides remains a challenge because of stringent valence balance demand on the stoichiometric ratio and the phase segregation originating from varying reactivity of various metal precursors [164].

5. Deactivation during the HDO Reaction

The main reasons for catalyst deactivation during the HDO reactions are coke formation, metal leaching, sintering, poisoning, and surface oxidation (attack from the water). Coke formation is the main factor of catalyst deactivation since the deposition on the active sites affects the catalyst activity and selectivity when it deposits inside the catalyst pores. The pore size, pore shape, and crystallite size determine the extent and nature of the coke deposit. Carbon is the undesired by-product of polymerization and polycondensation reactions.

The water produced under the HDO reaction conditions can attack P causing its oxidation towards phosphates that may block the active Ni sites and deactivate the catalyst. Additionally, water can induce phase transformation of supports, e.g., converting alumina to boehmite (AlOOH) [165]. According to density functional theory (DFT) calculations, oxygen on Ni_2P surface prefers to interact with phosphorus. These unphosphide $Ni^{\delta+}$ sites are responsible for phenol production during guaiacol HDO reactions. In addition, the P/Ni ratio of phosphide decreases during the HDO reaction, which is an indication of BrØnsted acid sites (i.e, P-OH) drop on the catalyst surface with a simultaneous increase in the phenol production by suppressing its dehydroxylation. To suppress the coke deposition, a proper adjustment of the operating conditions is needed. For example, hydrogen in the feed saturates the carbon precursor compounds (C_xH_y) formed under HDO. In addition, multi-step processes are better/preferred as they assist in minimizing the coking.

Another promising pathway to remove the coke deposits and reduce the deactivation of the catalysts is regeneration. Calcination under oxidant atmosphere at high temperature (500–600 °C) is used to remove the coke deposits. Regeneration cycles reduce the catalyst activity by producing a large amount of acid sites. Regeneration by oxidation is a more efficient reaction method where oxidation and/or reduction (with H_2) of catalyst. This methodology reduces the cost of biofuel by developing a highly active catalyst.

As reported by Liu et al. [166], O atoms interaction/adsorption with Ni_2P (001) originating from water can interact with both P and Ni atoms with a strong preference towards P atoms. In the resultant oxy-phosphide species, electron transfer from P to O takes place. Due to the ligand effect in the Ni_2P, P species may suppress the Ni-O interaction, thus increasing the oxidation resistance of Ni_2P compared to the Ni/SiO_2 catalyst. Kelun Li [167], who studied the HDO reaction of anisole over silica-supported Ni2P, MoP, and NiMoP reported that the higher affinity of Mo with oxygen leads to lower oxidation resistance. So, the deactivation of the NiMoP catalyst is decreased as the Ni/Mo ratio is increased.

6. Computational Approaches to Study the HDO Reaction over TMPs

Besides the experimental investigation of TMPs catalysts, the Density Function Theory (DFT) calculations have been widely adopted not only for mapping out possible reaction routes at the atomic level but also to deeply scrutinize the rate-controlling steps of reaction pathways. For instance, Wongnongwa and coworkers [168] have systemically studied the deoxygenation mechanisms of palmitic acid over Ni_2P catalyst via a multi-pronged approach of experimental, thermodynamics, and DFT studies. The DFT-based computations, based on butyric acid as model molecule of palmitic acid on Ni_2P (001) surface, showed that the rate controlling step of decarbonylation (DCO) route is the hydrogenolysis reaction step, i.e., conversion of butanal to propane as shown in Figure 13, whereas that of hydrodeoxygenation (HDO) is the transformation of butanol into butane. The energy barrier of the rate limiting steps for DCO and HDO routes are 1.52 and 1.99 eV, respectively. Considering the decarboxylation (DCO_2) route, the hydrogenolysis step, in which butyric acid is converted to propane, dominates the DCO_2 reaction, surmounting for a high activation barrier of 2.37 eV. These findings are in line with the experimental observations, which indicated that the DCO pathway is more favorable on Ni_2P catalyst than the HDO pathway with a ratio up to 4.13:1, while the DCO_2 route was found to be unlikely, which was borne out by the absence of the gaseous CO_2 product. Notably, the thermodynamic analysis indicated that the HDO route becomes more competitive when the high reaction temperature is considered.

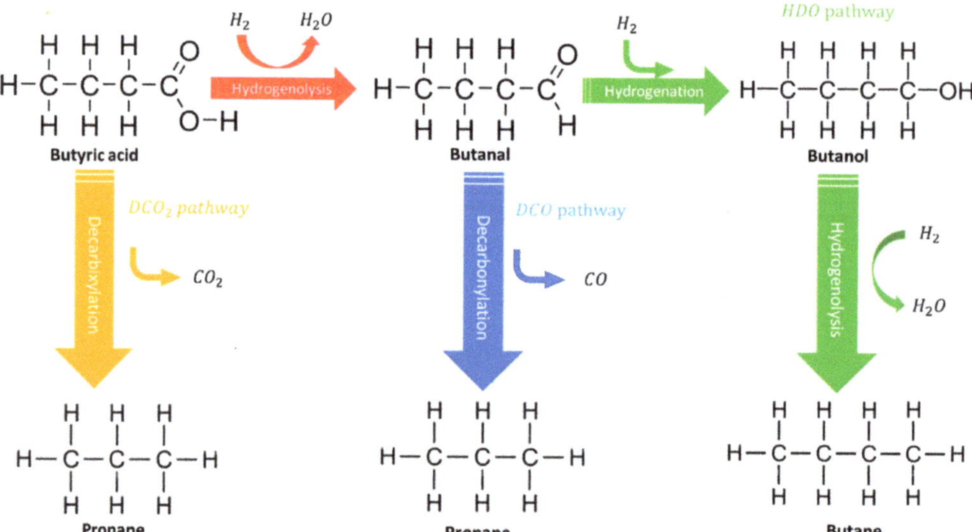

Figure 13. The proposed decarboxylation (DCO_2), decarbonylation (DCO) and hydrodeoxygenation (HDO) mechanisms for deoxygenation process. Adapted from Ref. [168].

It is well-known that the favorable adsorption of reactants and reaction intermediates and their activation could be easily determined by DFT computations, which, in many cases, represent the exact reaction routes during hydrodeoxygenation reactions. For example, Moon et al. [169] (Figure 14) reported that the active site of Ni_2P catalyst, for hydrodeoxygenation of guaiacol reaction, consists of P sites and threefold Ni_3 hollow sites, as confirmed by DFT study together with X-ray absorption fine structure spectroscopy (XAFS) and FTIR measurements, are responsible for adsorption of atomic H or OH groups. These findings indicated that the HDO of guaiacol is greatly affected by the extent of H and OH coverage on the P or Ni sites. In particular, the direct deoxygenation route facilitated the benzene formation on the more OH-covered surface as depicted in Figure 14, while a pre-hydrogenation route is favored on the reduced surface of Ni_2P surface.

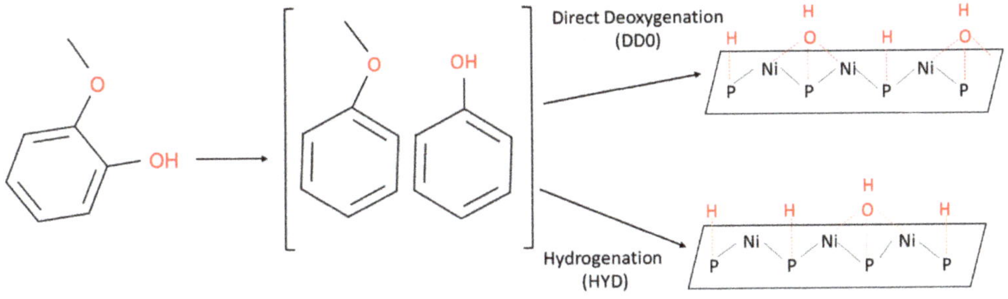

Figure 14. Suggested reaction pathways for guaiacol HDO on Ni_2P catalyst. Adapted from Ref. [169].

A fundamental understanding of the reaction mechanism, at the atomistic level, is of prime importance to unravel the rate-limiting steps for the hydrodeoxygenation reaction on TMPs catalysts. Jain et al. [170] investigated the phenol HDO on ternary phosphide catalytic surfaces, namely FeMoP, RuMoP, and NiMoP, using experimental and computational approaches to determine products distribution and C–O bond scission pathways. The results revealed that the FeMoP surface promotes a direct deoxygenation route, thanks to the lower kinetic barrier for cleaving the C-O bond, whereas the NiMoP and RuMoP favor the hydrogenation of the aromatic ring first, followed by the scission of the C-O bond. These results were found to be in accord with product distribution observed from experiments. He et al. [171] performed systematic DFT calculations for investigating the surface chemistry of selected 1st row TM phosphides, particularly VP, TiP, CrP, Co_2P, Ni_2P, and Fe_2P, in the model reaction of phenol deoxygenation. The results revealed that Ni_2P and Fe_2P exhibited the strongest adsorption energies of phenol, which favored parallel orientation, thanks to the TM ensembles present in these two surfaces. Notably, VP and Fe_2P presented lower kinetic barriers for C-O bond cleavage step, as shown in Figure 15, as compared to other the TMPs surfaces. Considering the selective hydrogenation step of phenyl fragment to benzene, the early 1st row TMPs, i.e., VP, CrP and TiP, demonstrated moderate activation energies, whereas lower energy barriers were recorded on Ni_2P, Co_2P and Fe_2P, pointing out a more facile selective hydrogenation. It is noteworthy that the deoxygenation reaction of phenol towards benzene proceeded via direct scission of $C_{ring} - OH$ bond over all TMPs with the exception of TiP, which was revealed to follow indirect route involves the dehydrogenation of OH_{phenol} then C–O scission as depicted in Figure 15.

It should be noted that TMPs possessed an appropriate level of atomic hydrogen reactivity, carbon affinity, and oxophilicity necessary to efficiently drive recalcitrant phenolic oxygen while maintaining the aromaticity of products. Further insights on how the surface chemistry of TMPs determines their HDO performance have been provided by the work of He and Laursen in the guaiacol deoxygenation over Ni_2P and TiP surfaces [172]. It was reported that the activation barrier for the direct demethoxylation step, i.e., removal of OCH_3 group, is lower on Ni_2P (1.0 eV) than TiP (1.4 eV). The dehydroxylation step, which appears to be kinetically hindered in some experimental studies, was reported to be moderately activated with kinetic barriers of 1.3 and 1.2 eV over Ni_2P and TiP, respectively. These activation energies signify that TMP are potent catalysts for promoting the removal of both O-containing groups easily under mild process conditions. The parallel orientation of guaiacol on Ni_2P surface was reported to inhibit the removal of methoxy group; however, a direct removal of oxygen moieties is thermodynamically favorable when perpendicular geometry is considered. It is well established that the orientation of the aromatic ring plays a decisive role in determining the reaction mechanism since it dictates $C_{ring}-$ surface interactions, overhydrogenation, C-C bond scission and C=C activation [170]. Parallel geometry can enhance strong C_{ring} − surface interactions, which modifies the aromatic ring geometry. On the other hand, perpendicular orientation promotes a higher degree of C-O interaction with the catalytic surface and thus more facile cleavage of C-O bond if an adequate surface activity is present. Furthermore, Jia et al. [173] examined the C-OH bond cleavage step of cyclohexanol, being a key oxygenated by-product in guaiacol HDO reaction and precursor of cyclohexane, on Ni_3P and Ni_3P_2 surface termination of Ni_2P crystal. The results revealed that the kinetic barrier of C-OH bond scission on Ni_3P (1.02 eV) is greater than that of Ni_3P_2 (0.71 eV) terminated surface, signifying the latter could readily break the C-OH bond during HDO reaction of guaiacol. Wang and coworkers [174] studied the anisole HDO reaction mechanism on Fe_2P (101) surface, and it was reported that the reaction proceeds via direct scission of C_{aryl} − O bond of anisole to produce CH_3O^* and $C_6H_5^*$ reaction intermediate while preventing the formation of $C_6H_5 - OH - CH_3$ intermediate, which coincides well with the experimental findings. The results revealed that C_{aryl} − O bond of anisole could be readily cleaved, requiring an energy barrier of 0.75 eV. The activation barriers were also reported for hydrogenation of $C_6H_5^*$ (0.08 eV) and CH_3O^* (1.10 eV), to form benzene and methanol, respectively. These findings suggest that the formation of the methanol reaction step dominates the HDO reaction on the Fe_2P (101) surface.

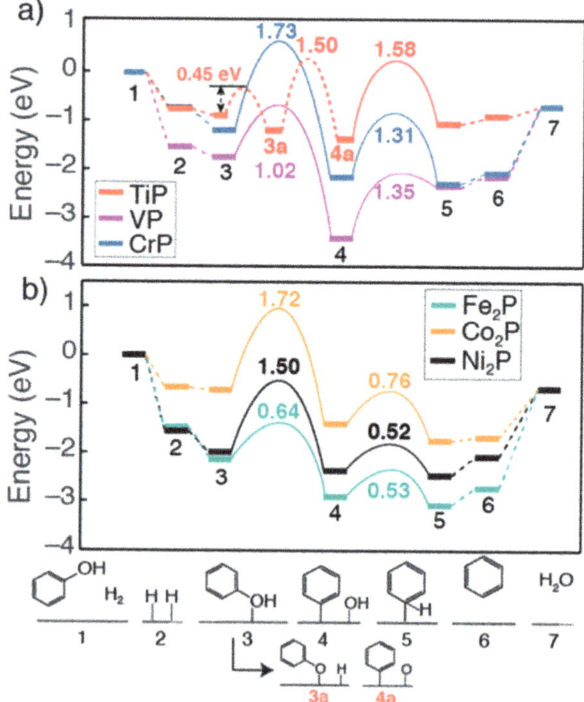

Figure 15. Potential energy landscape for phenol deoxygenation reaction over (**a**) TiP, VP and CrP and (**b**) Fe_2P, Co_2P and Ni_2P surfaces. Reprinted with permission from Ref. [171]. Copyright 2017 ACS Publication.

A few computational studies have reported the deoxygenation of furan derivatives on TMPs catalysts [174,175] For instance, Witzke et al. [175] investigated the hydrogenolysis mechanism of 2-Methyltetrahydrofuran (MTHF) over $Ni_2P(001)$, $Ni_{12}P_5(001)$, and $Ni(111)$ catalysts. DFT computations revealed that increasing the P:Ni ratio of the catalyst have resulted in decreasing the activation barrier for $^2C - O$ bond rupture, pertained to conversion of MTHF to 2-pentene-4-one intermediate pathway, as compared to that of ^{3} C $-$ O bond rupture related to hydrogenolysis pathway involving the formation of pentanal intermediate. The exposed Ni_3 hollow sites, present on Ni_2P and Ni surfaces, and Ni_4 hollow sites, existed on $Ni_{12}P_5$ surface, are the active sites for cleaving C $-$ O bond.

Regarding the sintering and coking, surface oxidation is deemed to be a possible source of catalyst deactivation [176,177]. Cecilia and coworkers [177] reported that phosphorus in Ni_2P catalyst could be oxidized, during the course of the reaction, to form phosphate, which is expected to cover the metallic active sites in TMPs catalysts, thereby resulting in catalyst deactivation. First-principles calculations revealed that atomic oxygen on a Ni_2P surface interact preferably with P surface atoms [178], which would be the convenient environment for oxidizing phosphorus to undesired phosphate species. It can be concluded that there is a pressing need for more computational studies to be considered for deactivation and regenerations of TMPs catalysts and HDO catalytic mechanisms at the atomistic level since these aspects are crucial before applying technology into industrial manufacture of biofuels. In this context, DFT computations should be adopted to deeply identify the detailed pathways and the rate-controlling steps for the HDO process.

Rensel et al. [178] studied the phenol HDO reaction on bimetallic $Fe_xMo_{2-x}P$ ($x = 1$ and 1.5) catalysts using both experimental and DFT computational approaches to elucidate the role of surface Lewis acidity on the reaction mechanism. The results indicated that the rate determining step is the $C_{ring} - O$ scission because it has the greatest activation barrier among the studied elementary reaction steps. The $C_{ring} - O$ bond cleavage takes place readily on Fe_1Mo_1P than $Fe_{1.5}Mo_{0.5}P$ surface, surmounting for the kinetic barrier of 0.39 eV and 0.77 eV, respectively, which was explained by the greater partial charge on metallic species of the former (+0.81 | e |) as compared to the latter (+0.63 | e |). This can be correlated to the improved Lewis acid character of the Fe_1Mo_1P as compared to $Fe_{1.5}Mo_{0.5}P$ catalyst.

Wanmolee and co-workers [133] studied the HDO of guaiacol on Cu_3P/SiO_2 catalysts. The most predominant facets based on XRD analysis, the guaiacol adsorption was investigated on (100) and (113) surfaces of Cu_3P. It was reported that the active sites for adsorbing the reactants, guaiacol and H_2, are Cu sites for the studied surfaces. The adsorption of guaiacol is more favorable via Cu-OH and Cu-C binding modes than Cu-OCH$_3$ mode because of the steric hindrance of the latter. The Cu_3P (113) surface presented an enhanced adsorption capability as compared to (100) for all reactants and reaction intermediates, which was correlated with the lesser stability of the former as confirmed by surface energy results. The Bader charge analysis revealed that the P sites are negatively charged while the Cu sites are positively charged, acting as Lewis acid sites.

7. Future Perspectives and Outlook

The up-to-date advancements of the HDO catalytic chemistry for converting pyrolysis bio-oils to fuels show that the process needs more time to be considered mature. TMPs are only one class of catalysts being successfully used for this reaction and with clear benefits compared to noble metal or pure transition metal catalysts. The complex electronic structure along with the bulk and surface properties of TMPs allow for a bifunctional catalytic chemistry to take place while it can be tuned by adding second metal or support. The increased activity of the TMPs towards oxygen drives the C–O bond breaking. The different electronic structure of the early TMP and the different extent of hybridization gives room for the rational design of catalysts. The above can be tailored by choosing the appropriate M, P precursors, and synthesis method to control the morphology and the size of the final TMP. The P precursors can be challenging as many are available (organic and inorganic). The role of P sites is often limited to that of spectator or binding provider. Paying more attention to the active site and the role of P in formulating that through a mechanistic approach would benefit the rational design of catalysts. In such a case, a better understanding will improve the TMPs stability for real life conditions/applications. In that direction, it would be worth understanding the TMPs behavior under other reactions/conditions where the

Nanomaterials **2022**, *12*, 1435

reactants, temperature, pressure vary. Having a closer look at the different reaction steps, the breaking of C–C/C–O bonds needs to be tailored as it leads to the loss of atoms in biomass; from that aspect, an atom-economical approach needs to be followed. In such an attempt, the OCH_3 (methoxy) groups can be used for chemical production so to utilize the carbon atoms efficiently. Additionally, OCH_3/OH groups can be used as H_2 carriers to self-support the HDO reaction, while dealing with the water produced during HDO is worth more attention as the latter can facilitate structure modifications or can co-catalyze some reactions steps (e.g., breaking of C–O bond). Another aspect is the facet-controlled growth of the TMPs and the study of the different exposed facets. Succeeding to grow TMPs that expose their most catalytically active facet can further add to the design of a catalyst. When it comes to the multi-elemental TMPs, such as ternary TMPs, strict control of the preparation conditions is needed (pH, P source, temperature, complexing agent) as in these systems, the activity of two metals is combined with P and thus control of the final stoichiometry is a challenging task. Multi-elemental TMPs catalysts have been rarely studied for HDO reactions. The key reason behind the importance of multi-elemental TMPs is the necessity of both hydrogenation and base/acid active sites. While many multi-elemental TMPs catalysts presented promising activity in HDO reactions, some key unresolved questions regarding the mechanism remain. Particularly, there is a pressing need to unravel the role of the interface between the different TMP phases in the reaction mechanism. In most studies to date, the density of interfacial sites has not been explicitly controlled, so it is difficult to discern the role proximity between distinct types of active sites plays in the reaction mechanism. It is importance to resolve these questions since the rational design of multi-elemental TMPs catalysts would consider tailoring the interface between the different TMPs phases. When these key challenges of TMPs catalysts can be overcome in the next decade, we foresee a promising potential for the industrial application of TMPs and multi-elemental TMPs for converting pyrolysis bio-oils to fuels. When it comes to the HDO reaction itself and how the plethora of TMP compositions impact the activity, more insightful studies are needed for the different process parameters, such as H_2 pressure, H_2/feed ratio, contact time, and how these parameters impact, from a mechanistic point of view the products selectivity/reaction pathway.

Additionally, the development of a cost-effective and easily scalable synthesis protocol still remains one of the barriers to the commercial applicability of TMPs and multi-elemental TMPs and catalysts. Although some of the discussed TMPs catalysts exhibited promising bi-functional catalytic activity for HDO reactions, most of the reported preparation methods are still restricted to lab scale. Therefore, developing a scalable and abundant preparation strategy plays an essential role in the realization of the industrial application of TMPs catalysts.

Author Contributions: L.I.A.-A.: Investigation, Methodology, Formal Analysis, Writing—original draft preparation. O.E.: Investigation, Formal Analysis, writing—original draft preparation. K.A.A.: Conceptualization, writing—review and editing, supervision, project administration. N.S.: Conceptualization, writing—review and editing, supervision, project administration. K.P.: Conceptualization, Writing—original draft preparation, writing—review & editing, Supervision, Resources, Project administration, Funding acquisition and evaluation of the findings. All authors have read and agreed to the published version of the manuscript.

Funding: This research was funded by Khalifa University grant number [RC2-2018-024 and CIRA-2020-077]. The APC was funded by [RC2-2018-024].

Data Availability Statement: Not Applicable.

Conflicts of Interest: The authors declare no conflict of interest.

References

1. Alvarez-Galvan, M.C.; Campos-Martin, J.M.; Fierro, J.L.G. Transition Metal Phosphides for the Catalytic Hydrodeoxygenation of Waste Oils into Green Diesel. *Catalysts* **2019**, *9*, 293. [CrossRef]
2. Jovičić, N.; Antonović, A.; Matin, A.; Antolović, S.; Kalambura, S.; Krička, T. Biomass Valorization of Walnut Shell for Liquefaction Efficiency. *Energies* **2022**, *15*, 495. [CrossRef]
3. Moretti, L.; Arpino, F.; Cortellessa, G.; Di Fraia, S.; Di Palma, M.; Vanoli, L. Reliability of Equilibrium Gasification Models for Selected Biomass Types and Compositions: An Overview. *Energies* **2022**, *15*, 61. [CrossRef]
4. Shi, Y.; Xing, E.; Wu, K.; Wang, J.; Yang, M.; Wu, Y. Recent progress on upgrading of bio-oil to hydrocarbons over metal/zeolite bifunctional catalysts. *Catal. Sci. Technol.* **2017**, *7*, 2385–2415. [CrossRef]
5. Kumar, R.; Strezov, V.; Lovell, E.; Kan, T.; Weldekidan, H.; He, J.; Dastjerdi, B.; Scott, J. Bio-oil upgrading with catalytic pyrolysis of biomass using Copper/zeolite-Nickel/zeolite and Copper-Nickel/zeolite catalysts. *Bioresour. Technol.* **2019**, *279*, 404–409. [CrossRef]
6. López-Renau, L.M.; García-Pina, L.; Hernando, H.; Gómez-Pozuelo, G.; Botas, J.A.; Serrano, D.P. Enhanced bio-oil upgrading in biomass catalytic pyrolysis using KH-ZSM-5 zeolite with acid-base properties. *Biomass Convers. Biorefin.* **2021**, *11*, 2311–2323. [CrossRef]
7. Charisiou, N.D.; Siakavelas, G.; Tzounis, L.; Dou, B.; Sebastian, V.; Hinder, S.J.; Baker, M.A.; Polychronopoulou, K.; Goula, M.A. Ni/Y_2O_3–ZrO_2 catalyst for hydrogen production through the glycerol steam reforming reaction. *Int. J. Hydrogen Energy* **2020**, *45*, 10442. [CrossRef]
8. Polychronopoulou, K.; Charisiou, N.D.; Siakavelas, G.I.; AlKhoori, A.A.; Sebastian, V.; Hinder, S.J.; Baker, M.A.; Goula, M.A. Ce–Sm–x Cu cost-efficient catalysts for H_2 production through the glycerol steam reforming reaction. *Sustain. Energy Fuels* **2019**, *3*, 673–691. [CrossRef]
9. Charisiou, N.D.; Siakavelas, G.I.; Dou, B.; Sebastian, V.; Hinder, S.J.; Baker, M.A.; Polychronopoulou, K.; Goula, M.A. Nickel supported on $AlCeO_3$ as a highly selective and stable catalyst for hydrogen production via the glycerol steam reforming reaction. *Catalysts* **2019**, *9*, 411. [CrossRef]
10. Polychronopoulou, K.; Bakandritsos, A.; Tzitzios, V.; Fierro, J.L.G.; Efstathiou, A.M. Absorption-enhanced reforming of phenol by steam over supported Fe catalysts. *J. Catal.* **2006**, *241*, 132–148. [CrossRef]
11. Polychronopoulou, K.; Fierro, J.L.G.; Efstathiou, A.M. The phenol steam reforming reaction over MgO-based supported Rh catalysts. *J. Catal.* **2004**, *228*, 417–432. [CrossRef]
12. Musso, M.; Veiga, S.; Perdomo, F.; Rodríguez, T.; Mazzei, N.; Decarlini, B.; Portugau, P.; Bussi, J. Hydrogen production via steam reforming of small organic compounds present in the aqueous fraction of bio-oil over Ni-La-Me catalysts (Me = Ce, Ti, Zr). *Biomass Convers. Biorefin.* **2022**, 1–17. [CrossRef]
13. Papageridis, K.N.; Charisiou, N.D.; Douvartzides, S.L.; Sebastian, V.; Hinder, S.J.; Baker, M.A.; Al Khoori, S.; Polychronopoulou, K.; Goula, M.A. Effect of operating parameters on the selective catalytic deoxygenation of palm oil to produce renewable diesel over Ni supported on Al_2O_3, ZrO_2 and SiO_2 catalysts. *Fuel Process. Technol.* **2020**, *209*, 106547. [CrossRef]
14. Papageridis, K.N.; Charisiou, N.D.; Douvartzides, S.; Sebastian, V.; Hinder, S.J.; Baker, M.A.; AlKhoori, A.A.; AlKhoori, S.I.; Polychronopoulou, K.; Goula, M.A. Continuous selective deoxygenation of palm oil for renewable diesel production over Ni catalysts supported on Al_2O_3 and La_2O_3–Al_2O_3. *RSC Adv.* **2021**, *11*, 8569–8584. [CrossRef]
15. Zhang, Q.; Chang, J.; Wang, T.; Xu, Y. Review of Biomass Pyrolysis Oil Properties and Upgrading Research. *Energy Convers. Manag.* **2007**, *48*, 87–92. [CrossRef]
16. Czernik, S.; Bridgwater, A.V. Overview of Applications of Biomass Fast Pyrolysis Oil. *Energy Fuels* **2004**, *18*, 590–598. [CrossRef]
17. Furimsky, E. Catalytic hydrodeoxygenation. *Appl. Catal. A Gen.* **2000**, *199*, 147–190. [CrossRef]
18. Sunarno, S.; Zahrina, I.; Nanda, W.R.; Amri, A. Upgrading of pyrolysis oil via catalytic co-pyrolysis of treated palm oil empty fruit bunch and plastic waste. *Biomass Convers. Biorefin.* **2022**, 1–9. [CrossRef]
19. Yun, G.; Takagaki, A.; Kikuchia, R.; Oyama, S.T. Hydrodeoxygenation of gamma-valerolactone on transition metal phosphide catalysts. *Catal. Sci. Technol.* **2017**, *7*, 281–292. [CrossRef]
20. Chen, J.; Xu, Q. Hydrodeoxygenation of biodiesel-related fatty acid methyl esters to diesel-range alkanes over zeolite-supported ruthenium catalysts. *Catal. Sci. Technol.* **2016**, *6*, 7239–7251. [CrossRef]
21. Satyarthi, J.K.; Chiranjeevi, T.; Gokak, D.T.; Viswanathan, P.S. An overview of catalytic conversion of vegetable oils/fats into middle distillates. *Catal. Sci. Technol.* **2013**, *3*, 70–80. [CrossRef]
22. Şenol, O.İ.; Ryymin, E.M.; Viljava, T.R.; Krause, A.O.I. Effect of hydrogen sulphide on the hydrodeoxygenation of aromatic and aliphatic oxygenates on sulphided catalysts. *J. Mol. Catal. A Chem.* **2007**, *277*, 107–112. [CrossRef]
23. Ryymin, E.M.; Honkela, M.L.; Viljava, T.R.; Krause, A.O.I. Insight to sulfur species in the hydrodeoxygenation of aliphatic esters over sulfided NiMo/γ-Al_2O_3 catalyst. *Appl. Catal. A Gen.* **2009**, *358*, 42–48. [CrossRef]
24. Dhandapani, B.; Clair, T.S.; Oyama, S.T. Simultaneous hydrodesulfurization, hydrodeoxygenation, and hydrogenation with molybdenum carbide. *Appl. Catal. A Gen.* **1998**, *168*, 219–228. [CrossRef]
25. Elliott, D.C.; Neuenschwander, G.G.; Hart, T.R.; Hu, J.; Solana, A.E.; Cao, C. *Hydrogenation of Bio-Oil for Chemicals and Fuels Production*; Pacific Northwest National Lab. (PNNL), Environmental Molecular Sciences Lab. (EMSL): Richland, WA, USA, 2006.
26. Wang, J.-J.; Li, X.-P.; Cui, B.-F.; Zhang, Z.; Hu, X.-F.; Ding, J.; Deng, Y.-D.; Han, X.-P.; Hu, W.-B. A review of non-noble metal-based electrocatalysts for CO_2 electroreduction. *Rare Met.* **2021**, *40*, 3019–3037. [CrossRef]

27. Lin, Z.; Chen, R.; Qu, Z.; Chen, J.G. Hydrodeoxygenation of biomass-derived oxygenates over metal carbides: From model surfaces to powder catalysts. *Green Chem.* **2018**, *20*, 2679–2696. [CrossRef]

28. Boullosa-Eiras, S.; Lodeng, R.; Bergem, H.; Stocker, M.; Hannevold, L.; Blekkana, E.A. Catalytic hydrodeoxygenation (HDO) of phenol over supported molybdenum carbide, nitride, phosphide and oxide catalysts. *Catal. Today* **2014**, *223*, 44–53. [CrossRef]

29. Wang, W.; Yang, Y.; Bao, J.; Luo, H. Characterization and catalytic properties of Ni–Mo–B amorphous catalysts for phenol hydrodeoxygenation. *Catal. Commun.* **2009**, *11*, 100–105. [CrossRef]

30. Chen, X.; Liang, C. Transition metal silicides: Fundamentals, preparation and catalytic applications. *Catal. Sci. Technol.* **2019**, *9*, 4785–4820. [CrossRef]

31. Wen, C.; Chen, L.; Xie, Y.; Bai, J.; Liang, X. MOF-derived carbon-containing Fe doped porous CoP nanosheets towards hydrogen evolution reaction and hydrodesulfurization. *Int. J. Hydrogen Energy* **2021**, *46*, 33420–33428. [CrossRef]

32. Oyama, S.T.; Wang, X.; Lee, Y.-K.; Bando, K.; Requejo, F.G. Effect of Phosphorus Content in Nickel Phosphide Catalysts Studied by XAFS and Other Techniques. *J. Catal.* **2002**, *210*, 207–217. [CrossRef]

33. Abu, I.I.; Smith, K.J. HDN and HDS of model compounds and light gas oil derived from Athabasca bitumen using supported metal phosphide catalysts. *Appl. Catal. A Gen.* **2007**, *328*, 58–67. [CrossRef]

34. Li, W.; Dhandapani, B.; Oyama, S.T. Molybdenum Phosphide: A Novel Catalyst for Hydrodenitrogenation. *Chem. Lett.* **1998**, *27*, 207–208. [CrossRef]

35. Yin, P.; Yang, Y.S.; Chen, L.F.; Xu, M.; Chen, C.Y.; Zhao, X.J.; Zhang, X.; Yan, H.; Wei, M. DFT Study on the Mechanism of the Water Gas Shift Reaction over NixPy Catalysts: The Role of P. *J. Phys. Chem. C* **2020**, *124*, 6598–6610. [CrossRef]

36. Hughes, M. Silica Supported Transition Metal Phosphides-Alternative Materials for the Water–Gas Shift Reaction. PhD. Thesis, University of Glasgow, Glasgow, UK, 2014.

37. Zhang, F.; Zhang, J.; Li, J.; Jin, X.; Li, Y.; Wu, M.; Kang, X.; Hu, T.; Wang, X.; Ren, W.; et al. Modulating charge transfer dynamics for g-C$_3$N$_4$ through a dimension and interface engineered transition metal phosphide co-catalyst for efficient visible-light photocatalytic hydrogen generation. *J. Mater. Chem. A* **2019**, *7*, 6939–6945. [CrossRef]

38. Zeng, D.; Lu, Z.; Gao, X.; Wu, B.; Ong, W. Hierarchical flower-like ZnIn$_2$S$_4$ anchored with well-dispersed Ni$_{12}$P$_5$ nanoparticles for high-quantum-yield photocatalytic H$_2$ evolution under visible light. *Catal. Sci. Technol.* **2019**, *9*, 4010–4016. [CrossRef]

39. Liu, S.; Huang, J.; Su, H.; Tang, G.; Liu, Q.; Sun, J.; Xu, J. Multiphase phosphide cocatalyst for boosting efficient photocatalytic H2 production and enhancing the stability. *Ceram. Int.* **2021**, *47*, 1414–1420. [CrossRef]

40. Boakye, F.O.; Fan, M.; Zhang, F.; Tang, H.; Zhang, R.; Zhang, H. Growth of branched heterostructure of nickel and iron phosphides on carbon cloth as electrode for hydrogen evolution reaction under wide pH ranges. *J. Solid State Electrochem.* **2022**, 1–11. [CrossRef]

41. Kolny-Olesiak, J. Recent Advances in the colloidal synthesis of ternarty transition metal phosphides. *Z. Naturforsch.* **2019**, *74*, 709–711. [CrossRef]

42. Liyanage, D.R.; Li, D.; Cheek, Q.B.; Baydoun, H.; Brock, S.L. Synthesis and oxygen evolution reaction (OER) catalytic performance of Ni$_{2-x}$Ru$_x$ P nanocrystals: Enhancing activity by dilution of the noble metal. *J. Mater. Chem. A* **2017**, *5*, 17609–17618. [CrossRef]

43. Mutinda, S.I.; Li, D.; Kay, J.; Brock, S.L. Synthesis and characterization of Co$_{2-x}$ Rh$_x$P nanoparticles and their catalytic activity towards the oxygen evolution reaction. *J. Mater. Chem. A* **2018**, *6*, 12142–12152. [CrossRef]

44. Ren, T.; Li, M.; Chu, Y.; Chen, J. Thioetherification of Isoprene and Butanethiol on Transition Metal Phosphides. *J. Energy Chem.* **2018**, *27*, 930–939. [CrossRef]

45. Lee, Y.K.; Oyama, S.T. Bifunctional nature of a SiO$_2$-supported Ni$_2$P catalyst for hydrotreating: EXAFS and FTIR studies. *J. Catal.* **2006**, *239*, 376–389. [CrossRef]

46. Golubevaa, M.A.; Zakharyana, E.M.; Maximov, A.L. Transition Metal Phosphides (Ni, Co, Mo, W) for Hydrodeoxygenation of Biorefinery Products (a Review). *Pet. Chem.* **2020**, *60*, 1109–1128. [CrossRef]

47. Oyama, S.T. Novel catalysts for advanced hydroprocessing: Transition metal phosphides. *J. Catal.* **2003**, *216*, 343. [CrossRef]

48. Brock, S.L.; Senevirathne, K. Recent developments in synthetic approaches to transition metal phosphide nanoparticles for magnetic and catalytic applications. *J. Solid State Chem.* **2008**, *181*, 1552–1559. [CrossRef]

49. Zhu, Y.; Lu, P.; Li, F.; Ding, Y.; Chen, Y. Metal-Rich Porous Copper Cobalt Phosphide Nanoplates as a High-Rate and Stable Battery-Type Cathode Material for Battery–Supercapacitor Hybrid Devices. *ACS Appl. Energy Mater.* **2021**, *4*, 3962–3974. [CrossRef]

50. Alexander, A.M.; Hargreaves, J.S. Alternative catalytic materials: Carbides, nitrides, phosphides and amorphous boron alloys. *Chem. Soc. Rev.* **2010**, *39*, 4388–4401. [CrossRef]

51. Chen, J.-H.; Whitmire, K.H. A Structural Survey of the Binary Transition Metal Phosphides and Arsenides of the D-Block Elements. *Coord. Chem. Rev.* **2018**, *355*, 271–327. [CrossRef]

52. Bussell, M.E. New methods for the preparation of nanoscale nickel phosphide catalysts for heteroatom removal reactions. *React. Chem. Eng.* **2017**, *2*, 628–635. [CrossRef]

53. Pu, Z.; Liu, T.; Amiinu, I.S.; Cheng, R.; Wang, P.; Zhang, C.; Ji, P.; Hu, W.; Liu, J.; Mu, S. Transition-Metal Phosphides: Activity Origin, Energy-Related Electrocatalysis Applications, and Synthetic Strategies. *Adv. Funct. Mater.* **2020**, *30*, 2004009. [CrossRef]

54. Shi, Y.; Li, M.; Yu, Y.; Zhang, B. Recent advantages in nanostructured transition metal phospides: Synthesis and energy-related applications. *Energy Environ. Sci.* **2020**, *13*, 4564–4582. [CrossRef]

55. Jun RE, N.; Wang, J.G.; Li, J.F.; Li, Y.W. Density functional theory study on crystal nickel phosphides. *J. Fuel Chem. Technol.* **2007**, *35*, 458–464.

56. Pei, Y.; Cheng, Y.; Chen, J.; Smith, W.; Dong, P.; Ajayan, P.M.; Ye, M.; Shen, J. Recent developments of transition metal phosphides as catalysts in the energy conversion field. *J. Mater. Chem. A* **2018**, *6*, 23220. [CrossRef]

57. Yu, S.H.; Chua, D.H. Toward high-performance and low-cost hydrogen evolution reaction electrocatalysts: Nanostructuring cobalt phosphide (CoP) particles on carbon fiber paper. *ACS Appl. Mater. Interfaces* **2018**, *10*, 14777–14785. [CrossRef]

58. Wu, C.; Yang, Y.; Dong, D.; Zhang, Y.; Li, J. In situ coupling of CoP polyhedrons and carbon nanotubes as highly efficient hydrogen evolution reaction electrocatalyst. *Small* **2017**, *13*, 1602873. [CrossRef]

59. Zeng, Y.; Wang, Y.; Huang, G.; Chen, C.; Huang, L.; Chen, R.; Wang, S. Porous CoP nanosheets converted from layered double hydroxides with superior electrochemical activity for hydrogen evolution reactions at wide pH ranges. *Chem. Commun.* **2018**, *54*, 1465–1468. [CrossRef]

60. Ye, C.; Wang, M.Q.; Chen, G.; Deng, Y.H.; Li, L.J.; Luo, H.Q.; Li, N.B. One-step CVD synthesis of carbon framework wrapped Co 2 P as a flexible electrocatalyst for efficient hydrogen evolution. *J. Mater. Chem. A* **2017**, *5*, 7791–7795. [CrossRef]

61. Huang, H.B.; Luo, S.H.; Liu, C.L.; Yi, T.F.; Zhai, Y.C. High-Surface-Area and Porous Co2P Nanosheets as Cost-Effective Cathode Catalysts for Li–O2 Batteries. *ACS Appl. Mater. Interfaces* **2018**, *10*, 21281–21290. [CrossRef]

62. Liu, M.; Zhang, R.; Zhang, L.; Liu, D.; Hao, S.; Du, G.; Asiri, A.M.; Kong, R.; Sun, X. Energy-efficient electrolytic hydrogen generation using a Cu_3P nanoarray as a bifunctional catalyst for hydrazine oxidation and water reduction. *Inorg. Chem. Front.* **2017**, *4*, 420–423. [CrossRef]

63. Fan, M.; Chen, Y.; Xie, Y.; Yang, T.; Shen, X.; Xu, N.; Yu, H.; Yan, C. Half-cell and full-cell applications of highly stable and binder-free sodium ion batteries based on Cu_3P nanowire anodes. *Adv. Funct. Mater.* **2016**, *26*, 5019–5027. [CrossRef]

64. Jiang, Y.; Lu, Y.; Lin, J.; Wang, X.; Shen, Z. A hierarchical MoP nanoflake array supported on Ni foam: A bifunctional electrocatalyst for overall water splitting. *Small Methods* **2018**, *2*, 1700369. [CrossRef]

65. Lan, K.; Wang, X.; Yang, H.; Iqbal, K.; Zhu, Y.; Jiang, P.; Tang, Y.; Yang, Y.; Gao, W.; Li, R. Ultrafine MoP Nanoparticles Well Embedded in Carbon Nanosheets as Electrocatalyst with High Active Site Density for Hydrogen Evolution. *ChemElectroChem* **2018**, *5*, 2256–2262. [CrossRef]

66. Zhang, R.; Wang, X.; Yu, S.; Wen, T.; Zhu, X.; Yang, F.; Sun, X.; Wang, X.; Hu, W. Ternary $NiCo_2Px$ nanowires as pH-universal electrocatalysts for highly efficient hydrogen evolution reaction. *Adv. Mater.* **2017**, *29*, 1605502. [CrossRef]

67. Liu, C.; Zhang, G.; Yu, L.; Qu, J.; Liu, H. Oxygen doping to optimize atomic hydrogen binding energy on NiCoP for highly efficient hydrogen evolution. *Small* **2018**, *14*, 1800421. [CrossRef]

68. Xuan, C.; Wang, J.; Xia, W.; Peng, Z.; Wu, Z.; Lei, W.; Xia, K.; Xin, H.L.; Wang, D. Porous structured Ni–Fe–P nanocubes derived from a prussian blue analogue as an electrocatalyst for efficient overall water splitting. *ACS Appl. Mater. Interfaces* **2017**, *9*, 26134–26142. [CrossRef]

69. Ma, Y.Y.; Wu, C.X.; Feng, X.J.; Tan, H.Q.; Yan, L.K.; Liu, Y.; Kang, Z.-H.; Wang, E.-B.; Li, Y.G. Highly efficient hydrogen evolution from seawater by a low-cost and stable CoMoP@ C electrocatalyst superior to Pt/C. *Energy Environ. Sci.* **2017**, *10*, 788–798. [CrossRef]

70. Asnavandi, M.; Suryanto, B.H.R.; Yang, W.; Bo, X.; Zhao, C. Dynamic Hydrogen Bubble Templated NiCu Phosphide Electrodes for pH-Insensitive Hydrogen Evolution Reactions. *ACS Sustain. Chem. Eng.* **2018**, *6*, 2866–2871. [CrossRef]

71. Park, J.; Koo, B.; Yoon, K.Y.; Hwang, Y.; Kang, M.; Park, J.G.; Hyeon, T. Generalized synthesis of metal phosphide nanorods via thermal decomposition of continuously delivered metal−phosphine complexes using a syringe pump. *J. Am. Chem. Soc.* **2005**, *127*, 8433–8440. [CrossRef]

72. Guan, Q.; Sun, C.; Li, R.; Li, W. The synthesis and investigation of ruthenium phosphide catalysts. *Catal. Commun.* **2011**, *14*, 114–117. [CrossRef]

73. Berenguer, A.; Sankaranarayanan, T.M.; Gómez, G.; Moreno, I.; Coronado, J.M.; Pizarro, P.; Serrano, D.P. Evaluation of transition metal phosphides supported on ordered mesoporous materials as catalysts for phenol hydrodeoxygenation. *Green Chem.* **2016**, *18*, 1938–1951. [CrossRef]

74. Shu, Y.; Oyama, S.T. Synthesis, characterization, and hydrotreating activity of carbon-supported transition metal phosphides. *Carbon* **2005**, *43*, 1517–1532. [CrossRef]

75. Korányi, T.I.; Vít, Z.; Poduval, D.G.; Ryoo, R.; Kim, H.S.; Hensen, E.J.M. SBA-15-supported nickel phosphide hydrotreating catalysts. *J. Catal.* **2008**, *253*, 119–131. [CrossRef]

76. D'Aquino, A.I.; Danforth, S.J.; Clinkingbeard, T.R.; Ilic, B.; Pullan, L.; Reynolds, M.A.; Murray, B.D.; Bussel, B.E. Highly-active nickel phosphide hydrotreating catalysts prepared in situ using nickel hypophosphite precursors. *J. Catal.* **2016**, *335*, 204–214. [CrossRef]

77. Wu, S.-K.; Lai, P.-C.; Lin, Y.-C.; Wan, H.-P.; Lee, H.-T.; Chang, Y.-H. Atmospheric Hydrodeoxygenation of Guaiacol over Alumina-, Zirconia-, and Silica-Supported Nickel Phosphide Catalysts. *ACS Sustain. Chem. Eng.* **2013**, *1*, 349–358. [CrossRef]

78. Gonçalves, V.O.O.; de Souza, P.M.; Cabioc'h, T.; da Silva, V.T.; Noronha, F.B.; Richard, F. Effect of P/Ni Ratio on the Performance of Nickel Phosphide Phases Supported on Zirconia for the Hydrodeoxygenation of m-Cresol. *Catal. Commun.* **2019**, *119*, 33–38. [CrossRef]

79. Moon, J.-S.; Lee, Y.-K. Support Effects of Ni_2P Catalysts on the Hydrodeoxygenation of Guaiacol: In Situ XAFS Studies. *Top. Catal.* **2015**, *58*, 211–218. [CrossRef]

80. Wang, Y.; Liu, F.; Han, H.; Xiao, L.; Wu, W. Metal Phosphide:A Highly Efficient Catalyst for the Selective Hydrodeoxygenation of Furfural to 2-Methylfuran. *ChemistrySelect* **2018**, *3*, 7926–7933. [CrossRef]

81. Oyama, S.T.; Gott, T.; Zhao, H.; Lee, Y.K. Transition metal phosphide hydroprocessing catalysts: A review. *Catal. Today* **2009**, *143*, 94–107. [CrossRef]

82. Prins, R.; Bussell, M.E. Metal phosphides: Preparation, characterization and catalytic reactivity. *Catal. Lett.* **2012**, *142*, 1413–1436. [CrossRef]

83. Rodriguez, J.A.; Kim, J.Y.; Hanson, J.C.; Sawhill, S.J.; Bussell, M.E. Physical and chemical properties of MoP, Ni₂P, and MoNiP hydrodesulfurization catalysts: Time-resolved X-ray diffraction, density functional, and hydrodesulfurization activity studies. *J. Phys. Chem. B* **2003**, *107*, 6276–6285. [CrossRef]

84. Liu, P.; Rodriguez, J.A.; Asakura, T.; Gomes, J.; Nakamura, K. Desulfurization reactions on Ni₂P (001) and α-Mo₂C (001) surfaces: Complex role of P and C sites. *J. Phys. Chem. B* **2005**, *109*, 4575–4583. [CrossRef]

85. Zhou, R.; Zhang, J.; Chen, Z.; Han, X.; Zhong, C.; Hu, W.; Deng, Y. Phase and Composition Controllable Synthesis of Nickel Phosphide-Based Nanoparticles via a Low-Temperature Process for Efficient Electrocatalytic Hydrogen Evolution. *Electrochim. Acta* **2017**, *258*, 866–875. [CrossRef]

86. Liu, P.; Rodriguez, J.A. Catalysts for hydrogen evolution from the [NiFe] hydrogenase to the Ni₂P (001) surface: The importance of ensemble effect. *J. Am. Chem. Soc.* **2005**, *127*, 14871–14878. [CrossRef] [PubMed]

87. Oyama, S.T.; Lee, Y.K. Mechanism of hydrodenitrogenation on phosphides and sulfides. *J. Phys. Chem. B* **2005**, *109*, 2109–2119. [CrossRef] [PubMed]

88. Robinson, A.M.; Hensley, J.E.; Medlin, J.W. Bifunctional Catalysts for Upgrading of Biomass-Derived Oxygenates: A Review. *ACS Catal.* **2016**, *6*, 5026–5043. [CrossRef]

89. Lu, X.; Baker, M.A.; Anjum, D.H.; Basina, G.; Hinder, S.J.; Papawassiliou, W.; Pell, A.J.; Karagianni, M.; Papavassiliou, G.; Shetty, D.; et al. Ni2P Nanoparticles Embedded in Mesoporous SiO₂ for Catalytic Hydrogenation of SO₂ to Elemental S. *ACS Appl. Nano Mater.* **2021**, *4*, 5665–5676. [CrossRef]

90. Lu, X.; Baker, M.A.; Anjum, D.H.; Papawassiliou, W.; Pell, A.J.; Fardis, M.; Papavassiliou, G.; Hinder, S.J.; Gaber, S.A.A.; Gaber, D.A.A.; et al. Nickel Phosphide Nanoparticles for Selective Hydrogenation of SO₂ to H₂S. *ACS Appl. Nano Mater.* **2021**, *4*, 6568–6582. [CrossRef]

91. Yao, Z.W.; Wang, L.; Dong, H. A new approach to the synthesis of molybdenum phosphide via internal oxidation and reduction route. *J. Alloys Compd.* **2009**, *473*, L10–L12. [CrossRef]

92. Aitken, J.A.; Ganzha-Hazen, V.; Brock, S.L. Solvothermal syntheses of Cu₃P via reactions of amorphous red phosphorus with a variety of copper sources. *J. Solid State Chem.* **2005**, *178*, 970–975. [CrossRef]

93. Cecilia, J.A.; Infantes-Molina, A.; Rodríguez-Castellón, E.; Jiménez-López, A. Dibenzothiophene hydrodesulfurization over cobalt phosphide catalysts prepared through a new synthetic approach: Effect of the support. *Appl. Catal. B Environ.* **2009**, *92*, 100–113. [CrossRef]

94. Zhang, G.; Li, B.; Liu, M.; Yuan, S.; Niu, L. Liquid Phase Synthesis of CoP Nanoparticles with High Electrical Conductivity for Advanced Energy Storage. *J. Nanomater.* **2017**, *2017*, 9728591. [CrossRef]

95. Jiang, J.; Wang, C.; Li, W.; Yang, Q. One-Pot Synthesis of Carbon-Coated Ni5P4 Nanoparticles and CoP Nanorods for High-Rate and High-Stability Lithium-ion Batteries. *J. Mater. Chem. A* **2015**, *3*, 23345–23351. [CrossRef]

96. Ha, D.; Moreau, L.M.; Bealing, C.R.; Zhang, H.; Hennig, R.G.; Robinson, R.D. The structural evolution and diffusion during the chemical transformation from cobalt to cobalt phosphide nanoparticles. *J. Mater. Chem.* **2011**, *21*, 11498–11510. [CrossRef]

97. D'Accriscio, F.; Schrader, E.; Sassoye, C.; Selmane, M.; André, R.F.; Lamaison, S.; Wakerley, D.; Fontecave, M.; Mougel, V.; le Corre, G.; et al. A Single Molecular Stoichiometric P-Source for Phase-Selective Synthesis of Crystalline and Amorphous Iron Phosphide Nanocatalysts. *ChemNanoMat* **2020**, *6*, 1208–1219. [CrossRef]

98. Li, H.; Lu, S.; Sun, J.; Pei, J.; Liu, D.; Xue, Y.; Mao, J.; Zhu, W.; Zhuang, Z. Phase controlled synthesis of nickel phosphide nanocrystals and their electro-catalytic performance for hydrogen evolution reaction. *Chem. Eur. J.* **2018**, *24*, 11748–11754. [CrossRef]

99. Sun, S.; Zhou, X.; Cong, B.; Hong, W.; Chen, G. Tailoring the d-Band Centers Endows (NixFe1-x)2P Nanosheets with Efficient Oxygen Evolution Catalysis. *ACS Catal.* **2020**, *10*, 9086–9097. [CrossRef]

100. Ye, E.; Zhang, S.Y.; Lim, S.H.; Bosman, M.; Zhang, Z.; Win, K.Y.; Han, M.Y. Ternary cobalt–iron phosphide nanocrystals with controlled compositions, properties, and morphologies from nanorods and nanorice to split nanostructures. *Chem.–A Eur. J.* **2011**, *17*, 5982–5988. [CrossRef]

101. Li, D.; Arachchige, M.P.; Kulikowski, B.; Lawes, G.; Seda, T.; Brock, S.L. Control of Composition and Size in Discrete Co x Fe2–x P Nanoparticles: Consequences for Magnetic Properties. *Chem. Mater.* **2016**, *28*, 3920–3927. [CrossRef]

102. Mendoza-Garcia, A.; Zhu, H.; Yu, Y.; Li, Q.; Zhou, L.; Su, D.; Kramer, M.J.; Sun, S. Controlled Anisotropic Growth of Co-Fe-P from Co-Fe-O Nanoparticles. *Angew. Chem.* **2015**, *127*, 9778–9781. [CrossRef]

103. Perera, S.C.; Tsoi, G.; Wenger, L.E.; Brock, S.L. Synthesis of MnP nanocrystals by treatment of metal carbonyl complexes with phosphines: A new, versatile route to nanoscale transition metal phosphides. *J. Am. Chem. Soc.* **2003**, *125*, 13960–13961. [CrossRef] [PubMed]

104. Li, Y.; Yang, X.; Zhu, L.; Zhang, H.; Chen, B. Hydrodeoxygenation of phenol as a bio-oil model compound over intimate contact noble metal–Ni₂P/SiO₂ catalysts. *RSC Adv.* **2015**, *5*, 80388–80396. [CrossRef]

105. Bui, P.; Cecilia, J.A.; Oyama, S.T.; Takagaki, A.; Infantes-Molina, A.; Zhao, H. Dan Li, Rodríguez-Castellón, E.; López, A.J. Studies of the synthesis of transition metal phosphides and their activity in the hydrodeoxygenation of a biofuel model compound. *J. Catal.* **2012**, *294*, 184–198. [CrossRef]

106. Koike, N.; Hosokai, S.; Takagaki, A.; Nishimura, S.; Kikuchi, R.; Ebitani, K.; Suzuki, Y.; Oyama, S.T. Upgrading of pyrolysis bio-oil using nickel phosphide catalysts. *J. Catal.* **2016**, *333*, 115–126. [CrossRef]

107. Rodríguez-Aguado, E.; Infantes-Molina, A.; Cecilia, J.A.; Ballesteros-Plata, D.; López-Olmo, R.; Rodríguez-Castellón, E. CoxPy Catalysts in HDO of Phenol and Dibenzofuran: Effect of P Content. *Top. Catal.* **2017**, *60*, 1094–1107. [CrossRef]

108. Wu, S.-K.; Lai, P.-C.; Lin, Y.-C. Atmospheric Hydrodeoxygenation of Guaiacol over Nickel Phosphide Catalysts: Effect of Phosphorus Composition. *Catal. Lett.* **2014**, *144*, 878–889. [CrossRef]

109. Gutiérrez-Rubio, S.; Berenguer, A.; Přech, J.; Opanasenko, M.; Ochoa-Hernández, C.; Pizarro, P.; Čejka, J.; Serrano, D.P.; Coronado, J.M.; Moreno, I. Guaiacol Hydrodeoxygenation over Ni2P Supported on 2D-Zeolites. *Catal. Today* **2020**, *345*, 48–58. [CrossRef]

110. Rensel, D.J.; Rouvimov, S.; Gin, M.E.; Hicks, J.C. Highly Selective Bimetallic FeMoP Catalyst for C–O Bond Cleavage of Aryl Ethers. *J. Catal.* **2013**, *305*, 256–263. [CrossRef]

111. Zhang, Z.; Tang, M.; Chen, J. Effects of P/Ni ratio and Ni content on performance of γ-Al$_2$O$_3$-supported nickel phosphides for deoxygenation of methyl laurate to hydrocarbons. *Appl. Surf. Sci.* **2016**, *360*, 353–364. [CrossRef]

112. Yang, Y.; Chen, J.; Shi, H. Deoxygenation of Methyl Laurate as a Model Compound to Hydrocarbons on Ni2P/SiO2, Ni2P/MCM-41, and Ni2P/SBA-15 Catalysts with Different Dispersions. *Energy Fuels* **2013**, *27*, 3400–3409. [CrossRef]

113. Yang, Y.; Ochoa-Hernández, C.; de la Peña O'Shea, V.A.; Coronado, J.M.; Serrano, D.P. Ni2P/SBA-15 As a Hydrodeoxygenation Catalyst with Enhanced Selectivity for the Conversion of Methyl Oleate Into n-Octadecane. *ACS Catal.* **2012**, *2*, 592–598. [CrossRef]

114. Chen, J.; Yang, Y.; Shi, H.; Li, M.; Chu, Y.; Pan, Z.; Yu, X. Regulating Product Distribution in Deoxygenation of Methyl Laurate on Silica-Supported Ni–Mo Phosphides: Effect of Ni/Mo Ratio. *Fuel* **2014**, *129*, 1–10. [CrossRef]

115. Xin, H.; Guo, K.; Li, D.; Yang, H.; Hu, C. Production of high-grade diesel from palmitic acid over activated carbon-supported nickel phosphide catalysts. *Appl. Catal. B Environ.* **2016**, *187*, 375–385. [CrossRef]

116. Liu, Y.; Yao, L.; Xin, H.; Wang, G.; Li, D.; Hu, C. The production of diesel-like hydrocarbons from palmitic acid over HZSM-22 supported nickel phosphide catalysts. *Appl. Catal. B Environ.* **2015**, 504–514. [CrossRef]

117. Le, T.A.; Ly, H.V.; Kim, J.; Kim, S.S.; Choi, J.H.; Woo, H.C.; Othman, M.R. Hydrodeoxygenation of 2-furyl methyl ketone as a model compound in bio-oil from pyrolysis of Saccharina Japonica Alga in fixed-bed reactor. *Chem. Eng. J.* **2014**, *250*, 157–163. [CrossRef]

118. Inocencio, C.V.M.; de Souza, P.M.; Rabelo-Neto, R.C.; da Silva, V.T.; Noronha, F.B. A systematic study of the synthesis of transition metal phosphides and their activity for hydrodeoxygenation of phenol. *Catal. Today* **2021**, *381*, 133–142. [CrossRef]

119. Peroni, M. Transition Metal Phosphides Catalysts for the Hydroprocessing of Triglycerides to Green Fuel. Ph.D. Thesis, Technical University of Munich, Munich, Germany, 2017.

120. Vereshchagin, O.S.; Pankin, D.V.; Smirnov, M.B.; Vlasenko, N.S.; Shilovskikh, V.V.; Britvin, S.N. Raman spectroscopy: A promising tool for the characterization of transition metal phosphides. *J. Alloys Compd.* **2021**, *853*, 156468. [CrossRef]

121. Layman, K.A.; Bussell, M.E. Infrared Spectroscopic Investigation of CO Adsorption on Silica-Supported Nickel Phosphide Catalysts. *J. Phys. Chem. B* **2004**, *108*, 10930–10941. [CrossRef]

122. Ballinger, T.H.; Yates, J.T., Jr. IR spectroscopic detection of Lewis acid sites on alumina using adsorbed carbon monoxide. Correlation with aluminum-hydroxyl group removal. *Langmuir* **1991**, *7*, 3041–3045. [CrossRef]

123. Wang, X.; Clark, P.; Oyama, S.T. Synthesis, characterization, and hydrotreating activity of several iron group transition metal phosphides. *J. Catal.* **2002**, *208*, 321–331. [CrossRef]

124. Feng, Z.; Liang, C.; Wu, W.; Wu, Z.; van Santen, R.A.; Li, C. Carbon monoxide adsorption on molybdenum phosphides: Fourier transform infrared spectroscopic and density functional theory studies. *J. Phys. Chem. B* **2003**, *107*, 13698–13702. [CrossRef]

125. Singh, K.P.; Bae, E.J.; Yu, J.S. Fe–P: A new class of electroactive catalyst for oxygen reduction reaction. *J. Am. Chem. Soc.* **2015**, *137*, 3165–3168. [CrossRef] [PubMed]

126. Du, H.; Gu, S.; Liu, R.; Li, C.M. Highly active and inexpensive iron phosphide nanorods electrocatalyst towards hydrogen evolution reaction. *Int. J. Hydrogen Energy* **2015**, *40*, 14272–14278. [CrossRef]

127. Burns, A.W.; Layman, K.A.; Bale, D.H.; Bussell, M.E. Understanding the relationship between composition and hydrodesulfurization properties for cobalt phosphide catalysts. *Appl. Catal. A Gen.* **2008**, *343*, 68–76. [CrossRef]

128. Cecilia, J.A.; Infantes-Molina, A.; Rodríguez-Castellón, E.; Jiménez-López, A. A novel method for preparing an active nickel phosphide catalyst for HDS of dibenzothiophene. *J. Catal.* **2009**, *263*, 4–15. [CrossRef]

129. Chen, J.; Shi, H.; Li, L.; Li, K. Deoxygenation of methyl laurate as a model compound to hydrocarbons on transition metal phosphide catalysts. *Appl. Catal. B Environ.* **2014**, *144*, 870–884. [CrossRef]

130. Bekaert, E.; Bernardi, J.; Boyanov, S.; Monconduit, L.; Doublet, M.L.; Ménétrier, M. Direct Correlation between the 31P MAS NMR Response and the Electronic Structure of Some Transition Metal Phosphides. *J. Phys. Chem. C* **2008**, *112*, 20481–20490. [CrossRef]

131. Zuzaniuk, V.; Prins, R. Synthesis and characterization of silica-supported transition-metal phosphides as HDN catalysts. *J. Catal.* **2003**, *219*, 85–96. [CrossRef]

132. Papawassiliou, W.; Carvalho, J.P.; Panopoulos, N.; al Wahedi, Y.; Wadi, V.K.S.; Lu, X.; Polychronopoulou, K.; Lee, J.B.; Lee, S.; Kim, C.Y.; et al. Crystal and electronic facet analysis of ultrafine Ni2P particles by solid-state NMR nanocrystallography. *Nat. Commun.* **2021**, *12*, 4334. [CrossRef]

133. Wanmolee, W.; Sosa, N.; Junkaew, A.; Youngjan, S.; Geantet, C.; Afanasiev, P.; Puzenat, E.; Laurenti, D.; Faungnawakij, K.; Khemthong, P. Phase speciation and surface analysis of copper phosphate on high surface area silica support by in situ XAS/XRD and DFT: Assessment for guaiacol hydrodeoxygenation. *Appl. Surf. Sci.* **2022**, *574*, 151577. [CrossRef]

134. Infantes-Molina, A.; Gralberg, E.; Cecilia, J.; Finocchio, E.; Rodríguez-Castellón, E. Nickel and cobalt phosphides as effective catalysts for oxygen removal of dibenzofuran: Role of contact time, hydrogen pressure and hydrogen/feed molar ratio. *Catal. Sci. Technol.* **2015**, *5*, 3403–3415. [CrossRef]

135. Mendes, M.J.; Santos, O.A.A.; Jordão, E.; Silva, A.M. Hydrogenation of oleic acid over ruthenium catalysts. *Appl. Catal. A Gen.* **2001**, *217*, 253–262. [CrossRef]

136. Pestman, R.; Koster, R.M.; Pieterse, J.A.Z.; Ponec, V. Reactions of Carboxylic Acids on Oxides: 1. Selective Hydrogenation of Acetic Acid to Acetaldehyde. *J. Catal.* **1997**, *168*, 255–264. [CrossRef]

137. He, Z.; Wang, X. Hydrodeoxygenation of model compounds and catalytic systems for pyrolysis bio-oils upgrading. *Catal. Sustain. Energy* **2013**, *1*, 28–52. [CrossRef]

138. Whiffen, V.M.L.; Smith, K.J. Hydrodeoxygenation of 4-Methylphenol over Unsupported MoP, MoS$_2$, and MoOx Catalysts. *Energy Fuels* **2010**, *24*, 4728–4737. [CrossRef]

139. Kanda, Y.; Chiba, T.; Aranai, R.; Yasuzawa, T.; Ueno, R.; Toyao, T.; Kato, K.; Obora, Y.; Shimizu, K.; Uemichi, Y. Catalytic Activity of Rhodium Phosphide for Selective Hydrodeoxygenation of Phenol. *Chem. Lett.* **2019**, *48*, 471–474. [CrossRef]

140. Feitosa, L.F.; Berhault, G.; Laurenti, D.; da Silva, V.T. Effect of the nature of the carbon support on the guaiacol hydrodeoxygenation performance of nickel phosphide: Comparison between carbon nanotubes and a mesoporous carbon support. *Ind. Eng. Chem. Res.* **2019**, *58*, 16164–16181. [CrossRef]

141. Yu, Z.; Wang, A.; Liu, S.; Yao, Y.; Sun, Z.; Li, X.; Wang, Y.; Camaioni, D.M.; Lercher, J.A. Hydrodeoxygenation of phenolic compounds to cycloalkanes over supported nickel phosphides. *Catal. Today* **2019**, *319*, 48–56. [CrossRef]

142. Wang, W.; Zhang, K.; Liu, H.; Qiao, Z.; Yang, Y.; Ren, K. Hydrodeoxygenation of p-cresol on unsupported Ni–P catalysts prepared by thermal decomposition method. *Catal. Commun.* **2013**, *41*, 41–46. [CrossRef]

143. Jing, Y.; Wang, Y. Catalytic Hydrodeoxygenation of Lignin-Derived Feedstock into Arenes and Phenolics. *Front. Chem. Eng.* **2020**, *10*, 10. [CrossRef]

144. Zhao, H.Y.; Li, D.; Bui, P.; Oyamaa, S.T. Hydrodeoxygenation of guaiacol as model compound for pyrolysis oil on transition metal phosphide hydroprocessing catalysts. *Appl. Catal. A Gen.* **2011**, *391*, 305–310. [CrossRef]

145. Hurff, S.J.; Klein, M.T. Reaction pathway analysis of thermal and catalytic lignin fragmentation by use of model compounds. *Ind. Eng. Chem. Fundam.* **1983**, *22*, 426–430. [CrossRef]

146. Bredenberg, J.S.; Huuska, M.; Toropainen, P. Hydrogenolysis of differently substituted methoxyphenols. *J. Catal.* **1989**, *120*, 401–408. [CrossRef]

147. Huuska, M.; Rintala, J. Effect of catalyst acidity on the hydrogenolysis of anisole. *J. Catal.* **1985**, *94*, 230–238. [CrossRef]

148. Ferrari, M.; Maggi, R.; Delmon, B.; Grange, P. Influences of the hydrogen sulfide partial pressure and of a nitrogen compound on the hydrodeoxygenation activity of a CoMo/carbon catalyst. *J. Catal.* **2001**, *198*, 47–55. [CrossRef]

149. Gonçalves, V.O.; de Souza, P.M.; Cabioc'h, T.; da Silva, V.T.; Noronha, F.B.; Richard, F. Hydrodeoxygenation of m-cresol over nickel and nickel phosphide based catalysts. Influence of the nature of the active phase and the support. *Appl. Catal. B Environ.* **2017**, *219*, 619–628. [CrossRef]

150. Shi, H.; Chen, J.X.; Yang, Y.; Tian, S.S. Catalytic deoxygenation of methyl laurate as a model compound to hydrocarbons on nickel phosphide catalysts: Remarkable support effect. *Fuel Process. Technol.* **2014**, *118*, 161–170. [CrossRef]

151. Peroni, M.; Lee, I.; Huang, X.; Baráth, E.; Gutiérrez, O.Y.; Lercher, J.A. Deoxygenation of palmitic acid on unsupported transition metal phosphides. *ACS Catal.* **2017**, *7*, 6331–6341. [CrossRef]

152. Lee, S.P.; Chen, Y.W. Selective hydrogenation of furfural on Ni–P, Ni–B, and Ni–P–B ultrafine materials. *Ind. Eng. Chem. Res.* **1999**, *38*, 2548–2556. [CrossRef]

153. Zhang, J.; Matsubara, K.; Yun, G.N.; Zheng, H.; Takagaki, A.; Kikuchi, R.; Oyama, S.T. Comparison of phosphide catalysts prepared by temperature-programmed reduction and liquid-phase methods in the hydrodeoxygenation of 2-methylfuran. *Appl. Catal. A Gen.* **2017**, *548*, 39–46. [CrossRef]

154. Yu, Z.; Meng, F.; Wang, Y.; Sun, Z.; Liu, Y.; Shi, C.; Wang, W.; Wang, A. Catalytic transfer hydrogenation of levulinic acid to γ-valerolactone over Ni$_3$P-CePO$_4$ catalysts. *Ind. Eng. Chem. Res.* **2020**, *59*, 7416–7425. [CrossRef]

155. Wilson, C.L. Reactions of furan compounds. Part V. Formation of furan from furfuraldehyde by the action of nickel or cobalt catalysts: Importance of added hydrogen. *J. Chem. Soc.* **1945**, 61–63. [CrossRef]

156. Kijeński, J.K.; Winiarek, P.; Paryjczak, T.; Lewicki, A.; Mikołajska, A. Platinum deposited on monolayer supports in selective hydrogenation of furfural to furfuryl alcohol. *Appl. Catal. A* **2002**, *233*, 171–182. [CrossRef]

157. Sitthisa, S.; Pham, T.; Prasomsri, T.; Sooknoi, T.; Mallinson, R.G.; Resasco, D.E. Conversion of furfural and 2-methylpentanal on Pd/SiO$_2$ and Pd–Cu/SiO$_2$ catalysts. *J. Catal.* **2011**, *280*, 17–27. [CrossRef]

158. Movick, W.J.; Yun, G.; Vargheese, V.; Bando, K.K.; Kikuchi, R.; Oyama, S.T. Hydrodeoxygenation of benzofuran on novel CoPdP catalysts supported on potassium ion exchanged ultra-stable Y-zeolites. *J. Catal.* **2021**, *403*, 160–172. [CrossRef]

159. Cecilia, J.A.; Infantes-Molina, A.; Rodríguez-Castellón, E.; Jiménez-López, A.; Oyama, S.T. Oxygen-removal of dibenzofuran as a model compound in biomass derived bio-oil on nickel phosphide catalysts: Role of phosphorus. *Appl. Catal. B Environ.* **2013**, *136–137*, 140–149. [CrossRef]

160. Liu, T.; Li, A.; Wang, C.; Zhou, W.; Liu, S.; Guo, L. Interfacial Electron Transfer of Ni_2P–NiP_2 Polymorphs Inducing Enhanced Electrochemical Properties. *Adv. Mater.* **2018**, *30*, 1803590. [CrossRef]
161. Bonita, Y.; Hicks, J.C. Periodic Trends from Metal Substitution in Bimetallic Mo-Based Phosphides for Hydrodeoxygenation and Hydrogenation Reactions. *J. Phys. Chem. C* **2018**, *122*, 13322–13332. [CrossRef]
162. Cho, A.; Shin, J.; Takagaki, A.; Kikuchi, R.; Oyama, S.T. Ligand and ensemble effects in bimetallic NiFe phosphide catalysts for the hydrodeoxygenation of 2-methyltetrahydrofuran. *Top. Catal.* **2012**, *55*, 969–980. [CrossRef]
163. Kang, Q.; Li, M.; Shi, J.; Lu, Q.; Gao, F. A Universal Strategy for Carbon-Supported Transition Metal Phosphides as High-Performance Bifunctional Electrocatalysts towards Efficient Overall Water Splitting. *ACS Appl. Mater. Interfaces* **2020**, *12*, 19447–19456. [CrossRef]
164. Weigold, H. Behaviour of Co-Mo-Al_2O_3 catalysts in the hydrodeoxygenation of phenols. *Fuel* **1982**, *61*, 1021–1026. [CrossRef]
165. Liu, P.; Rodriguez, J.A.; Takahashi, Y.; Nakamura, K. Water-gas-shift reaction on a Ni2P(001) catalyst: Formation of oxy-phosphides and highly active reaction sites. *J. Catal.* **2009**, *262*, 294–303. [CrossRef]
166. Li, K.; Wang, R.; Chen, J. Hydrodeoxygenation of Anisole over Silica-Supported Ni_2P, MoP, and NiMoP Catalysts. *Energy Fuels* **2011**, *25*, 854–863. [CrossRef]
167. Wongnongwa, Y.; Jungsuttiwong, S.; Pimsuta, M.; Khemthong, P.; Kunaseth, M. Mechanistic and thermodynamic insights into the deoxygenation of palm oils using Ni_2P catalyst: A combined experimental and theoretical study. *Chem. Eng. J.* **2020**, *399*, 125586. [CrossRef]
168. Moon, J.S.; Kim, E.G.; Lee, Y.K. Active sites of Ni_2P/SiO_2 catalyst for hydrodeoxygenation of guaiacol: A joint XAFS and DFT study. *J. Catal.* **2014**, *311*, 144–152. [CrossRef]
169. Jain, V.; Bonita, Y.; Brown, A.; Taconi, A.; Hicks, J.C.; Rai, N. Mechanistic insights into hydrodeoxygenation of phenol on bimetallic phosphide catalysts. *Catal. Sci. Technol.* **2018**, *8*, 4083–4096. [CrossRef]
170. He, Y.; Laursen, S. The surface and catalytic chemistry of the first row transition metal phosphides in deoxygenation. *Catal. Sci. Technol.* **2018**, *8*, 5302–5314. [CrossRef]
171. He, Y.; Laursen, S. Trends in the Surface and Catalytic Chemistry of Transition-Metal Ceramics in the Deoxygenation of a Woody Biomass Pyrolysis Model Compound. *ACS Catal.* **2017**, *7*, 3169–3180. [CrossRef]
172. Jia, Z.; Ji, N.; Diao, X.; Li, X.; Zhao, Y.; Lu, X.; Liu, Q.; Liu, C.; Chen, G.; Ma, L.; et al. Highly Selective Hydrodeoxygenation of Lignin to Naphthenes over Three-Dimensional Flower-like Ni_2P Derived from Hydrotalcite. *ACS Catal.* **2022**, *12*, 1338–1356. [CrossRef]
173. Wang, S.; Xu, D.; Chen, Y.; Zhou, S.; Zhu, D.; Wen, X.; Yang, Y.; Li, Y. Hydrodeoxygenation of anisole to benzene over an Fe2P catalyst by a direct deoxygenation pathway. *Catal. Sci. Technol.* **2020**, *10*, 3015–3023. [CrossRef]
174. Witzke, M.E.; Almithn, A.; Conrad, C.L.; Triezenberg, M.D.; Hibbitts, D.D.; Flaherty, D.W. In Situ Methods for Identifying Reactive Surface Intermediates during Hydrogenolysis Reactions: C-O Bond Cleavage on Nanoparticles of Nickel and Nickel Phosphides. *J. Am. Chem. Soc.* **2019**, *141*, 16671–16684. [CrossRef] [PubMed]
175. Witzke, M.E.; Almithn, A.; Conrad, C.L.; Hibbitts, D.D.; Flaherty, D.W. Mechanisms and Active Sites for C-O Bond Rupture within 2-Methyltetrahydrofuran over Ni, $Ni12P5$, and Ni_2P Catalysts. *ACS Catal.* **2018**, *8*, 7141–7157. [CrossRef]
176. Lan, X.; Hensen, E.J.M.; Weber, T. Hydrodeoxygenation of Guaiacol over Ni_2P/SiO_2–Reaction Mechanism and Catalyst Deactivation. *Appl. Catal. A Gen.* **2018**, *550*, 57–66. [CrossRef]
177. Li, D.; Baydoun, H.; Verani, C.N.; Brock, S.L. Efficient Water Oxidation Using CoMnP Nanoparticles. *J. Am. Chem. Soc.* **2016**, *138*, 4006–4009. [CrossRef]
178. Rensel, D.J.; Kim, J.; Jain, V.; Bonita, Y.; Rai, N.; Hicks, J.C. Composition-directed FeXMo2-XP bimetallic catalysts for hydrodeoxygenation reactions. *Catal. Sci. Technol.* **2017**, *7*, 1857–1867. [CrossRef]

 *nanomaterials*

Article

Propane Steam Reforming over Catalysts Derived from Noble Metal (Ru, Rh)-Substituted LaNiO$_3$ and La$_{0.8}$Sr$_{0.2}$NiO$_3$ Perovskite Precursors

Theodora Ramantani, Georgios Bampos, Andreas Vavatsikos, Georgios Vatskalis and Dimitris I. Kondarides *

Department of Chemical Engineering, University of Patras, 26504 Patras, Greece;
ramantani@chemeng.upatras.gr (T.R.); geoba@chemeng.upatras.gr (G.B.); up1019056@upnet.gr (A.V.);
up1047645@upnet.gr (G.V.)
* Correspondence: dimi@chemeng.upatras.gr; Tel.: +30-2610969527; Fax: +30-2610991527

Abstract: The propane steam reforming (PSR) reaction was investigated over catalysts derived from LaNiO$_3$ (LN), La$_{0.8}$Sr$_{0.2}$NiO$_3$ (LSN), and noble metal-substituted LNM$_x$ and LSNM$_x$ (M = Ru, Rh; x = 0.01, 0.1) perovskites. The incorporation of foreign cations in the A and/or B sites of the perovskite structure resulted in an increase in the specific surface area, a shift of XRD lines toward lower diffraction angles, and a decrease of the mean primary crystallite size of the parent material. Exposure of the as-prepared samples to reaction conditions resulted in the in situ development of new phases including metallic Ni and La$_2$O$_2$CO$_3$, which participate actively in the PSR reaction. The LN-derived catalyst exhibited higher activity compared to LSN, and its performance for the title reaction did not change appreciably following partial substitution of Ru for Ni. In contrast, incorporation of Ru and, especially, Rh in the LSN perovskite matrix resulted in the development of catalysts with significantly enhanced catalytic performance, which improved by increasing the noble metal content. The best results were obtained for the LSNRh$_{0.1}$-derived sample, which exhibited excellent long-term stability for 40 hours on stream as well as high propane conversion (X_{C3H8} = 92%) and H$_2$ selectivity (S_{H2} = 97%) at 600 °C.

Keywords: propane; steam reforming; hydrogen production; perovskite; ruthenium; rhodium; La$_2$O$_2$CO$_3$; stability

Citation: Ramantani, T.; Bampos, G.; Vavatsikos, A.; Vatskalis, G.; Kondarides, D.I. Propane Steam Reforming over Catalysts Derived from Noble Metal (Ru, Rh)-Substituted LaNiO$_3$ and La$_{0.8}$Sr$_{0.2}$NiO$_3$ Perovskite Precursors. *Nanomaterials* **2021**, *11*, 1931. https://doi.org/10.3390/nano11081931

Academic Editor: Ioannis V. Yentekakis

Received: 29 June 2021
Accepted: 24 July 2021
Published: 27 July 2021

Publisher's Note: MDPI stays neutral with regard to jurisdictional claims in published maps and institutional affiliations.

1. Introduction

The continuously increasing energy demand associated with global population growth and the rapid evolution of the industrial sectors has led to the search for alternative energy systems that are affordable, reliable, and sustainable, with low environmental impact [1,2]. Hydrogen (H$_2$) appears to be the most promising future energy carrier since, when combined with fuel cells, it makes it possible to produce electricity for mobile, stationary, and industrial applications with minimal pollutant emissions [3–5]. Hydrogen is mainly produced via hydrocarbon conversion employing steam reforming (SR), dry reforming (DR), autothermal reforming (ATR), and partial oxidation (POX) reactions [6]. Currently, the most common industrial process for hydrogen production is the steam reforming of natural gas, which accounts for ca. 85% of the total H$_2$ produced worldwide [3,4,7]. Other fuels that can be used for this purpose include ethanol, methanol, gasoline, ammonia, and dimethyl ether [8,9] as well as light alkanes such as propane and butane [10–12]. The use of propane as a source of hydrogen has attracted significant attention in recent years because of its favorable physical properties (liquefaction at room temperature and ca. 9 bar), which facilitate its safe handling, storage, and transportation as well as its relatively low cost, abundance, and availability through the existing distribution network of liquefied petroleum gas (LPG) [13–17].

Propane steam reforming (PSR) is a strongly endothermic reaction (1) requiring high temperatures (>700 °C) to succeed high H_2 yields, and proceeds in parallel with the moderately exothermic water-gas shift (WGS) reaction (2):

$$C_3H_8 + 3\,H_2O \leftrightarrow 3\,CO + 7\,H_2 \qquad \Delta H^\circ = 499\ kJ\ mol^{-1} \qquad (1)$$

$$CO + H_2O \leftrightarrow CO_2 + H_2 \qquad \Delta H^\circ = -41\ kJ\ mol^{-1} \qquad (2)$$

By-products such as methane and ethylene may also be produced under reaction conditions via the CO/CO_2 methanation [18] and propane decomposition reactions [19], respectively. The PSR reaction can be carried out by employing Ni-based catalysts, which have been extensively studied because of the good activity and low cost of nickel, compared to precious metals [19]. However, the application of high temperatures results in the sintering of Ni particles and the concomitant deterioration of catalytic performance [20,21]. In addition, exposure of Ni-based catalysts to reforming reaction conditions results in carbon deposition on their surface via the Boudouard reaction or/and the decomposition of propane and by-product hydrocarbons such as methane and ethylene, which further accelerate the catalyst deactivation [11,15,22–25]. Consequently, research efforts currently focus on the development of novel catalytic systems, with improved resistance against coke deposition and metal particle agglomeration [21]. In this respect, noble metals such as Rh, Ru, Pt, and Ir have been studied either alone or in combination with Ni because of their lower tendency to form coke [14,26–30]. The addition of promoters such as alkali metals in the form of oxides has also been shown to suppress coke deposition, but results in generally inferior catalytic activity [10,22].

Recently, perovskite-derived catalysts have attracted significant attention for the production of H_2 or syngas via reforming reactions [3] because of their high activity, carbon tolerance, thermal stability, and low cost [23,31]. Perovskites are a class of crystalline oxides described by the general formula ABO_3, where the A-site (12-fold coordination) is generally occupied by an alkaline-earth or alkali metal cation with a larger size, and the B-site (6-fold coordination) is occupied by a transition metal ion with a smaller radius [3,32,33]. Generally, the A-sites provide the basicity and high thermal stability of the whole structure. When perovskites are exposed to a reductive environment, the B-site cations migrate (exsolve) spontaneously from the bulk to the surface, forming highly dispersed metal nanoparticles. The exsolved metal particles, which are confined on the catalyst surface, exhibit superior reactivity and stability due to their enhanced resistance to coking and metal sintering [32,33]. Removal of the active metal from the host lattice results in the rearrangement of the nonreducible metals and the formation of oxygen vacancies [32]. The physicochemical and catalytic properties of perovskite-derived catalysts can be greatly influenced by the partial substitution of the A- and/or B-sites by other metal cations, which results in the formation of solids described by the formula $A_{1-x}A'_xB_{1-y}B'_yO_3$ [31,34]. As a general trend, A-site substitution enhances the oxygen mobility in the perovskite structure by generating oxygen vacancies, which suppress the carbon deposition, while B-site substitution tends to increase the activity and stability of the derived catalysts due to bimetallic synergy effects [3,33].

The performance of perovskite-derived catalysts for hydrogen production has been studied by several authors. The majority of these studies focused on the dry reforming of methane [3,35–43] and to a lesser extent on the reformation of other feedstocks such as acetic acid [44,45], biomass tar [46], bio-oil [47], diesel [48–50], ethane [51], ethanol [52–54], glycerol [55,56], phenol [57], and toluene [58,59]. The propane reforming over perovskite-derived catalysts has rarely been reported in the literature [60–62]. For example, Lim et al. [60] investigated the autothermal reforming of propane over Ce-modified $Ni/LaAlO_3$ catalysts and reported the increased thermal stability and low carbon deposition over the Ce-doped sample. Yeyongchaiwat et al. [62] studied the oxidative reforming of C_3H_8 over $Pr_2Ni_{0.75}Cu_{0.25}Ga_{0.05}O_4$ perovskite whereas Sudhakaran et al. [61] investigated the dry reforming of propane over the $SrNiO_3$ catalyst. The observed enhancement of CO selectivity and minimal carbon formation was attributed to the strong basicity of the perovskite [61].

In the present work, the propane steam reforming reaction was investigated over catalysts derived from $LaNiO_3$-based perovskites promoted with Sr and/or noble metals (Ru, Rh). The $LaNiO_3$ perovskite was chosen as the parent material because of its well-known activity and stability for reforming reactions [3,63] originating from its transformation into metallic nickel (Ni^0) and lanthanum oxide (La_2O_3) when exposed to reductive conditions at high temperatures [31]. The effects of partial substitution of La by Sr in the A-sites and/or of Ni by Ru or Rh in the B-sites of $LaNiO_3$ have been investigated in an attempt to systematically study the effects of the nature and composition of the A- and B-sites on the catalytic performance for the title reaction. Strontium was chosen because of its ability to increase the number of vacancies that facilitates the mobility of oxygen toward the surface of the solid, to improve the adsorption and desorption of CO_2, and to enhance the resistance of the derived catalysts against carbon deposition [64–66]. On the other hand, partial substitution of Ni by small amounts of Ru or Rh aimed at further improving the catalytic activity and resistance against deactivation, because of the excellent affinity of these metals for the rupture of C–C and C–H bonds and their ability to minimize coke accumulation [32,52,67,68]. Results obtained show that catalysts derived from $LSNRu_x$ and, especially, $LSNRh_x$ perovskites are characterized by high activity and selectivity toward H_2 as well as by long-term stability for the PSR reaction. It is anticipated that the present work may contribute toward understanding the effects of the nature and composition of the A and B sites on the catalytic performance of perovskite-derived catalysts, thereby facilitating the design of efficient catalytic materials for the steam reforming of propane and other similar reactions.

2. Materials and Methods

2.1. Synthesis of Perovskite-Type Oxides

A series of perovskites including $LaNiO_3$ (LN), $La_{0.8}Sr_{0.2}NiO_3$ (LSN), and noble metal-substituted oxides denoted in the following as LNM_x and $LSNM_x$ (M = Rh or Rh; x = 0.01 or 0.1) were prepared with the combustion synthesis method [69]. Citric acid monohydrate (Merck, Darmstadt, Germany) was used as a fuel whereas $Sr(NO_3)_2$ (Sigma-Aldrich, St. Louis, MO, USA), $La(NO_3)_3 \cdot 6H_2O$ (Alfa Aesar, Karlsruhe, Germany), $Ni(NO_3)_2 \cdot 6H_2O$ (Alfa Aesar, Karlsruhe, Germany), $Ru(NO)(NO_3)_3$ (Alfa Aesar, Karlsruhe, Germany), and N_3O_9Rh (Alfa Aesar, Karlsruhe, Germany) were used as metal precursor salts. In a typical synthesis, stoichiometric quantities of the respective metal nitrates were dissolved in triple-distilled water followed by the addition of citric acid, NH_4NO_3 (extra-oxidant), and, finally, ammonia solution for the neutralization of the excess citric acid. The resulting solution was heated until ignition at ca. 400 °C and the powder obtained was calcined at 900 °C for 5 h and then ground in a mortar. The notations and nominal formulas of the materials thus prepared are listed in Table 1.

2.2. Physicochemical Characterization

The specific surface areas (SSAs) of the freshly prepared and the used materials were determined with the BET method using a Micromeritics (Gemini III 2375) instrument (Norcross, GA, USA). X-ray diffraction (XRD) patterns were obtained on a Bruker D8 instrument (Bruker, Karlsruhe, Germany) and were analyzed by employing JCPDS data files. The average crystallite size of the resulting phases was estimated by applying the Scherrer equation. Details on the above methods and procedures can be found elsewhere [70]. Scanning electron microscopy (SEM) images were obtained on the JEOL JSM 6300 instrument (JEOL, Akishima, Tokyo, Japan), equipped with X-ray Energy Dispersive Spectrometer, EDS (ISIS Link 300, Oxford Instruments, Oxford, UK).

Table 1. Notation, nominal composition, BET specific surface area, and main phases detected with XRD for the as-prepared (fresh) perovskites and the used catalysts.

Notation	Nominal Formula	Specific Surface Area ($m^2\ g^{-1}$)		Phases Detected with XRD	
		Fresh	Used	Fresh	Used
LN	$LaNiO_3$	3	42	$LaNiO_3$ (rhombohedral, rh)	$La_2O_2CO_3$ Ni
LSN	$La_{0.8}Sr_{0.2}NiO_3$	5	18	$Sr_{0.5}La_{1.5}NiO_4$ NiO	La_2SrO_x $La_2O_2CO_3$ Ni
$LNRu_{0.01}$	$LaNi_{0.99}Ru_{0.01}O_3$	4	56	$LaNiO_3$ (rh)	$La_2O_2CO_3$ Ni
$LNRu_{0.1}$	$LaNi_{0.9}Ru_{0.1}O_3$	6	71	$LaNiO_3$ (cubic)	$La_2O_2CO_3$ La_2O_3 Ni
$LSNRu_{0.01}$	$La_{0.8}Sr_{0.2}Ni_{0.99}Ru_{0.01}O_3$	6	55	$Sr_{0.5}La_{1.5}NiO_4$ $LaNiO_3$ (rh) $SrNiO_3$ NiO	$La_2O_2CO_3$ $SrCO_3$ SrC_2 Ni
$LSNRu_{0.1}$	$La_{0.8}Sr_{0.2}Ni_{0.9}Ru_{0.1}O_3$	8	59	$Sr_{0.5}La_{1.5}NiO_4$ $LaNiO_3$ (rh) NiO	$La_2O_2CO_3$ $SrCO_3$ La_2O_3 Ni
$LSNRh_{0.01}$	$La_{0.8}Sr_{0.2}Ni_{0.99}Rh_{0.01}O_3$	6	66	$Sr_{0.5}La_{1.5}NiO_4$ NiO	$La_2O_2CO_3$ $SrCO_3$ Ni
$LSNRh_{0.1}$	$La_{0.8}Sr_{0.2}Ni_{0.9}Rh_{0.1}O_3$	6	29	$Sr_{0.5}La_{1.5}NiO_4$ NiO	$La_2O_2CO_3$ $SrCO_3$ La_2O_3 Ni

2.3. Catalytic Performance Tests

The catalytic performance of the synthesized materials for the propane steam reforming reaction was evaluated using a fixed bed reactor operating at near atmospheric pressure. The reactor consists of a 35-cm long quartz tube (6 mm O.D.) with an expanded 6-cm long section in the middle (12 mm O.D.) where the catalyst was placed. The flow of the inlet gases is controlled by employing mass-flow controllers (MKS Instruments, Andover, MA, USA). An HPLC pump (type Marathon; Spark-Holland, Emmen, The Netherlands) was used to feed water into a vaporizer with a set temperature of 200 °C and the steam produced was mixed with the gas stream coming from the mass-flow controllers. The reaction temperature was measured in the middle of the catalyst bed using a K-type thermocouple placed within a quartz capillary well running through the cell. The reactor is placed in an electric furnace, the temperature of which is controlled using a second K-type thermocouple placed between the reactor and the walls of the furnace. A pressure indicator was used to measure the pressure drop in the catalyst bed. A cold trap was placed at the exit of the reactor to condense water before the introduction of the sample to the analysis system. Separation and analysis of hydrocarbons such as C_3H_8, C_3H_6, C_2H_6, C_2H_4, and CH_4 was accomplished with a gas chromatograph (GC-14A, Shimadzu, Kyoto, Japan) equipped with a Carbosieve column and an FID detector, whereas analysis of Ar, CO, CO_2, and CH_4 was made using a Carboxen column and a TCD detector using He as the carrier gas. A separate GC-TCD system (GC-8A, Shimadzu, Kyoto, Japan), operated with N_2 as the carrier gas, was used to measure the amount of produced H_2. Determination of the response factors of the GC detectors was performed using gas streams of known composition.

In a typical experiment, 100 mg of a fresh perovskite sample (particle size: 0.18 mm < d < 0.25 mm) was placed in the reactor and heated at 750 °C under He flow. The flow was then switched to the reaction mixture (200 cm^3 min^{-1}) consisting of 2.3% C_3H_8, 22.9% H_2O, and 0.7% Ar, which was used as an internal standard (He balance). The reactor effluent was analyzed using the gas chromatographs described above. The reaction temperature was then stepwise decreased down to 450 °C and similar measurements were obtained. The propane conversion was calculated using the following equation

$$X_{C3H8} = \frac{[C_3H_8]_{in} - [C_3H_8]_{out}}{[C_3H_8]_{in}} \times 100 \tag{3}$$

where $[C_3H_8]_{in}$ and $[C_3H_8]_{out}$ are the inlet and outlet concentrations of C_3H_8, respectively.

The selectivities toward carbon-containing products (e.g., CO, CO_2, CH_4, C_2H_4, C_2H_6) were evaluated according to:

$$S_i = \frac{[C_i] \times n}{\sum [C_i] \times n} \times 100 \tag{4}$$

where n is the number of carbon atoms in product i and C_i is its concentration at the reactor effluent. Hydrogen selectivity is defined as the concentration of H_2 in the effluent gas over the sum of the concentrations of all reaction products containing hydrogen, multiplied by the number of hydrogen atoms (y) in each product:

$$S_{H2} = \frac{[H_2]}{[H_2] + \sum (\frac{y}{2} \times [C_xH_y]_i)} \times 100 \tag{5}$$

2.4. Temperature-Programmed Oxidation Experiments

The amount of carbon deposits accumulated on the catalyst surface following exposure to reaction conditions was estimated with temperature-programmed oxidation (TPO) experiments employing the apparatus and following the procedure described elsewhere [71]. Briefly, the "used" catalyst sample was placed in a quartz microreactor, exposed to a 3% O_2/He mixture at room temperature for 10 min, and then heated linearly (β = 10 °C min^{-1}) up to 750 °C under the same flow. Analysis of gases at the reactor effluent was accomplished by employing an online mass spectrometer (Omnistar/Pfeiffer Vacuum). The transient-MS signals at m/z = 2 (H_2), 18 (H_2O), 28 (CO), 32 (O_2), and 44 (CO_2) were continuously recorded. Responses of the mass spectrometer were calibrated against self-prepared mixtures of accurately known composition.

3. Results and Discussion

3.1. Physicochemical Characteristics of the as-Prepared Perovskite Samples

The specific surface areas of the as-prepared (fresh) perovskite samples were measured with the BET method and the results obtained are listed in Table 1. It was observed that the SSA of $LaNiO_3$ (LN) was very low (3 m^2 g^{-1}). This is typical for perovskite-type oxide structures where SSA corresponds mainly to the external area of non-porous particles [72]. Partial substitution of La by Sr resulted in an increase of the SSA to 5 m^2 g^{-1} for the $La_{0.8}Sr_{0.2}NiO_3$ (LSN) sample. This is in agreement with the results of previous studies, which showed that partial substitution of the A-sites of a perovskite generally results in an increase of the SSA induced by structural disorder and delay in the crystallite growth [52,73,74]. A further increase of the SSA up to 8 m^2 g^{-1} was observed upon partial substitution of Ni by Ru or Rh in the B-sites of the LN and LSN perovskites, which is more significant for the samples containing larger amounts of noble metals (Table 1).

The XRD patterns of the freshly prepared samples are shown in Figure 1, and the various phases detected in each case are listed in Table 1. It was observed that the diffractogram obtained for LN (Figure 1A, trace a) contained only reflections attributable to the rhombohedral $LaNiO_3$ perovskite structure (JCPDS Card No. 34-1028). Substitution of 1% at. of Ni by Ru in the B-sites of LN ($LNRu_{0.01}$ sample) resulted in a shift of the XRD peaks toward lower angles (trace b). This shift, which becomes more pronounced upon further increase of the Ru content (LNRu$_{0.1}$ sample, trace c), indicates the preservation of the LaNiO$_3$ perovskite structure but in the cubic system (JCPDS Card No. 33-710).

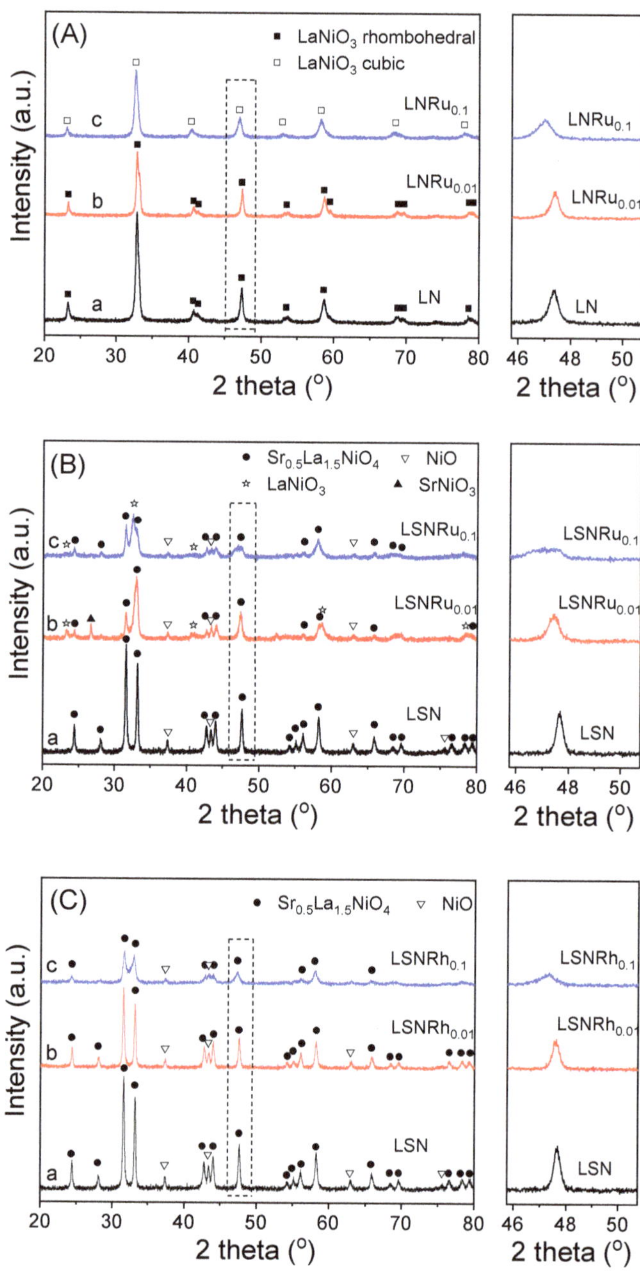

Figure 1. X-ray diffraction patterns of the as-prepared perovskite samples: (**A**) LNRu$_x$, (**B**) LSNRu$_x$, and (**C**) LSNRh$_x$. In the magnified pattern on the right of each graph is shown the peak attributed to the (200) reflection.

The sample obtained following partial substitution of 1% at. Ni by Ru in the LSN structure (LSNRu$_{0.01}$ sample) was also characterized by the presence of Sr$_{0.5}$La$_{1.5}$NiO$_4$ and NiO phases, whereas smaller peaks due to LaNiO$_3$ (cubic) and SrNiO$_3$ (hexagonal) could also be observed (trace b). Increasing the Ru content resulted in the formation of the same phases except for SrNiO$_3$, which did not appear in the XRD pattern of the LSNRu$_{0.1}$ sample (trace c). The presence of the NiO phase in the diffraction patterns of all LSNRu$_x$ samples (Figure 1B) indicates that a portion of nickel is located outside the perovskite structure. Regarding the LSNRh$_x$ samples, the XRD patterns presented in Figure 1C show that the partial substitution of Ni by Rh in the B-sites of the perovskite resulted in the formation of the same main phases observed for LSN (Sr$_{0.5}$La$_{1.5}$NiO$_4$ and NiO). The peaks attributed to Sr$_{0.5}$La$_{1.5}$NiO$_4$ were much broader for the LSNRh$_{0.1}$ sample (trace c), compared to LSNRh$_{0.01}$ (trace b), indicating the development of smaller crystallites. The calculated cell parameters of the crystalline phases detected by XRD for the as-prepared perovskite samples are presented in Table S1. It was observed that the incorporation of Ru or Rh in the LSN samples resulted in an increase in the *a* cell parameter and a decrease in the *c* cell parameter of the Sr$_{0.5}$La$_{1.5}$NiO$_4$ tetragonal structure.

It is of interest to note that the XRD patterns presented in Figure 1 do not contain any peaks attributable to Ru or Rh species, indicating that noble metals are incorporated in the perovskite structure. It is known that the formation of perovskites may be restricted depending on the size of the cations present in the A and/or B sites of the material. In general, the radius of cation A should be larger than 0.09 nm, whereas the radius of cation B should be larger than 0.051 nm [75]. The partial substitution of A by an alkaline earth metal A$'$ and/or the partial substitution of B by another transition metal B$'$ may result in the destruction of the perovskite matrix [75,76]. This can be avoided if the radii of A (r_A), B (r_B), and oxygen (r_O) ions in the perovskite structure obey the restriction $0.75 < t < 1.0$, where t is the tolerance factor expressed by [75]:

$$t = \frac{(r_A + r_O)}{\sqrt{2}(r_B + r_O)} \tag{6}$$

Based on the above, and considering the sizes of the Ru and Rh cations, it may be concluded that the noble metals used as promoters in this work can only be incorporated in the B-sites of the LN and LSN perovskite structures.

As already mentioned, the partial substitution of Ni by Ru or Rh resulted in a shift in the XRD reflections of the LN and LSN perovskites toward lower angles (2θ), which is accompanied by a decrease in the mean primary size of the perovskite crystallites, estimated by the Scherrer equation. This was the case, for example, for the (200) reflection which can be clearly discerned in the XRD patterns of all samples investigated (see expanded sections of the XRD patterns in Figure 1). The dependence of the (200) diffraction angle (2θ) and the mean crystallite size on the type (Ru or Rh) and content ($x = 0, 0.01, 0.1$) of the noble metal in the LNRu$_x$, LSNRu$_x$, and LSNRh$_x$ samples are shown in Figure 2. It was observed that in all cases, partial substitution of Ni by Ru or Rh resulted in a progressive shift of the diffraction peak toward lower angles and in the development of smaller crystallites. Qualitatively similar results have been reported by other authors [48,69,77]. For example, Mota et al. [48] found that substitution of Ru for Co in LaCo$_{1-x}$Ru$_x$O$_3$ resulted in a shift of XRD peaks toward lower 2θ values accompanied by an alteration in the perovskite structure from rhombohedral to orthorhombic. This has been attributed to the larger ionic radius of Ru^{3+} (0.68 Å) compared to Co^{3+} (0.61 Å) and the concomitant decrease of the tolerance factor. Similar results were obtained by Ivanova et al. [77] over LaNi$_x$Co$_{1-x}$O$_3$ and LaFe$_x$Co$_{1-x}$O$_3$ perovskites as well as by Safakas et al. [69], who also observed a change in the perovskite structure from orthorhombic to rhombohedral upon partial substitution of Fe by Co in La$_{0.8}$Sr$_{0.2}$Co$_x$Fe$_{1-x}$O$_{3-\delta}$. In the present work, a change in the LaNiO$_3$ structure from rhombohedral to cubic was observed following partial substitution of Ni by Ru (Figure 1A). The shift toward lower 2θ angles observed for the LNRu$_x$, LSNRu$_x$, and LSNRh$_x$ samples (Figure 2) indicates a high degree of incorporation of the noble metals in the B-sites of the perovskite structure [49] and can be explained considering that the ionic radii of Ru^{3+} (0.68 Å) and Rh^{3+} (0.67 Å) are larger than that of Ni^{3+} (0.60 Å).

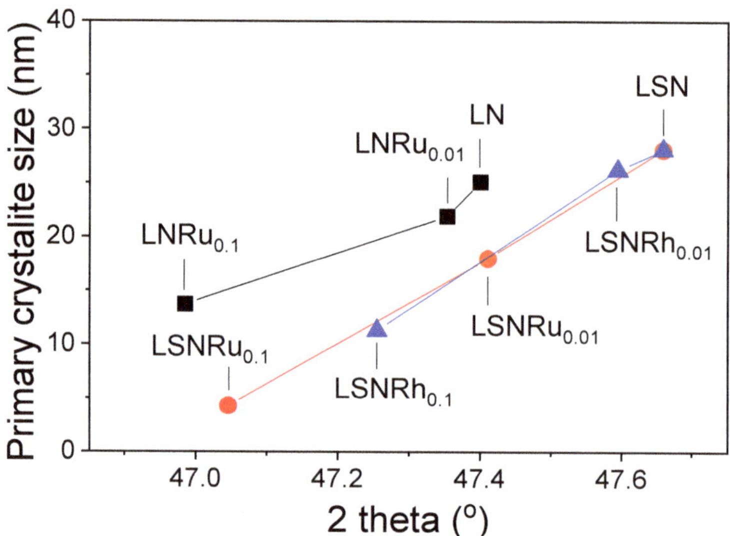

Figure 2. Dependence of the mean primary size of the perovskite crystallites, and of the angle corresponding to the (200) reflection on the type (Ru or Rh) and content ($x = 0, 0.01, 0.1$) of the B-site cation in LNRu$_x$, LSNRu$_x$, and LSNRh$_x$ samples (data extracted from the XRD patterns shown in Figure 1).

Regarding the decrease in the primary crystallite size of the perovskites following substitution of Ni by Ru or Rh (Figure 2), it can be attributed to structural disorder and delay in the crystallite growth above-mentioned, and is reflected to the higher SSA of the noble metal-containing samples (Table 1). Qualitatively similar results have been reported by Mota et al. [48,49] for LaCo$_{1-x}$Ru$_x$O$_3$ perovskites.

A representative SEM image obtained for the fresh LSNRh$_{0.1}$ perovskite sample is shown in Figure S1, together with elemental mapping results, which confirm the presence and uniform distribution of La, Sr, Ni, and Rh in the material. Moreover, the elemental analysis showed that the wt.% content of La (47.42%), Sr (7.34%), Ni (22.04%), and Rh (4.16%) is in agreement with the nominal composition of this sample.

3.2. Catalytic Performance Tests

Results of catalytic performance tests performed under PSR reaction conditions using the noble metal-free LN and LSN perovskite samples are shown in Figure 3, where the conversion of propane (X_{C3H8}) and the selectivities to reaction products (S_i) are plotted as functions of reaction temperature. It was observed that the LN sample exhibited considerable activity for the title reaction, with X_{C3H8} increasing progressively from 26% at 450 °C to ca. 100% at 750 °C (Figure 3A). The selectivity toward CO_2 decreased with increasing temperature while S_{CO} followed the opposite trend, indicating the occurrence of the reverse water gas shift (RWGS) reaction (reverse of Equation (2)), which is thermodynamically favored at higher temperatures [78].

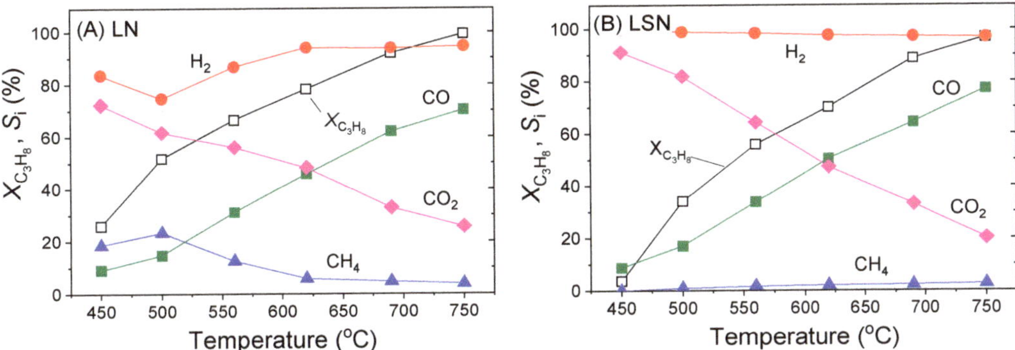

Figure 3. Conversion of propane ($X_{C_3H_8}$) and selectivities to the indicated reaction products obtained over (**A**) the LaNiO$_3$ (LN) and (**B**) the La$_{0.8}$Sr$_{0.2}$NiO$_3$ (LSN) samples. Experimental conditions: mass of catalyst: 100 mg; particle diameter: 0.18 < d < 0.25 mm; feed composition: 2.3% C$_3$H$_8$, 22.9% H$_2$O, 0.7% Ar (balance He); total flow rate: 200 cm^3 min^{-1}.

Production of H$_2$ exceeds 95% at temperatures higher than 550 °C. However, at lower temperatures, S_{H2} is decreased due to the production of methane. Selectivity to methane passes through a maximum of ca. 20% at around 500 °C and progressively decreases with increasing temperature due to the onset of the CH$_4$ steam reforming reaction. In addition to the above-mentioned products, negligible amounts of C$_2$H$_4$ and C$_2$H$_6$ were also detected at the reactor effluent (not shown for clarity). The propane conversion curve obtained for the LSN sample (Figure 3B) shifted toward higher temperatures, compared to LN, indicating that the partial substitution of La with Sr in the A-sites of the perovskite negatively affects the catalytic activity. This is in general agreement with results of previous studies showing that substitution of Sr into the A-sites of LaNiO$_3$ [64,79] or LaCoO$_3$ [80] perovskites usually decreases the CH$_4$ reforming activity. On the other hand, LSN is highly selective toward H$_2$, with S_{H2} exceeding 97% over the entire temperature range investigated (Figure 3B).

Results of similar catalytic performance tests obtained for the Ru-substituted LaNiO$_3$ perovskites (LNRu$_{0.01}$ and LNRu$_{0.1}$ samples) are shown in Figure 4. It was observed that partial substitution of Ni by Ru resulted in slightly lower propane conversions compared to unpromoted LN (Figure 4A). This can be attributed, at least in part, to the different crystal structures obtained following the addition of Ru (i.e., cubic for LNRu$_{0.1}$ vs. rhombohedral for LN (see Figure 1A)). Regarding selectivities to reaction products, the results presented in Figure 4B,C show that the presence of Ru does not appreciably affect the product distribution.

Results obtained following the substitution of Ni by Ru or Rh in the B-sites of LSN are presented in Figure 5. Regarding the sample with the lower Ru content (LSNRu$_{0.01}$), it was observed that the propane conversion curve was slightly improved compared to pristine LSN (Figure 5A). An increase in the Ru content (LSNRu$_{0.1}$ sample) resulted in a moderate increase in propane conversion in the whole temperature range investigated. Regarding selectivities to reaction products, both the LSNRu$_{0.01}$ and LSNRu$_{0.1}$ catalysts yielded higher amounts of methane compared to LSN (Figure 5B), whereas S_{CO} and S_{CO2} were not significantly affected (Figure 5C). On the other hand, the substitution of Ni by a small amount of Rh (1% at.) in the B-sites of the LSN perovskite (LSNRh$_{0.01}$ sample) resulted in a significant increase of propane conversion in the whole temperature range investigated, compared to the pristine LSN (Figure 5A). For example, X_{C3H8} increased from ca. 4% to 38% at 450 °C and from 70 to 86% at 620 °C. This was accompanied by a decrease in the selectivity toward H$_2$ and the production of larger amounts of CH$_4$, which was maximized at ca. 550 °C (Figure 5B). The selectivity toward CO$_2$ followed the opposite trend than that of CH$_4$, whereas S_{CO} was not appreciably affected by the presence of Rh (Figure 5C). Further increase in the Rh content (LSNRh$_{0.1}$ sample) resulted in an even higher propane conversion, which exceeded 90% at temperatures above ca. 550 °C

(Figure 5A). Regarding selectivity to reaction products, the LSNRh$_{0.1}$ sample exhibited superior selectivity toward H$_2$, with S_{H2} being higher than 91% over the entire temperature range investigated and close to 100% above ca. 620 °C (Figure 5B). The selectivity toward CO$_2$ was also very high and comparable to that of LSN (Figure 5C).

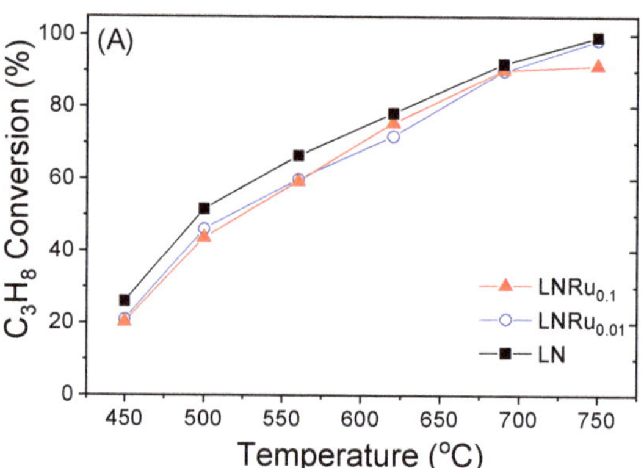

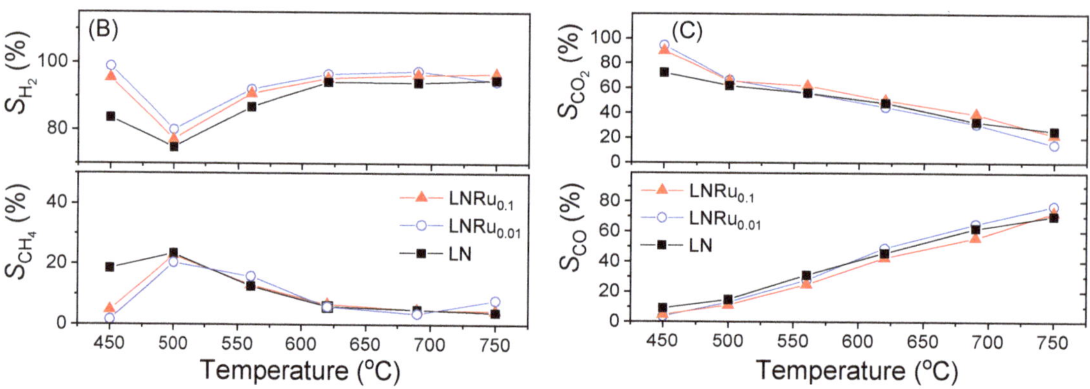

Figure 4. (**A**) Conversion of propane, (**B**) selectivities to H$_2$ and CH$_4$, and (**C**) selectivities to CO$_2$ and CO obtained over the LNRu$_x$ catalysts. Experimental conditions: same as in Figure 3.

3.3. Physicochemical Characteristics of the Used Catalyst Samples

The "used" catalyst samples obtained following the catalytic performance tests presented in Figure 5 were characterized with the use of BET and XRD techniques in order to investigate the effects of exposure to the reaction conditions on the physicochemical properties of the materials.

As shown in Table 1, the SSAs of the used samples were considerably larger than those of the fresh perovskites. For example, the SSA of the LNRu$_{0.1}$ sample increased from 6 m^2 g^{-1} to 71 m^2 g^{-1}. As discussed below, this increase in SSA was due to the destruction of the perovskite structures and the formation of new phases under reaction conditions.

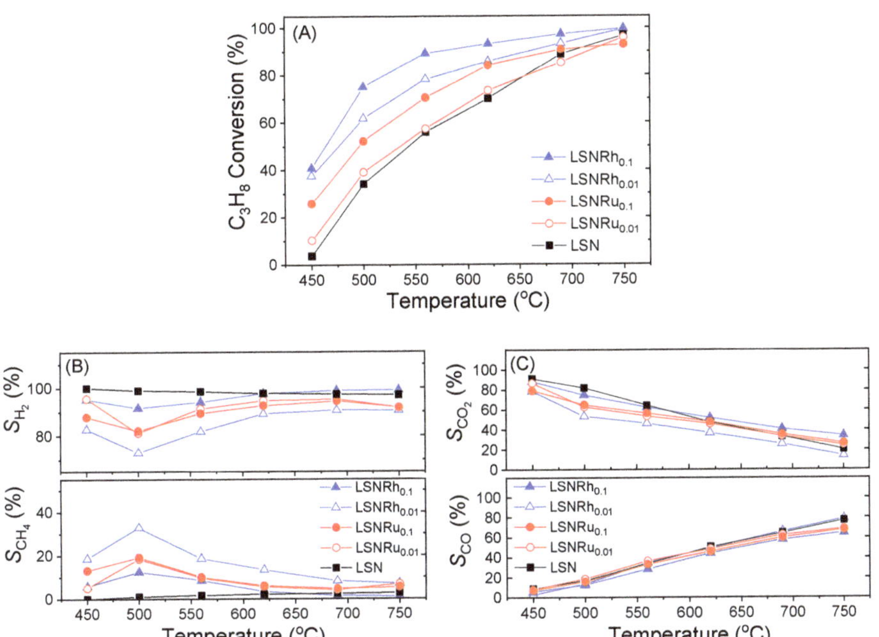

Figure 5. (**A**) Conversion of propane, (**B**) selectivities to H_2 and CH_4, and (**C**) selectivities to CO_2 and CO obtained over the $LSNRh_x$ and $LSNRu_x$ catalysts. Experimental conditions: same as in Figure 3.

The XRD profiles of the used catalysts are presented in Figure 6 and the various phases identified for each sample are listed in Table 1. It was observed that the XRD pattern of LN (Figure 6A, trace a) consisted of peaks attributed to hexagonal $La_2O_2CO_3$ (JCPDS Card No. 37-804) and cubic Ni (JCPDS Card No. 1-1260). Qualitatively similar results were obtained for the used $LNRu_{0.01}$ (trace b) and $LNRu_{0.1}$ (trace c) catalysts. In the latter case, reflections attributable to hexagonal La_2O_3 (JCPDS Card No. 5-602) could also be discerned. This implies that, under the reducing atmosphere existing under reaction conditions, the parent $LaNiO_3$ perovskite is in situ transformed into metallic Ni and La_2O_3. Because of the high affinity between La_2O_3 and CO_2, lanthanum oxycarbonate ($La_2O_2CO_3$) is formed according to [66]:

$$La_2O_3 + CO_2 \rightarrow La_2O_2CO_3 \tag{7}$$

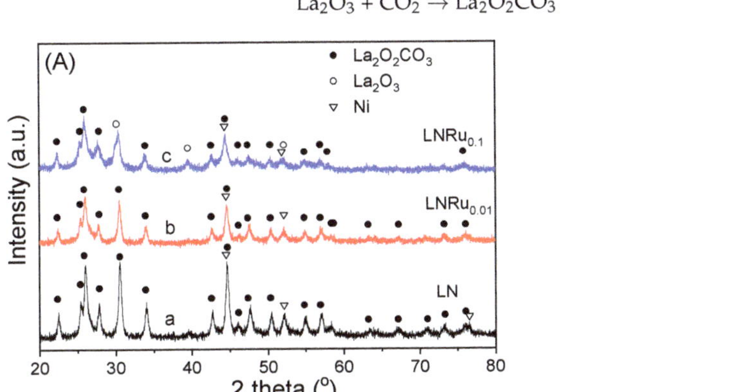

Figure 6. *Cont.*

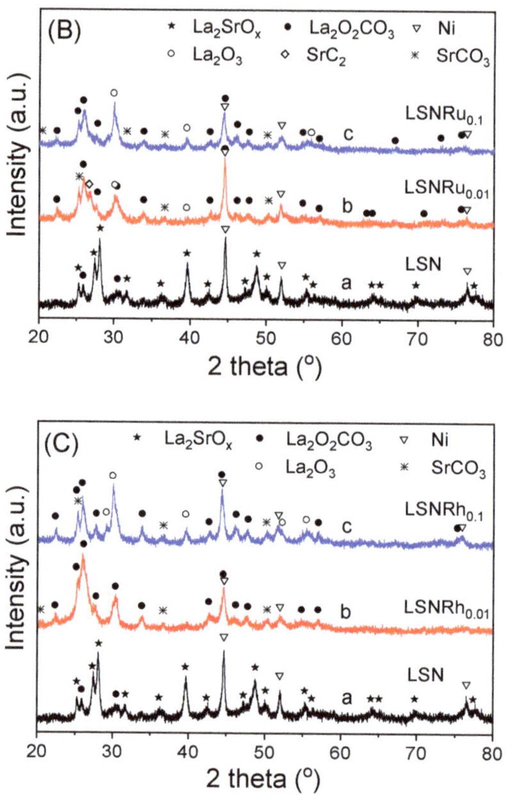

Figure 6. X-ray diffraction patterns obtained for the (**A**) LNRu$_x$, (**B**) LSNRu$_x$, and (**C**) LSNRh$_x$ catalyst samples following exposure to reaction conditions.

The Ni^0 crystallites, which provide the active sites for propane reforming, were well dispersed and interacted strongly with the La$_2$O$_3$/La$_2$O$_2$CO$_3$ support, thereby inhibiting metal sintering [3]. The absence of peaks attributable to Ru provides evidence that the exsolved noble metal was also well dispersed on the catalyst surface. As shown in Figure 6B, the XRD pattern of the used LSN sample (trace a) was dominated by reflections attributed to the La$_2$SrO$_x$ structure (JCPDS Card No. 42-343) as well as hexagonal La$_2$O$_3$ and cubic Ni. Interestingly, the La$_2$SrO$_x$ phase was absent in the diffraction profiles of the used LSNRu$_{0.01}$ (trace b) and LSNRu$_{0.1}$ (trace c) catalysts that consisted mainly of La$_2$O$_2$CO$_3$, Ni as well as orthorhombic SrCO$_3$ (JCPDS Card No. 5-418), La$_2$O$_3$, and cubic SrC$_2$ (JCPDS Card No. 1-1022). This indicates that the presence of Ru in the perovskite structure facilitates the conversion of La$_2$SrO$_x$ to La$_2$O$_3$ and SrO, which may interact with CO$_2$ to form La$_2$O$_2$CO$_3$ (Equation (7)) and SrCO$_3$:

$$SrO + CO_2 \rightarrow SrCO_3 \tag{8}$$

Qualitatively similar results were obtained for the used LSNRh$_x$ catalyst samples (Figure 6C). In particular, the presence of Rh leads in the development of the La$_2$O$_2$CO$_3$, Ni^0, SrCO$_3$, and La$_2$O$_3$ phases under reaction conditions. As a general trend, the presence of carbonate phases in the used LSNRu$_x$ and LSNRh$_x$ catalyst samples seems to be related to improved catalytic performance for the title reaction. For example, LSN, which is the least active among the samples investigated, was the only one that contained small amounts of La$_2$O$_2$CO$_3$ after exposure to the reaction conditions (Figure 6). The cell parameters estimated for the La$_2$O$_2$CO$_3$ structure detected in the used samples are presented in

Table S1. It was observed that in the case of the LSNRu$_x$ and LSNRh$_x$ catalysts, the *a* cell parameter of La$_2$O$_2$CO$_3$ decreased and the *c* cell parameter increased, compared to the used LSN sample. The EDS mapping analysis performed for the used LSNRh$_{0.1}$ sample (Figure S2) showed intense signals for the carbon and La elements originating from La$_2$O$_2$CO$_3$. It is important to note that no XRD peaks attributable to graphite were detected over the used catalyst samples, indicating that the amount of accumulated carbon deposited was relatively small.

3.4. Long-Term Stability Tests

The stability of representative catalyst samples derived from LN, LSN, and LSNRh$_{0.1}$ was studied at 600 °C for 40 h. The results obtained are presented in Figure 7, where the conversion of propane (X_{C3H8}) and the selectivities toward H$_2$, CO, CO$_2$, and CH$_4$ are plotted as functions of time-on-stream. Dashed vertical lines indicate shutting down of the system overnight, where the catalysts were kept at room temperature under He flow. It was observed that the conversion of propane over the LN sample increased substantially during the first three hours-on-stream from an initial value of 38% to ca. 70% (Figure 7A). Prolonged exposure to the reaction mixture resulted in a further increase of X_{C3H8}, which stabilized at ca. 77% after about 25 hours. This behavior can be attributed to the progressive reduction of the as-prepared LaNiO$_3$ perovskite (Figure 1A) to the catalytically active Ni^0 and La$_2$O$_2$CO$_3$ species (Figure 6A), in agreement with the results of previous studies [31,33,63]. Selectivities to reaction products also varied during the first few hours-on-stream and then remained practically stable. Selectivity toward H$_2$ exceeded 97%, whereas S_{CO} and S_{CO2} had values around 50% (Figure 7A).

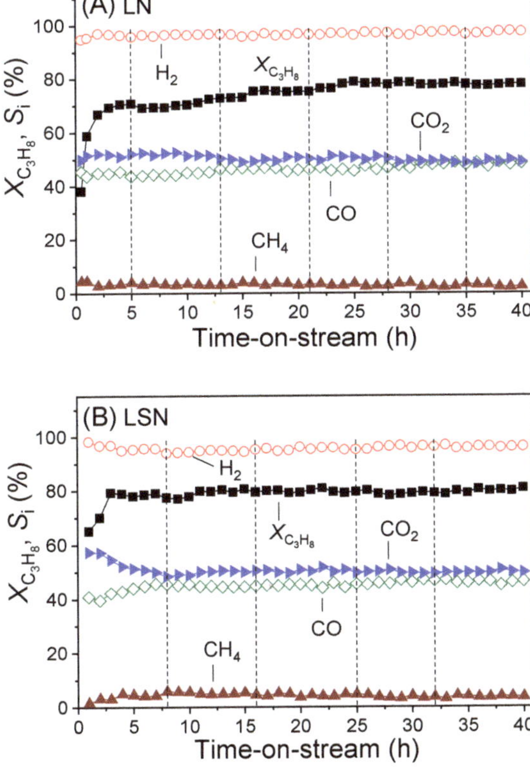

Figure 7. *Cont.*

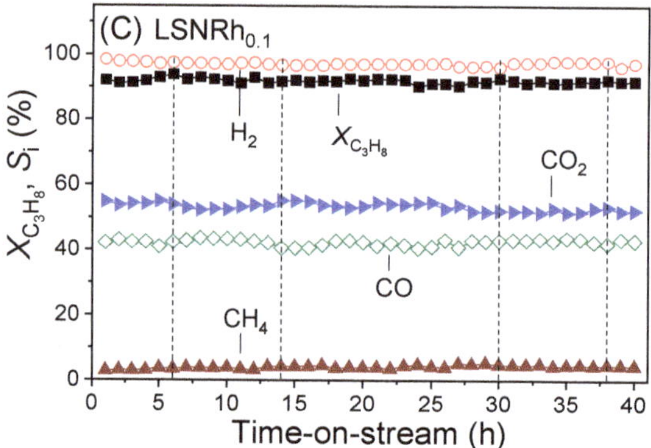

Figure 7. Long-term stability tests obtained at $T = 600\ °C$ over the (**A**) LN, (**B**) LSN, and (**C**) LSNRh$_{0.1}$ catalysts. Experimental conditions: same as in Figure 3. Vertical lines indicate shutting down of the system overnight, where the catalyst was kept at room temperature under He flow.

The performance of the LSN catalyst (Figure 7B) was qualitatively similar to that of LN (i.e., X_{C3H8} increased during the first few hours-on-stream and then remained stable at ca. 80% throughout the run). The selectivity toward H$_2$ remained practically constant at ca. 96%. It is of interest to note that the in situ activation of the LSN sample and the stabilization of catalytic performance (Figure 7B) occurred after a shorter time of exposure to the reaction mixture compared to the LN sample (Figure 7A). This can be attributed to the ability of the Sr dopant to decrease the reduction temperature of the parent material [81] and to facilitate the formation of La$_2$O$_2$CO$_3$ according to:

$$La_2O_3 + SrCO_3 \rightarrow La_2O_2CO_3 + SrO \tag{9}$$

Finally, the LSNRh$_{0.1}$ catalyst (Figure 7C) exhibited excellent stability and was characterized by the highest values of both propane conversion ($X_{C3H8} = 92\%$) and selectivity toward H$_2$ ($S_{H2} = 97\%$) compared to the LN and LSN samples. The fact that X_{C3H8} and S_i acquired stable values at relatively very short time periods indicates that the presence of Rh in the structure of the parent material has a beneficial effect on the reduction processes that lead to the in situ formation of the catalytically active Ni^0 and La$_2$O$_2$CO$_3$ phases. Regarding the higher activity and H$_2$ selectivity of the LSNRh$_{0.1}$ catalyst compared to LN and LSN, these can be attributed to the presence of well-dispersed Rh crystallites on the catalyst surface, which act synergistically with Ni^0 species [72].

The excellent stability of the LN, LSN, and LSNRh$_{0.1}$ catalysts (Figure 7) can be related to the formation of La$_2$O$_2$CO$_3$ under reaction conditions. It is well known that this species can prevent catalyst deactivation by reacting with carbon deposits according to [82,83]:

$$La_2O_2CO_3 + C \rightleftarrows La_2O_3 + 2\,CO \tag{10}$$

Similar results have been reported for conventionally prepared Ni/La$_2$O$_3$ catalysts, which transform into Ni/La$_2$O$_3$/La$_2$O$_2$CO$_3$ under reduction conditions [84]. It may also be noted that pre-synthesized La$_2$O$_2$CO$_3$ either alone [85,86] or in combination with Al$_2$O$_3$ [87] has also been used as catalyst supports for reforming reactions. Compared to these materials, the perovskite-derived catalysts have the advantage of comprising very well dispersed and strongly bound metal crystallites on the support, rendering them more active and resistant against carbon deposition.

3.5. Carbon Accumulation on the Catalyst Surface

After the completion of the stability tests presented in Figure 7, TPO experiments were carried out to quantify the amount of reactive carbon-containing species that accumulated on the catalyst surfaces. The results obtained are presented in Figure 8, where the CO_2 concentration at the reactor effluent is plotted as a function of temperature. It was observed that the TPO profile of the LN-derived catalyst is characterized by an intense peak centered at 530 °C and two shoulders at ca. 560 and 600 °C (trace a). The features appearing at temperatures lower than ca. 550 °C can be attributed to carbon species with amorphous filamentous morphology (Cα) characterized by a higher reactivity due to structural disorder [88]. The high-temperature shoulder at ca. 600 °C can be assigned to the oxidation of graphite-like carbonaceous species, C_β, which are produced from C_α and are more difficult to remove from the catalyst surface ([88] and refs. therein). The amount of accumulated carbon on the LN catalyst was calculated from the area below the TPO curve and was found to be 57 mg C g_{cat}^{-1}.

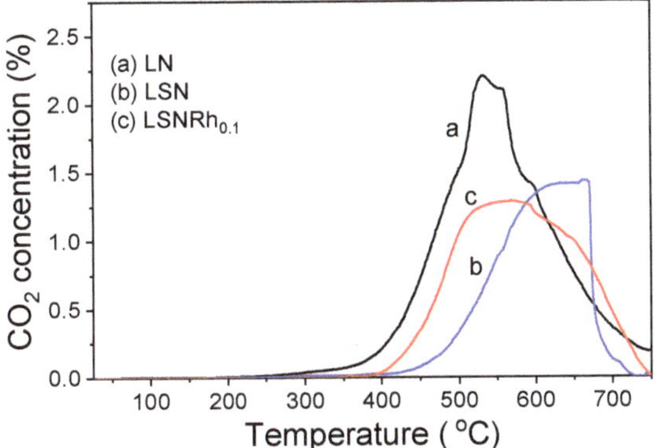

Figure 8. Temperature-programmed oxidation (TPO) profiles obtained for the indicated catalysts following exposure to the long-term stability tests shown in Figure 7.

The TPO profile of the LSN catalyst consists of a peak with a broad maximum located above ca. 600 °C (trace b). The amount of the accumulated carbon (35 mg C g_{cat}^{-1}) was much lower than that obtained for the LN sample, indicating that the co-existence of La and Sr at the A-sites of the parent perovskite material decreased the tendency of the derived catalyst toward carbon deposition. As discussed above, this can be related to the ability of $SrCO_3$ to facilitate the formation of $La_2O_2CO_3$ under reaction conditions. Finally, the LSNRh$_{0.1}$ catalyst presented a TPO peak with a broad maximum located at ca. 500–650 °C (trace c), which corresponded to 42.6 mg C g_{cat}^{-1}. Although the amount of CO_2 produced was higher than that obtained for LSN (35 mg C g_{cat}^{-1}), the conversion of propane was also considerably higher (i.e., 92% for LSNRh$_{0.1}$ vs. 80% for LSN (Figure 7)). In addition, the CO_2 peak started evolving at lower temperatures for the LSNRh$_{0.1}$ sample, indicating that the presence of Rh resulted in the accumulation of filamentous carbonate species, which can be more easily oxidized. This can be attributed to the Rh-induced enhancement of the reducibility of Ni, the inhibition of the sintering of the metallic particles, and the promotion of the gasification rate of carbon deposits due to Ni–Rh interactions [89–91].

After the TPO measurements, the used samples were characterized with XRD, and the results obtained are shown in Figure 9. It was observed that the dominant phase in the XRD patterns of all three samples, namely LN (trace a), LSN (trace b), and LSNRh$_{0.1}$ (trace c), was $La_2O_2CO_3$. Nickel was clearly oxidized toward NiO only in the case of LSN, whereas it mainly existed in its metallic phase in the cases of the LN and LSNRh$_{0.1}$ samples. The TPO did not result in the reappearance of the initial perovskite structures. In addition, the $La_2O_2CO_3$ phase did not decompose, implying that the CO_2 that evolved during TPO should be mainly attributed to the oxidation of accumulated carbon and not to the decomposition of carbonate phases such as $La_2O_2CO_3$. This is rather expected since temperatures higher than ca. 900 °C are typically necessary for the decomposition of the

$La_2O_2CO_3$ phase [81]. However, the possibility that the TPO peaks presented in Figure 8 are, at least in part, due to the elimination of CO_2 from surface oxycarbonates and/or strongly adsorbed CO_2 cannot be excluded [66,92]. Further investigation of this issue is beyond the scope of the present work and will be the subject of our future studies.

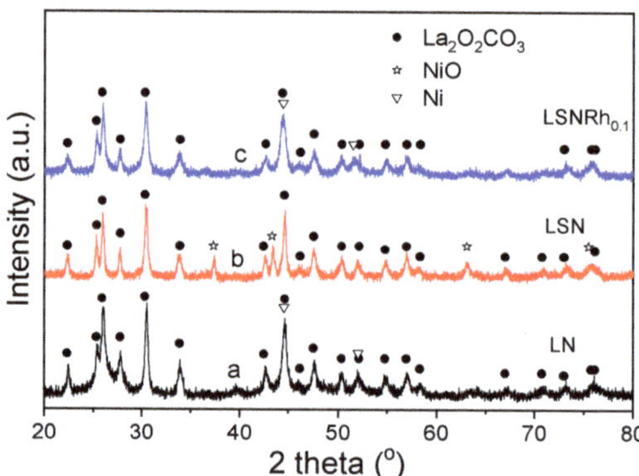

Figure 9. X-ray diffraction patterns of (**a**) LN, (**b**) LSN, and (**c**) LSNRh$_{0.1}$ samples obtained after the TPO experiments presented in Figure 8.

4. Conclusions

The effects of partial substitution of La by Sr and of Ni by Ru or Rh in the LaNiO$_3$ structure were investigated in an attempt to systematically study the effects of the nature and composition of the A- and B-sites of the perovskite-derived catalysts for the propane steam reforming (PSR) reaction. The physicochemical characteristics of the as-prepared LaNiO$_3$ (LN), La$_{0.8}$Sr$_{0.2}$NiO$_3$ (LSN), and noble metal-substituted LNM$_x$ and LSNM$_x$ perovskites (M = Ru or Rh; x = 0.01 or 0.1) as well as the used catalysts obtained following exposure to reaction conditions were studied by employing the BET method and the XRD technique. Results obtained can be summarized as follows:

1. Incorporation of noble metals in the matrix of LN and LSN perovskites resulted in an increase in the specific surface area (SSA), a shift of the XRD lines toward lower angles, and a decrease in the mean primary crystallite size of the materials. These modifications of the physicochemical characteristics of the perovskites, which are more pronounced for samples with higher noble metal content, have been attributed to the distortion of the perovskite structure induced by the incorporation of Ru or Rh in the matrix;

2. Exposure of the perovskite samples to PSR reaction conditions resulted in the destruction of the perovskite structure and the development of new phases, accompanied by a considerable increase in the specific surface areas of the materials. Specifically, the in situ reduction of the parent perovskites resulted in the exsolution of Ni (as well as Rh or Ru) and the formation of well-dispersed metal nanoparticles on the resulting support, which consisted mainly of La$_2$O$_2$CO$_3$ produced from the interaction of La$_2$O$_3$ with CO$_2$. The main phases detected with XRD for the LNRu$_x$-derived samples included metallic Ni and La$_2$O$_2$CO$_3$ whereas the LNSRu$_x$ and LNSRh$_x$-derived catalysts also contained La$_2$SrO$_x$, La$_2$O$_3$, and SrCO$_3$. The presence of noble metals in the derived catalysts was confirmed by SEM/EDS measurements;

3. The LN-derived catalyst exhibited higher activity compared to LSN, and its performance for the title reaction did not change appreciably following partial substitution of Ru in the B-sites of the perovskite. In contrast, incorporation of Ru and, especially, Rh in the perovskite matrix resulted in the development of catalysts with significantly improved catalytic performance, which increased with an increase in the noble metal content;

4. Results of long-term stability test obtained at 600 °C using the LN, LSN, and LSNRh$_{0.1}$ samples showed that all catalysts were characterized by high stability for 40 hours-on-stream, which was reflected in the accumulation of relatively small amounts of carbon deposits on the catalyst surfaces. This has been attributed to the in situ formation of La$_2$O$_2$CO$_3$ under reaction conditions, which facilitates the oxidation of accumulated carbon, thereby preventing catalyst deactivation; and

5. Best results were obtained for the LSNRh$_{0.1}$-derived catalyst, which was characterized by high activity (X_{C3H8} = 92%) and selectivity toward H$_2$ (S_{H2} = 97%) at 600 °C as well as excellent stability for 40 hours-on-stream. This has been attributed to the presence of well dispersed Rh crystallites on the catalyst surface, which act synergistically with Ni^0 species.

Supplementary Materials: The following are available online at https://www.mdpi.com/article/10.3390/nano11081931/s1. Figure S1. SEM image (A), EDS mapping results showing the distribution of (B) La, (C) Sr, (D) Ni, (E) Rh and (F) La (green spots), Ni (blue spots) and Rh (red spots) elements and corresponding EDS spectrum (G) over the as-prepared LSNRh$_{0.1}$ perovskite sample; Figure S2. SEM image (A), EDS mapping results showing the distribution of (B) Carbon, (C) La, (D) Sr, (E) Ni and (F) Rh elements and corresponding EDS spectrum (G) over the "used" LSNRh0.1 perovskite sample; Table S1: Cell parameters of the crystalline phases detected by XRD for the as-prepared perovskite samples and the derived (used) catalysts.

Author Contributions: T.R. and D.I.K. conceived and designed the experiments; T.R., A.V. and G.V. performed the experiments; T.R., G.B. and D.I.K. analyzed the results and wrote the paper. All authors have read and agreed to the published version of the manuscript.

Funding: This research was co-financed by the European Union and Greek national funds through the Operational Program Competitiveness, Entrepreneurship and Innovation, under the call RESEARCH-CREATE-INNOVATE (project code: T1EDK-02442).

Data Availability Statement: Not applicable.

Conflicts of Interest: The authors declare no conflict of interest. The funders had no role in the design of the study; in the collection, analyses, or interpretation of data; in the writing of the manuscript, or in the decision to publish the results.

References

1. Rusman, N.A.A.; Dahari, M. A review on the current progress of metal hydrides material for solid-state hydrogen storage applications. *Int. J. Hydrog. Energy* **2016**, *41*, 12108–12126. [CrossRef]
2. Owusu, P.A.; Asumadu-Sarkodie, S. A review of renewable energy sources, sustainability issues and climate change mitigation. *Cogent Eng.* **2016**, *3*, 1167990. [CrossRef]
3. Bian, Z.; Wang, Z.; Jiang, B.; Hongmanorom, P.; Zhong, W.; Kawi, S. A review on perovskite catalysts for reforming of methane to hydrogen production. *Renew. Sustain. Energy Rev.* **2020**, *134*, 110291. [CrossRef]
4. Sazali, N. Emerging technologies by hydrogen: A review. *Int. J. Hydrogen Energy* **2020**, *45*, 18753–18771. [CrossRef]
5. Panagiotopoulou, P.; Papadopoulou, C.; Matralis, H.; Verykios, X. Production of Renewable Hydrogen by Reformation of Biofuels. *Adv. Bioenergy* **2016**, 109–130.
6. Kalamaras, C.M.; Efstathiou, A.M. Hydrogen Production Technologies: Current State and Future Developments. *Conf. Pap. Energy* **2013**, *2013*, 690627. [CrossRef]
7. Lee, D.-Y.; Elgowainy, A. By-product hydrogen from steam cracking of natural gas liquids (NGLs): Potential for large-scale hydrogen fuel production, life-cycle air emissions reduction, and economic benefit. *Int. J. Hydrog. Energy* **2018**, *43*, 20143–20160. [CrossRef]
8. Ashcroft, A.T.; Cheetham, A.K.; Foord, J.S.; Green, M.L.H.; Grey, C.P.; Murrell, A.J.; Vernon, P.D.F. Selective oxidation of methane to synthesis gas using transition metal catalysts. *Nature* **1990**, *344*, 319–321. [CrossRef]
9. Ashcroft, A.T.; Cheetham, A.K.; Green, M.L.H.; Vernon, P.D.F. Partial oxidation of methane to synthesis gas using carbon dioxide. *Nature* **1991**, *352*, 225–226. [CrossRef]
10. Silva, P.P.; Ferreira, R.A.R.; Noronha, F.B.; Hori, C.E. Hydrogen production from steam and oxidative steam reforming of liquefied petroleum gas over cerium and strontium doped LaNiO$_3$ catalysts. *Catal. Today* **2017**, *289*, 211–221. [CrossRef]
11. Im, Y.; Lee, J.H.; Kwak, B.S.; Do, J.Y.; Kang, M. Effective hydrogen production from propane steam reforming using M/NiO/YSZ catalysts (M = Ru, Rh, Pd, and Ag). *Catal. Today* **2018**, *303*, 168–176. [CrossRef]

12. Rakib, M.A.; Grace, J.R.; Lim, C.J.; Elnashaie, S.S.E.H.; Ghiasi, B. Steam reforming of propane in a fluidized bed membrane reactor for hydrogen production. *Int. J. Hydrog. Energy* **2010**, *35*, 6276–6290. [CrossRef]

13. Park, K.S.; Son, M.; Park, M.-J.; Kim, D.H.; Kim, J.H.; Park, S.H.; Choi, J.-H.; Bae, J.W. Adjusted interactions of nickel nanoparticles with cobalt-modified MgAl2O4-SiC for an enhanced catalytic stability during steam reforming of propane. *Appl. Catal. A Gen.* **2018**, *549*, 117–133. [CrossRef]

14. Karakaya, C.; Karadeniz, H.; Maier, L.; Deutschmann, O. Surface Reaction Kinetics of the Oxidation and Reforming of Propane over Rh/Al$_2$O$_3$ Catalysts. *ChemCatChem* **2017**, *9*, 685–695. [CrossRef]

15. Do, J.Y.; Lee, J.H.; Park, N.-K.; Lee, T.J.; Lee, S.T.; Kang, M. Synthesis and characterization of Ni$_{2-x}$Pd$_x$MnO$_4$/γ-Al$_2$O$_3$ catalysts for hydrogen production via propane steam reforming. *Chem. Eng. J.* **2018**, *334*, 1668–1678. [CrossRef]

16. Barzegari, F.; Kazemeini, M.; Farhadi, F.; Rezaei, M.; Keshavarz, A. Preparation of mesoporous nanostructure NiO–MgO–SiO$_2$ catalysts for syngas production via propane steam reforming. *Int. J. Hydrog. Energy* **2020**, *45*, 6604–6620. [CrossRef]

17. Do, J.Y.; Kwak, B.S.; Park, N.-K.; Lee, T.J.; Lee, S.T.; Jo, S.W.; Cha, M.S.; Jeon, M.-K.; Kang, M. Effect of acidity on the performance of a Ni-based catalyst for hydrogen production through propane steam reforming: K-AlSi$_x$O$_y$ support with different Si/Al ratios. *Int. J. Hydrog. Energy* **2017**, *42*, 22687–22697. [CrossRef]

18. Rönsch, S.; Schneider, J.; Matthischke, S.; Schlüter, M.; Götz, M.; Lefebvre, J.; Prabhakaran, P.; Bajohr, S. Review on methanation— From fundamentals to current projects. *Fuel* **2016**, *166*, 276–296. [CrossRef]

19. Gambo, Y.; Adamu, S.; Abdulrasheed, A.A.; Lucky, R.A.; Ba-Shammakh, M.S.; Hossain, M.M. Catalyst design and tuning for oxidative dehydrogenation of propane—A review. *Appl. Catal. A Gen.* **2021**, *609*, 117914. [CrossRef]

20. Ou, Z.; Zhang, Z.; Qin, C.; Xia, H.; Deng, T.; Niu, J.; Ran, J.; Wu, C. Highly active and stable Ni/perovskite catalysts in steam methane reforming for hydrogen production. *Sustain. Energy Fuels* **2021**, *5*, 1845–1856. [CrossRef]

21. Yentekakis, I.V.; Goula, G. Biogas Management: Advanced Utilization for Production of Renewable Energy and Added-value Chemicals. *Front. Environ. Sci.* **2017**, *5*, 7. [CrossRef]

22. Kim, K.M.; Kwak, B.S.; Park, N.-K.; Lee, T.J.; Lee, S.T.; Kang, M. Effective hydrogen production from propane steam reforming over bimetallic co-doped NiFe/Al$_2$O$_3$ catalyst. *J. Ind. Eng. Chem.* **2017**, *46*, 324–336. [CrossRef]

23. Kokka, A.; Katsoni, A.; Yentekakis, I.V.; Panagiotopoulou, P. Hydrogen production via steam reforming of propane over supported metal catalysts. *Int. J. Hydrog. Energy* **2020**, *45*, 14849–14866. [CrossRef]

24. Malaibari, Z.O.; Croiset, E.; Amin, A.; Epling, W. Effect of interactions between Ni and Mo on catalytic properties of a bimetallic Ni-Mo/Al$_2$O$_3$ propane reforming catalyst. *Appl. Catal. A Gen.* **2015**, *490*, 80–92. [CrossRef]

25. Arvaneh, R.; Fard, A.A.; Bazyari, A.; Alavi, S.M.; Abnavi, F.J. Effects of Ce, La, Cu, and Fe promoters on Ni/MgAl$_2$O$_4$ catalysts in steam reforming of propane. *Korean J. Chem. Eng.* **2019**, *36*, 1033–1041. [CrossRef]

26. Yu, L.; Sato, K.; Nagaoka, K. Rh/Ce$_{0.25}$Zr$_{0.75}$O$_2$ Catalyst for Steam Reforming of Propane at Low Temperature. *ChemCatChem* **2019**, *11*, 1472–1479. [CrossRef]

27. Modafferi, V.; Panzera, G.; Baglio, V.; Frusteri, F.; Antonucci, P.L. Propane reforming on Ni–Ru/GDC catalyst: H$_2$ production for IT-SOFCs under SR and ATR conditions. *Appl. Catal. A Gen.* **2008**, *334*, 1–9. [CrossRef]

28. Aartun, I.; Silberova, B.; Venvik, H.; Pfeifer, P.; Görke, O.; Schubert, K.; Holmen, A. Hydrogen production from propane in Rh-impregnated metallic microchannel reactors and alumina foams. *Catal. Today* **2005**, *105*, 469–478. [CrossRef]

29. Kolb, G.; Zapf, R.; Hessel, V.; Löwe, H. Propane steam reforming in micro-channels—results from catalyst screening and optimisation. *Appl. Catal. A Gen.* **2004**, *277*, 155–166. [CrossRef]

30. Aartun, I.; Gjervan, T.; Venvik, H.; Görke, O.; Pfeifer, P.; Fathi, M.; Holmen, A.; Schubert, K. Catalytic conversion of propane to hydrogen in microstructured reactors. *Chem. Eng. J.* **2004**, *101*, 93–99. [CrossRef]

31. Royer, S.; Duprez, D.; Can, F.; Courtois, X.; Batiot-Dupeyrat, C.; Laassiri, S.; Alamdari, H. Perovskites as substitutes of noble metals for heterogeneous catalysis: Dream or reality. *Chem. Rev.* **2014**, *114*, 10292–10368. [CrossRef]

32. Wang, C.; Wang, Y.; Chen, M.; Liang, D.; Yang, Z.; Cheng, W.; Tang, Z.; Wang, J.; Zhang, H. Recent advances during CH$_4$ dry reforming for syngas production: A mini review. *Int. J. Hydrog. Energy* **2021**, *46*, 5852–5874. [CrossRef]

33. Bhattar, S.; Abedin, M.A.; Kanitkar, S.; Spivey, J.J. A review on dry reforming of methane over perovskite derived catalysts. *Catal. Today* **2021**, *365*, 2–23. [CrossRef]

34. Gomez-Cuaspud, J.; CA, P.; Schmal, M. Nanostructured La$_{0.8}$Sr$_{0.2}$Fe$_{0.8}$Cr$_{0.2}$O$_3$ Perovskite for the Steam Methane Reforming. *Catal. Lett.* **2016**, *146*, 2504–2515. [CrossRef]

35. Anil, C.; Modak, J.M.; Madras, G. Syngas production via CO$_2$ reforming of methane over noble metal (Ru, Pt, and Pd) doped LaAlO$_3$ perovskite catalyst. *Mol. Catal.* **2020**, *484*, 110805. [CrossRef]

36. Da Silva, B.C.; Bastos, P.H.C.; Junior, R.B.S.; Checca, N.R.; Fréty, R.; Brandão, S.T. Perovskite-type catalysts based on nickel applied in the Oxy-CO$_2$ reforming of CH$_4$: Effect of catalyst nature and operative conditions. *Catal. Today* **2020**, *369*, 19–30. [CrossRef]

37. Kim, W.Y.; Jang, J.S.; Ra, E.C.; Kim, K.Y.; Kim, E.H.; Lee, J.S. Reduced perovskite LaNiO$_3$ catalysts modified with Co and Mn for low coke formation in dry reforming of methane. *Appl. Catal. A Gen.* **2019**, *575*, 198–203. [CrossRef]

38. Nuvula, S.; Sagar, T.V.; Valluri, D.K.; Sai Prasad, P.S. Selective substitution of Ni by Ti in LaNiO$_3$ perovskites: A parameter governing the oxy-carbon dioxide reforming of methane. *Int. J. Hydrog. Energy* **2018**, *43*, 4136–4142. [CrossRef]

39. Wang, M.; Zhao, T.; Dong, X.; Li, M.; Wang, H. Effects of Ce substitution at the A-site of LaNi$_{0.5}$Fe$_{0.5}$O$_3$ perovskite on the enhanced catalytic activity for dry reforming of methane. *Appl. Catal. B Environ.* **2018**, *224*, 214–221. [CrossRef]

40. Wang, H.; Dong, X.; Zhao, T.; Yu, H.; Li, M. Dry reforming of methane over bimetallic Ni-Co catalyst prepared from La(Co$_x$Ni$_{1-x}$)$_{0.5}$Fe$_{0.5}$O$_3$ perovskite precursor: Catalytic activity and coking resistance. *Appl. Catal. B Environ.* **2019**, *245*, 302–313. [CrossRef]
41. Wei, T.; Jia, L.; Luo, J.-L.; Chi, B.; Pu, J.; Li, J. CO$_2$ dry reforming of CH$_4$ with Sr and Ni co-doped LaCrO$_3$ perovskite catalysts. *Appl. Surf. Sci.* **2020**, *506*, 144699. [CrossRef]
42. Yadav, P.K.; Das, T. Production of syngas from carbon dioxide reforming of methane by using LaNi$_x$Fe$_{1-x}$O$_3$ perovskite type catalysts. *Int. J. Hydrog. Energy* **2019**, *44*, 1659–1670. [CrossRef]
43. Bai, Y.; Wang, Y.; Yuan, W.; Sun, W.; Zhang, G.; Zheng, L.; Han, X.; Zhou, L. Catalytic performance of perovskite-like oxide doped cerium (La$_{2-x}$Ce$_x$CoO$_{4\pm y}$) as catalysts for dry reforming of methane. *Chin. J. Chem. Eng.* **2019**, *27*, 379–385. [CrossRef]
44. Junior, R.B.S.; Rabelo-Neto, R.C.; Gomes, R.S.; Noronha, F.B.; Fréty, R.; Brandão, S.T. Steam reforming of acetic acid over Ni-based catalysts derived from La$_{1-x}$Ca$_x$NiO$_3$ perovskite type oxides. *Fuel* **2019**, *254*, 115714. [CrossRef]
45. Li, L.; Jiang, B.; Tang, D.; Zhang, Q.; Zheng, Z. Hydrogen generation by acetic acid steam reforming over Ni-based catalysts derived from La$_{1-x}$Ce$_x$NiO$_3$ perovskite. *Int. J. Hydrog. Energy* **2018**, *43*, 6795–6803. [CrossRef]
46. Zhang, Z.; Ou, Z.; Qin, C.; Ran, J.; Wu, C. Roles of alkali/alkaline earth metals in steam reforming of biomass tar for hydrogen production over perovskite supported Ni catalysts. *Fuel* **2019**, *257*, 116032. [CrossRef]
47. Zhang, X.; Su, Y.; Pei, C.; Zhao, Z.-J.; Liu, R.; Gong, J. Chemical looping steam reforming of methane over Ce-doped perovskites. *Chem. Eng. Sci.* **2020**, *223*, 115707. [CrossRef]
48. Mota, N.; Navarro, R.M.; Alvarez-Galvan, M.C.; Al-Zahrani, S.M.; Fierro, J.L.G. Hydrogen production by reforming of diesel fuel over catalysts derived from LaCo$_{1-x}$Ru$_x$O$_3$ perovskites: Effect of the partial substitution of Co by Ru (x = 0.01–0.1). *J. Power Sources* **2011**, *196*, 9087–9095. [CrossRef]
49. Mota, N.; Alvarez-Galvan, M.C.; Al-Zahrani, S.M.; Navarro, R.M.; Fierro, J.L.G. Diesel fuel reforming over catalysts derived from LaCo$_{1-x}$Ru$_x$O$_3$ perovskites with high Ru loading. *Int. J. Hydrog. Energy* **2012**, *37*, 7056–7066. [CrossRef]
50. Navarro, R.M.; Alvarez-Galvan, M.C.; Villoria, J.A.; González-Jiménez, I.D.; Rosa, F.; Fierro, J.L.G. Effect of Ru on LaCoO$_3$ perovskite-derived catalyst properties tested in oxidative reforming of diesel. *Appl. Catal. B Environ.* **2007**, *73*, 247–258. [CrossRef]
51. Zhao, B.; Yan, B.; Yao, S.; Xie, Z.; Wu, Q.; Ran, R.; Weng, D.; Zhang, C.; Chen, J.G. LaFe$_{0.9}$Ni$_{0.1}$O$_3$ perovskite catalyst with enhanced activity and coke-resistance for dry reforming of ethane. *J. Catal.* **2018**, *358*, 168–178. [CrossRef]
52. Martínez, A.H.; Lopez, E.; Cadús, L.E.; Agüero, F.N. Elucidation of the role of support in Rh/perovskite catalysts used in ethanol steam reforming reaction. *Catal. Today* **2021**, *372*, 59–69. [CrossRef]
53. De Lima, S.M.; da Silva, A.M.; da Costa, L.O.O.; Assaf, J.M.; Jacobs, G.; Davis, B.H.; Mattos, L.V.; Noronha, F.B. Evaluation of the performance of Ni/La$_2$O$_3$ catalyst prepared from LaNiO$_3$ perovskite-type oxides for the production of hydrogen through steam reforming and oxidative steam reforming of ethanol. *Appl. Catal. A Gen.* **2010**, *377*, 181–190. [CrossRef]
54. Wang, M.; Yang, J.; Chi, B.; Pu, J.; Li, J. High performance Ni exsolved and Cu added La$_{0.8}$Ce$_{0.2}$Mn$_{0.6}$Ni$_{0.4}$O$_3$-based perovskites for ethanol steam reforming. *Int. J. Hydrogen Energy* **2020**, *45*, 16458–16468. [CrossRef]
55. Aman, D.; Radwan, D.; Ebaid, M.; Mikhail, S.; van Steen, E. Comparing nickel and cobalt perovskites for steam reforming of glycerol. *Mol. Catal.* **2018**, *452*, 60–67. [CrossRef]
56. Kamonsuangkasem, K.; Therdthianwong, S.; Therdthianwong, A.; Thammajak, N. Remarkable activity and stability of Ni catalyst supported on CeO$_2$-Al$_2$O$_3$ via CeAlO$_3$ perovskite towards glycerol steam reforming for hydrogen production. *Appl. Catal. B Environ.* **2017**, *218*, 650–663. [CrossRef]
57. Baamran, K.S.; Tahir, M. Ni-embedded TiO$_2$-ZnTiO$_3$ reducible perovskite composite with synergistic effect of metal/support towards enhanced H$_2$ production via phenol steam reforming. *Energy Convers. Manag.* **2019**, *200*, 112064. [CrossRef]
58. Takise, K.; Manabe, S.; Muraguchi, K.; Higo, T.; Ogo, S.; Sekine, Y. Anchoring effect and oxygen redox property of Co/La$_{0.7}$Sr$_{0.3}$AlO$_{3-\delta}$ perovskite catalyst on toluene steam reforming reaction. *Appl. Catal. A Gen.* **2017**, *538*, 181–189. [CrossRef]
59. Zhang, Z.; Qin, C.; Ou, Z.; Ran, J. Resistance of Ni/perovskite catalysts to H$_2$S in toluene steam reforming for H$_2$ production. *Int. J. Hydrog. Energy* **2020**, *45*, 26800–26811. [CrossRef]
60. Lim, S.-S.; Lee, H.-J.; Moon, D.-J.; Kim, J.-H.; Park, N.-C.; Shin, J.-S.; Kim, Y.-C. Autothermal reforming of propane over Ce modified Ni/LaAlO$_3$ perovskite-type catalysts. *Chem. Eng. J.* **2009**, *152*, 220–226. [CrossRef]
61. MSP, S.; Hossain, M.M.; Gnanasekaran, G.; Mok, Y.S. Dry Reforming of Propane over γ-Al$_2$O$_3$ and Nickel Foam Supported Novel SrNiO$_3$ Perovskite Catalyst. *Catalysts* **2019**, *9*, 68.
62. Yeyongchaiwat, J.; Matsumoto, H.; Ishihara, T. Oxidative reforming of propane with oxygen permeating membrane reactor using Pr$_2$Ni$_{0.75}$Cu$_{0.25}$Ga$_{0.05}$O$_4$ perovskite related mixed conductor. *Solid State Ion.* **2017**, *301*, 23–27. [CrossRef]
63. Batiot-Dupeyrat, C.; Gallego, G.A.S.; Mondragon, F.; Barrault, J.; Tatibouët, J.M. CO$_2$ reforming of methane over LaNiO$_3$ as precursor material. *Catal. Today* **2005**, *107*, 474–480. [CrossRef]
64. Sutthiumporn, K.; Kawi, S. Promotional effect of alkaline earth over Ni-La$_2$O$_3$ catalyst for CO$_2$ reforming of CH$_4$: Role of surface oxygen species on H$_2$ production and carbon suppression. *Int. J. Hydrog. Energy* **2011**, *36*, 14435–14446. [CrossRef]
65. Valderrama, G.; Goldwasser, M.R.; Navarro, C.U.D.; Tatibouët, J.M.; Barrault, J.; Batiot-Dupeyrat, C.; Martínez, F. Dry reforming of methane over Ni perovskite type oxides. *Catal. Today* **2005**, *107*, 785–791. [CrossRef]
66. Yang, E.H.; Noh, Y.S.; Hong, G.H.; Moon, D.J. Combined steam and CO$_2$ reforming of methane over La$_{1-x}$Sr$_x$NiO$_3$ perovskite oxides. *Catal. Today* **2018**, *299*, 242–250. [CrossRef]
67. Rostrup-Nielsen, J.R.; Bak Hansen, J.H. CO$_2$-reforming of methane over transition metals. *J. Catal.* **1993**, *144*, 38–49. [CrossRef]

68. Borges, R.P.; Moura, L.G.; Spivey, J.J.; Noronha, F.B.; Hori, C.E. Hydrogen production by steam reforming of LPG using supported perovskite type precursors. *Int. J. Hydrog. Energy* **2020**, *45*, 21166–21177. [CrossRef]

69. Safakas, A.; Bampos, G.; Bebelis, S. Oxygen reduction reaction on $La_{0.8}Sr_{0.2}Co_xFe_{1-x}O_{3-\delta}$ perovskite/carbon black electrocatalysts in alkaline medium. *Appl. Catal. B Environ.* **2019**, *244*, 225–232. [CrossRef]

70. Bampos, G.; Bebelis, S.; Kondarides, D.I.; Verykios, X. Comparison of the Activity of Pd–M (M: Ag, Co, Cu, Fe, Ni, Zn) Bimetallic Electrocatalysts for Oxygen Reduction Reaction. *Top. Catal.* **2017**, *60*, 1260–1273. [CrossRef]

71. Bampos, G.; Bika, P.; Panagiotopoulou, P.; Verykios, X.E. Reactive adsorption of CO from low CO concentrations streams on the surface of Pd/CeO_2 catalysts. *Appl. Catal. A Gen.* **2019**, *588*, 117305. [CrossRef]

72. Rivas, M.E.; Fierro, J.L.G.; Goldwasser, M.R.; Pietri, E.; Pérez-Zurita, M.J.; Griboval-Constant, A.; Leclercq, G. Structural features and performance of $LaNi_{1-x}Rh_xO_3$ system for the dry reforming of methane. *Appl. Catal. A Gen.* **2008**, *344*, 10–19. [CrossRef]

73. Merino, N.A.; Barbero, B.P.; Grange, P.; Cadús, L.E. $La_{1-x}Ca_xCoO_3$ perovskite-type oxides: Preparation, characterisation, stability, and catalytic potentiality for the total oxidation of propane. *J. Catal.* **2005**, *231*, 232–244. [CrossRef]

74. Zhan, H.; Li, F.; Gao, P.; Zhao, N.; Xiao, F.; Wei, W.; Zhong, L.; Sun, Y. Methanol synthesis from CO_2 hydrogenation over La-M-Cu-Zn-O (M = Y, Ce, Mg, Zr) catalysts derived from perovskite-type precursors. *J. Power Sources* **2014**, *251*, 113–121. [CrossRef]

75. Zhu, J.; Li, H.; Zhong, L.; Xiao, P.; Xu, X.; Yang, X.; Zhao, Z.; Li, J. Perovskite oxides: Preparation, characterizations, and applications in heterogeneous catalysis. *ACS Catal.* **2014**, *4*, 2917–2940. [CrossRef]

76. Chen, D.; Chen, C.; Baiyee, Z.M.; Shao, Z.; Ciucci, F. Nonstoichiometric Oxides as Low-Cost and Highly-Efficient Oxygen Reduction/Evolution Catalysts for Low-Temperature Electrochemical Devices. *Chem. Rev.* **2015**, *115*, 9869–9921. [CrossRef] [PubMed]

77. Ivanova, S.; Senyshyn, A.; Zhecheva, E.; Tenchev, K.; Stoyanova, R.; Fuess, H. Crystal structure, microstructure and reducibility of $LaNi_xCo_{1-x}O_3$ and $LaFe_xCo_{1-x}O_3$ Perovskites ($0 < x \leq 0.5$). *J. Solid State Chem.* **2010**, *183*, 940–950.

78. Pastor-Pérez, L.; Baibars, F.; Le Sache, E.; Arellano-García, H.; Gu, S.; Reina, T.R. CO_2 valorisation via Reverse Water-Gas Shift reaction using advanced Cs doped $Fe-Cu/Al_2O_3$ catalysts. *J. CO2 Util.* **2017**, *21*, 423–428. [CrossRef]

79. Rynkowski, J.; Samulkiewicz, P.; Ladavos, A.K.; Pomonis, P.J. Catalytic performance of reduced $La_{2-x}Sr_xNiO_4$ perovskite-like oxides for CO_2 reforming of CH_4. *Appl. Catal. A Gen.* **2004**, *263*, 1–9. [CrossRef]

80. Valderrama, G.; Urbina De Navarro, C.; Goldwasser, M.R. CO_2 reforming of CH_4 over Co-La-based perovskite-type catalyst precursors. *J. Power Sources* **2013**, *234*, 31–37. [CrossRef]

81. Valderrama, G.; Kiennemann, A.; Goldwasser, M.R. La-Sr-Ni-Co-O based perovskite-type solid solutions as catalyst precursors in the CO_2 reforming of methane. *J. Power Sources* **2010**, *195*, 1765–1771. [CrossRef]

82. Nezhad, P.D.K.; Bekheet, M.F.; Bonmassar, N.; Schlicker, L.; Gili, A.; Kamutzki, F.; Gurlo, A.; Doran, A.; Gao, Y.; Heggen, M.; et al. Mechanistic in situ insights into the formation, structural and catalytic aspects of the La_2NiO_4 intermediate phase in the dry reforming of methane over Ni-based perovskite catalysts. *Appl. Catal. A Gen.* **2021**, *612*, 117984. [CrossRef]

83. Bonmassar, N.; Bekheet, M.F.; Schlicker, L.; Gili, A.; Gurlo, A.; Doran, A.; Gao, Y.; Heggen, M.; Bernardi, J.; Klötzer, B.; et al. In Situ-Determined Catalytically Active State of $LaNiO_3$ in Methane Dry Reforming. *ACS Catal.* **2020**, *10*, 1102–1112. [CrossRef]

84. Zhang, Z.; Verykios, X.E. Mechanistic aspects of carbon dioxide reforming of methane to synthesis gas over Ni catalysts. *Catal. Lett.* **1996**, *38*, 175–179. [CrossRef]

85. Shi, Q.; Peng, Z.; Chen, W.; Zhang, N. $La_2O_2CO_3$ supported Ni-Fe catalysts for hydrogen production from steam reforming of ethanol. *J. Rare Earths* **2011**, *29*, 861–865. [CrossRef]

86. Shi, Q.J.; Li, B.; Chen, W.Q.; Liu, C.W.; Huang, B.W. Ethanol Steam Reforming over $La_2O_2CO_3$ Supported Ni-Ru Bimetallic Catalysts. *Adv. Mater. Res.* **2012**, *457–458*, 314–319.

87. Li, K.; Pei, C.; Li, X.; Chen, S.; Zhang, X.; Liu, R.; Gong, J. Dry reforming of methane over $La_2O_2CO_3$-modified Ni/Al_2O_3 catalysts with moderate metal support interaction. *Appl. Catal. B Environ.* **2020**, *264*, 118448. [CrossRef]

88. Da Silva, B.C.; Bastos, P.H.C.; Junior, R.B.S.; Checca, N.R.; Costa, D.S.; Fréty, R.; Brandão, S.T. Oxy-CO_2 reforming of CH_4 on Ni-based catalysts: Evaluation of cerium and aluminum addition on the structure and properties of the reduced materials. *Catal. Today* **2020**. In Press. [CrossRef]

89. Italiano, C.; Bizkarra, K.; Barrio, V.L.; Cambra, J.F.; Pino, L.; Vita, A. Renewable hydrogen production via steam reforming of simulated bio-oil over Ni-based catalysts. *Int. J. Hydrog. Energy* **2019**, *44*, 14671–14682. [CrossRef]

90. García-Diéguez, M.; Pieta, I.S.; Herrera, M.C.; Larrubia, M.A.; Alemany, L.J. RhNi nanocatalysts for the CO_2 and $CO_2 + H_2O$ reforming of methane. *Catal. Today* **2011**, *172*, 136–142. [CrossRef]

91. Villegas, L.; Guilhaume, N.; Mirodatos, C. Autothermal syngas production from model gasoline over Ni, Rh and Ni-Rh/Al_2O_3 monolithic catalysts. *Int. J. Hydrog. Energy* **2014**, *39*, 5772–5780. [CrossRef]

92. Barros, B.S.; Melo, D.M.A.; Libs, S.; Kiennemann, A. CO_2 reforming of methane over La_2NiO_4/α-Al_2O_3 prepared by microwave assisted self-combustion method. *Appl. Catal. A Gen.* **2010**, *378*, 69–75. [CrossRef]

nanomaterials

MDPI

Article

Support Effects on the Activity of Ni Catalysts for the Propane Steam Reforming Reaction

Aliki Kokka [1], Athanasia Petala [2] and Paraskevi Panagiotopoulou [1,*]

[1] School of Chemical and Environmental Engineering, Technical University of Crete, 73100 Chania, Greece; akokka@isc.tuc.gr
[2] Department of Chemical Engineering, University of Patras, 26504 Patras, Greece; natpetala@chemeng.upatras.gr
* Correspondence: ppanagiotopoulou@isc.tuc.gr; Tel.: +30-28210-37770

Abstract: The catalytic performance of supported Ni catalysts for the propane steam reforming reaction was investigated with respect to the nature of the support. It was found that Ni is much more active when supported on ZrO_2 or YSZ compared to TiO_2, whereas Al_2O_{3-} and CeO_2-supported catalysts exhibit intermediate performance. The turnover frequency (TOF) of C_3H_8 conversion increases by more than one order of magnitude in the order $Ni/TiO_2 < Ni/CeO_2 < Ni/Al_2O_3 < Ni/YSZ < Ni/ZrO_2$, accompanied by a parallel increase of the selectivity toward the intermediate methane produced. In situ FTIR experiments indicate that CH_x species produced via the dissociative adsorption of propane are the key reaction intermediates, with their hydrogenation to CH_4 and/or conversion to formates and, eventually, to CO, being favored over the most active Ni/ZrO_2 catalyst. Long term stability test showed that Ni/ZrO_2 exhibits excellent stability for more than 30 h on stream and thus, it can be considered as a suitable catalyst for the production of H_2 via propane steam reforming.

Keywords: propane steam reforming; H_2 production; Ni; TiO_2; CeO_2; YSZ; ZrO_2; Al_2O_3; drifts

Citation: Kokka, A.; Petala, A.; Panagiotopoulou, P. Support Effects on the Activity of Ni Catalysts for the Propane Steam Reforming Reaction. *Nanomaterials* **2021**, *11*, 1948. https://doi.org/10.3390/nano11081948

Academic Editors: Nikolaos Dimitratos and Alberto Villa

Received: 30 June 2021
Accepted: 27 July 2021
Published: 28 July 2021

Publisher's Note: MDPI stays neutral with regard to jurisdictional claims in published maps and institutional affiliations.

1. Introduction

The high energy efficiency of fuel cells has drawn considerable attention toward the development of hydrogen production technologies. Hydrogen can be produced from fossil fuels either via hydrocarbon pyrolysis or hydrocarbon reforming processes including steam reforming, partial oxidation or autothermal reforming [1–5]. Biomass processes consisting of biological (bio-photolysis, dark fermentation, photo fermentation) and thermochemical (pyrolysis, gasification, combustion, liquefaction) methods as well as water splitting processes such as electrolysis, thermolysis or photolysis can be alternatively applied for the production of H_2 from renewable energy sources [2,6]. However, the latter approaches are facing some major obstacles mostly related to high cost and low H_2 yields. Currently, steam reforming of light hydrocarbons, including natural gas, ethane, propane, butane and liquified petroleum gas (LPG), are considered among the most promising and economical routes for hydrogen production [7]. Propane, which is the main component of LPG, has many advantages such as high energy density, compressibility to a transportable liquid at normal temperature and well-developed infrastructure which enable its use worldwide [8–10]. Moreover, propane can be stored and transferred as LPG through a wide distribution network or in high pressure cylinders in order to be supplied in remote places (e.g., agricultural, inaccessible or camping areas) or for domestic uses (e.g., households) [3,7].

Under propane steam reforming conditions, the water-gas shift reaction occurs simultaneously at low temperatures contributing to H_2 and CO_2 production, whereas CO/CO_2 methanation may also run in parallel yielding CH_4 and H_2O. Methane can be also formed via hydrogenation of CH_x species derived by the dissociative adsorption of propane on the catalyst surface or through propane decomposition accompanied by ethylene production. In certain cases, the C_2H_4, CH_4 and CO thus produced are further decomposed leading to the formation of coke on the catalyst surface and consequently, to its progressive deactivation [11,12].

Supported noble metal (Rh, Ru, Pt, Pd) catalysts have been proposed to efficiently catalyze the production of H_2 via propane steam reforming exhibiting high resistance to coke formation [3,11,13,14]. However, the high cost and low availability of noble metals are major drawbacks restricting their use in practical applications [15–17]. Noble metals can be replaced by nickel, which is less expensive and able to convert propane with high H_2 yields. However, Ni is susceptible to coke formation and particles sintering considered to be responsible for catalyst deactivation [5,10,17]. The lifetime of Ni-based catalysts can be improved by optimization of the reaction conditions, catalyst promotion, improvement of the catalyst synthesis method as well as by the proper selection of the support [3,9,10,12,16,18,19].

Regarding the support material, it has been proposed that metal oxides characterized by high oxygen storage capacity, such as CeO_2, YSZ, TiO_2, ZrO_2 or CeO_2-ZrO_2, are able to suppress carbon deposition through the participation of their lattice oxygen in carbon removal [3,16]. Moreover, Ni catalysts supported on metal oxides or promoted metal oxides, which favor steam adsorption and mobility of surface hydroxyls, have been found to facilitate coke gasification [17,20].

Metal-support interactions have been also reported to impose a dramatic effect on both carbon deposition and metal particles sintering under conditions of hydrocarbons reforming [5,21,22]. It was demonstrated that stronger interactions between Ni and $MgAl_2O_4$ support resulted in high Ni dispersion and inhibition of the formation of large Ni clusters [22], whereas weak interactions between Ni and SiO_2 carrier were found to accelerate sintering and coke formation [21].

Therefore, the economic viability and practical applicability of H_2 production via propane steam reforming may be facilitated by the development of efficient Ni catalysts supported on suitable metal oxides, which will be able to possess both high activity and resistance against carbon deposition and particles sintering to realize economically viable reforming processes. In the present study the effect of the nature of the support (TiO_2, CeO_2, Al_2O_3, YSZ, ZrO_2) on the activity and selectivity of Ni-based catalysts for the propane steam reforming reaction was investigated. Mechanistic aspects related to the support influence on the reaction pathway were also studied and are discussed.

2. Materials and Methods

2.1. Catalyst Preparation and Characterization

The wet impregnation method was applied to prepare Ni (5 wt.%) catalysts supported on commercial metal oxide powders by using an aqueous solution of Ni ($Ni(NO_3)_2 \cdot 6H_2O$) as the metal precursor salt. The commercial metal oxide carriers were used as received and were (a) activated aluminum oxide (Al_2O_3), catalyst support, 99% (metals basis) (Alfa Aesar, Kandel, Germany), (b) AEROXIDE® TiO_2 P25 (TiO_2) (Evonik Industries AG, Essen, Germany), (c) cerium (IV) oxide (CeO_2), nanopowder, 99.5% min (Alfa Aesar, Kandel, Germany), (d) yttria-stabilized zirconia (YSZ) (8Y-SZ, Tosoh, Amsterdam, The Netherlands) and (e) zirconium (IV) oxide (ZrO_2) 99% (metal basis) (Alfa Aesar, Kandel, Germany). The resulting materials were dried at 110 °C for 24 h followed by reduction at 400 °C under H_2 flow for 2 h. The selection of reducing Ni catalysts at 400 °C for 2 h was based on previous studies which indicated that under these reducing conditions nickel oxide species are able to be completely converted to metallic nickel [23–25].

The X-ray diffraction patterns of the catalysts were recorded using an X-ray powder diffractometer (A Brucker D8 Advance instrument, Bruker, Karlsruhe, Germany) using Cu K_a radiation ($\lambda = 0.15406$ nm, 40 kV, 40 mA) and a scan rate of 0.025°/s over a range of 2θ between 20 and 80°. The diffraction pattern was identified by comparison with those supplied from the JCPDS data base, whereas the primary crystallite size of M_xO_y ($d_{M_xO_y}$) was estimated according to Scherrer's equation:

$$d_{M_xO_y} = \frac{0.9 \cdot \lambda}{B \cdot \cos\theta} \qquad (1)$$

where θ is the angle of diffraction corresponding to the peak broadening, B is the full-width at half maximum intensity (in radians) and $\lambda = 0.15406$ nm is the X-ray wavelength corresponding to CuK_a radiation.

The specific surface area (SSA) of the supported Ni catalysts were measured by N_2 adsorption at 77 K (B.E.T. technique) using a Gemini III 2375 instrument (Micromeritics, Norcross, GA, USA). Carbon monoxide chemisorption measurements at 25 °C were applied for the determination of Ni dispersion and mean particle size using a modified Sorptomatic 1900 apparatus (Fisons Instruments, Glaskow, UK) and assuming a CO:Me stoichiometry of 1:1, an atomic surface area of 6.5 $Å^2$ and spherical particles. CO chemisorption measurements were used instead of H_2 chemisorption in order to avoid overestimation of Ni dispersion due to hydrogen spillover effects, which have been previously found to occur over supported Ni catalysts [26,27]. Nickel particle size was calculated according to the following equation:

$$d_{Ni} = \frac{60000}{\rho_{Ni} \cdot S_{Ni}} \, [Å] \qquad (2)$$

where d_{Ni} is the mean crystallite diameter, ρ_{Ni} (= 8.9 $g \cdot cm^{-3}$) is the density of Ni and S_{Ni} [m^2/g_{Ni}] is the surface area per gram of Ni.

Transmission electron microscopy (TEM) images were obtained with a JEM-2100 system (JEOL, Akishima, Tokyo, Japan) operated at 200 kV (point resolution 0.23 nm) using an Erlangshen CCD Camera (Model 782 ES500W, Gatan Inc., Pleasanton, CA, USA). Samples were dispersed in water and spread onto a carbon-coated copper grid (200 mesh). Details related to the equipment and procedures used for catalyst characterization have been described in detail elsewhere [28].

2.2. Catalytic Performance Tests and Kinetic Measurements

The catalytic performance of the synthesized materials was studied in a tubular fixed-bed quartz reactor under atmospheric pressure using an apparatus which has been described in detail elsewhere [11]. The reaction conditions were as follows: temperature range 400–750 °C, $H_2O/C = 3.25$, and gas hourly space velocity (GHSV) = 55,900 h^{-1}. The reactor was loaded with 150 mg of catalyst (particle diameter: $0.15 < d_p < 0.25$ mm) and placed in an electric furnace, where it was reduced in situ at 300 °C for 1 h under $50\%H_2/He$ flow (60 cm^3 min^{-1}) to ensure that the Ni exists in its metallic phase prior to catalytic performance tests. Catalyst reduction was followed by heating at 750 °C under He and subsequent switch of the flow to the feed stream consisted of $4.5\%C_3H_8$ + $0.15\%Ar$ + $44\%H_2O$ (He balance). Argon was used as internal standard in order to account for the volume change. Water was fed through an HPLC pump (LD Class Pump, TELEDYNE SSI, PA, USA) into a vaporizer maintained at 180 °C and mixed with the gas stream coming from mass-flow controllers. A condenser immersed in an ice bath was placed at the exit of the reactor to condensate water prior to introduction of the gas stream to the analysis system. Reaction gases (He, $30\%C_3H_8-1\%Ar/He$, H_2) are supplied from high-pressure gas cylinders (Buse Gas, Bad Hönningen, Germany) and are of ultrahigh purity. Measurements of reactants' and products' concentrations were obtained by stepwise decreasing temperature up to 400 °C. The effluent from the reactor was analyzed using two gas chromatographs (Shimadzu, Kyoto, Japan) which were connected in parallel. The procedure used for gas phase analysis was described in our previous study [11]. The conversion of propane ($X_{C_3H_8}$) was calculated using the following expression:

$$X_{C_3H_8} = \frac{[Carbon]_{total,out}}{[Carbon]_{total,out} + [C_3H_8]_{out}} \times 100 \qquad (3)$$

where $[\text{Carbon}]_{\text{total,out}}$ is the sum of the concentrations of all carbon containing products:

$$[\text{Carbon}]_{\text{total,out}} = \frac{[CO] + [CO_2] + [CH_4]}{3} + 2 \times \frac{[C_2H_4] + [C_2H_6]}{3} \quad (4)$$

Selectivity toward reaction products containing carbon was defined using Equation (5). The factor n corresponds to the number of carbon atoms in the corresponding molecule (e.g., for CO is 1, for C_2H_4 is 2 etc.):

$$S_{Cn} = \frac{[C_n] \times n}{3 \times [\text{Carbon}]_{\text{total,out}}} \times 100 \quad (5)$$

Selectivity toward hydrogen production was defined as the concentration of hydrogen produced divided with the concentration of all products containing hydrogen according to Equation (6). The factor m represents the number of hydrogen atoms in the corresponding molecule (e.g., for CH_4 and C_2H_4 is 4).

$$S_{H_2}(\%) = \frac{[H_2]}{[H_2] + m/2 \times [C_nH_m]} \times 100 \quad (6)$$

The intrinsic reaction rates for propane steam reforming reaction were measured for low propane conversions ($X_{C_3H_8} < 10\%$) by varying W/F using the following expression:

$$R_{C_3H_8} = \frac{[C_3H_8]_{in} \cdot F_{in} - [C_3H_8]_{out} \cdot F_{out}}{W} \times 100 \quad (7)$$

where R_{C3H8} is the molar rate of C_3H_8 consumption (mol s^{-1} g$_{cat}{}^{-1}$), $[C_3H_8]_{in}$, $[C_3H_8]_{out}$, are the inlet and outlet concentrations (v/v) of C_3H_8, respectively, F_{in} and F_{out} are the total flow rates in the inlet and outlet of the reactor (mols^{-1}), respectively, and W is the mass of catalyst (g$_{cat}$).

Turnover frequencies (TOFs) of propane conversion were estimated following Equation (8) taking into account the measurements of both the reaction rates and nickel dispersions:

$$TOF = \frac{R_{C_8H_8} \cdot AW_{Ni}}{D_{Ni} \cdot X_{Ni}} \quad (8)$$

where AW_{Ni} is the atomic weight of nickel (g$_{Ni}$/mol$_{Ni}$), X_{Ni} is the nickel loading (g$_{Ni}$/g$_{cat}$) and D_{Ni} is the dispersion of nickel.

2.3. In Situ FTIR Spectroscopy

In situ Fourier transform infrared (FTIR) experiments were carried out using an iS20 FTIR spectrometer (Nicolet, Thermo Fischer Scientific, Waltham, MA, USA) equipped with an MCT detector, a KBr beam splitter and a diffuse reflectance (DRIFT) sampling system (Specac, Orpington, UK) accompanied by an environmental chamber suitable for the study of diffusely reflecting solid samples in a controlled atmosphere. A flow system equipped with mass flow controllers, a steam saturator and a set of valves used for controlling the gas stream interacted with the catalyst surface, was directly connected to the gas inlet of the environmental chamber.

In a typical experiment, the catalyst powder was placed in the sampling system and heated at 500 °C in flowing helium for 10 min and then reduced under hydrogen flow at 300 °C for 30 min. The flow was then switched to He and the temperature was increased at 500 °C. After remaining 10 min at this temperature the sample was cooled at 100 °C. While cooling, the background spectra were recorded at the desired temperatures. Finally, the flow was switched to the reaction mixture, which consisted of 0.5%C_3H_8 +5%H_2O (in He). Steam was introduced to the system via an independent He line passing through a saturator containing water maintained at 60 °C. The resulting gas mixture was fed to the DRIFT cell through stainless steel tubing maintained at 60 °C by means of heating tapes. A spectrum was collected at 100 °C after 15 min-on-stream followed by a stepwise increase of temperature up to 500 °C. During heating, spectra were recorded at selected temperatures after an equilibration for 15 min. In all experiments, the total flow through the DRIFT cell was 30 cm^3 min^{-1}. Reaction gases (He, 2%C_3H_8/He, H_2) are supplied from high-pressure gas cylinders (Buse Gas, Bad Hönningen, Germany) and are of ultrahigh purity.

3. Results

3.1. Catalyst Characterization

The XRD patterns of Ni/M$_x$O$_y$ catalysts are shown in Figure 1. The characteristic peaks located at the diffraction angles of 32.7°, 37.7°, 39.9°, 45.8° and 67.5° were appeared for Ni/Al$_2$O$_3$ which are attributed to (220), (311), (222), (400) and (440) facets of cubic Al$_2$O$_3$ (JCPDS Card No. 10-425), respectively. The XRD spectra recorded for Ni/CeO$_2$ catalyst consisted of peaks located at 2θ = 28.79°, 33.26°, 47.62°, 56.34°, 59.1°, 69.42°, 76.66°, 79.11° attributed to (111), (200), (220), (311), (222), (400), (331) and (420) planes of the cubic CeO$_2$ (JCPDS Card No. 2-1306), whereas in the case of Ni/ZrO$_2$ catalyst numerous peaks were detected on the XRD pattern. In particular, the peaks detected at 24.2°, 24.5°, 28.5°, 31.7°, 34.5°, 35.5°, 38.9°, 41.0°, 41.5°, 45.1°, 45.8°, 49.5°, 50.5°, 54.4°, 55.7°, 57.5°, 60.2°, 62.1°, 63.2°, 66.0°, 71.6°, 75.3° correspond to (011), (−110), (−111), (111), (002), (200), (021), (−211), (−121), (112), (211), (022), (−221), (202), (013), (212), (−302), (113), (311), (−321), (−104), (−140) planes of monoclinic ZrO$_2$ (JCPDS Card No. 13-307). When Ni was supported on YSZ, the XRD pattern was characterized by reflections at 30.4°, 35.1°, 50.4°, 59.9°, 62.9° and 74.0° attributed to (111), (200), (220), (311), (222) and (400) planes of YSZ (JCPDS Card No. 82-1246), respectively.

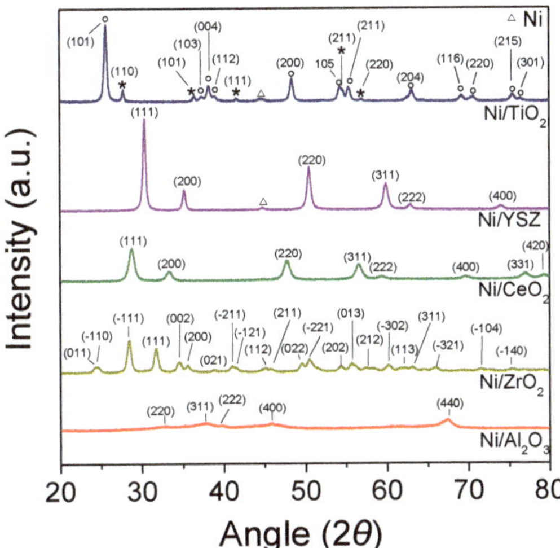

Figure 1. XRD patterns obtained from 5 wt.% Ni catalysts supported on the indicated commercial metal oxide carriers. The reflection planes of anatase (°) and rutile (*) TiO$_2$ phases are indicated in the diffractogram of Ni/TiO$_2$ catalyst.

Results obtained from Ni/TiO$_2$ catalyst showed that the sample consisted of TiO$_2$ in both its anatase and rutile form exhibiting peaks at 2θ = 25.6° (101), 37.2° (103), 38.2° (004), 38.9° (112), 48.4° (200), 54.3° (105), 55.4° (211), 63.1° (204), 69.3° (116), 70.7° (220), 75.4° (215) and 76.4° (301) for anatase (JCPDS Card No. 21-1272), and at 2θ = 27.6° (110), 36.3° (101), 41.6° (111), 44.3° (210), 54.6° (211), 56.9° (220) and 64.3° (310) for rutile (JCPDS Card No. 21-1276).

In the case of Ni/TiO$_2$ and Ni/YSZ catalysts an additional weak peak located 44.5° was appeared corresponding to Ni (111) plane (JCPDS Card No. 04-0850). The absence of peaks corresponding to metallic Ni for the rest catalysts investigated is due to the low Ni loading and/or particle size. The primary crystallite size of the supports was estimated according to Scherrer's formula at the diffraction angles corresponding to (440) plane for Al_2O_3, (-111) plane for ZrO_2, (111) plane for CeO_2, (111) plane for YSZ and (101) plane for TiO_2, and it was found to be 6.0 nm for Ni/Al_2O_3, 10.5 nm for Ni/CeO_2, 15.0 nm for Ni/ZrO_2, 20.9 nm for Ni/YSZ and 21.8 nm for Ni/TiO_2 (Table 1).

The SSAs of Ni catalysts supported on metal oxide (M_xO_y) carriers were estimated equal to 39 m^2/g for Ni/ZrO_2, 11 m^2/g for Ni/YSZ, 66 m^2/g for Ni/Al_2O_3, 39 m^2/g for Ni/CeO_2 and 41 m^2/g for Ni/TiO_2 (Table 1).

Table 1. Physicochemical properties of supported Ni (5 wt.%) catalysts and their apparent activation energies for propane steam reforming reaction.

Catalyst	SSA [a] (m^2/g)	d_{MxOy} [b] (nm)	D_{Ni} [c] (%)	d_{Ni} [c] (nm)	Activation Energy (kJ/mol)
5%Ni/ZrO$_2$	39	15.0	5.7	17.8	154
5%Ni/YSZ	11	20.9	4.7	21.4	140
5%Ni/Al$_2$O$_3$	66	6.0	4.0	25.5	121
5%Ni/CeO$_2$	39	10.5	11.9	8.5	102
5%Ni/TiO$_2$	41	21.8	2.8	36.1	127

[a] Specific surface area, estimated with the BET method. [b] Primary crystallite size of M_xO_y, estimated from XRD line broadening. [c] Dispersion and mean particle size of Ni, estimated from selective chemisorption measurements.

Results of Ni dispersion (D_{Ni}) and mean particle size (d_{Ni}) estimated from CO chemisorption meaurements are summarized in Table 1. Generally, low Ni dispersions were estimated for all the investigated catalysts, most possibly due to the high Ni content (5 wt.%) in agreement with previous studies [4,6]. Higher Ni dispersion of 11.9% and smaller particle size of 8.5 nm was found for Ni/CeO_2 catalyst, whereas Ni/TiO_2 exhibited the lowest value of Ni dispersion of 2.8% and the largest particle size of 36.1 nm.

Figure 2 shows representative TEM images and selected area electron diffraction (SAED) patterns obtained from Ni/YSZ, Ni/CeO_2 and Ni/TiO_2 catalysts. In all cases Ni particles appear as fairly homogeneously distributed spherical particles with average sizes of 20 nm for Ni/YSZ, 10 nm for Ni/CeO_2 and 30 nm for Ni/TiO_2, in agreement with those estimated according to CO chemisorption measurements (Table 1).

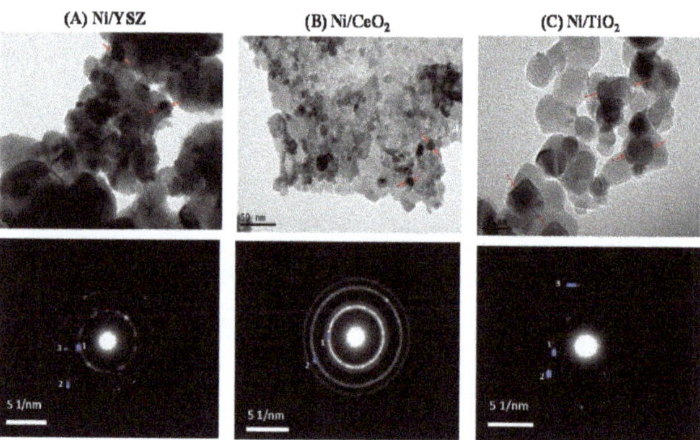

Figure 2. TEM images and Selected Area Electron Diffraction (SAED) patterns obtained for (**A**) 5%Ni/YSZ, (**B**) 5%Ni/CeO$_2$ and (**C**) 5%Ni/TiO$_2$ catalysts. Ni particles are indicated with red arrows.

It should be noted that, based on the results of Table 1, the mean particle size of Ni is similar to the average size of the corresponding metal oxide used as support for all the investigated catalysts. This may hinder distinguishing between Ni particles and the M_xO_y carrier in TEM images. Thus, SAED analysis was performed to calculate the d-spacing in an attempt to further discerned Ni particles from those of metal oxide support (Figure 2, Table 2). Results indicated that, in all cases, Ni particles are present in TEM images as evidenced by the appearance of the (111) plane of Ni ($d_{spacing}$ = 0.21 nm, JCPDS No 1-1258). The appearance of the (101) ($d_{spacing}$ = 0.35 nm, JCPDS No 1-562) and (103) ($d_{spacing}$ = 0.24 nm, JCPDS No 1-562) planes of TiO_2, the (111) ($d_{spacing}$ = 0.30 nm, JCPDS No 30-1468) and (200) ($d_{spacing}$ = 0.26 nm, JCPDS No 30-1468) planes of YSZ, as well as the (111) ($d_{spacing}$ = 0.31 nm, JCPDS No 1-800) and (200) ($d_{spacing}$ = 0.27 nm, JCPDS No 1-800) planes of CeO_2 also confirmed the presence of the metal oxides.

Table 2. Selected area electron diffraction (SAED) analysis of TEM images obtained for 5%Ni/YSZ, 5%Ni/CeO_2 and 5%Ni/TiO_2 catalysts.

Catalyst	Spot	d-Spacing (Å)	Formula	Crystallographic Plane (h k l)	JCPDS No
5%Ni/YSZ	1	3.0	$Y_{0.15}Zr_{0.85}O_{1.93}$	(111)	30-1468
	2	2.6	$Y_{0.15}Zr_{0.85}O_{1.93}$	(200)	30-1468
	3	2.1	Ni	(111)	1-1258
5%Ni/CeO_2	1	3.1	CeO_2	(111)	1-800
	2	2.7	CeO_2	(200)	1-800
	3	2.1	Ni	(111)	1-1258
5%Ni/TiO_2	1	3.5	TiO_2	(101)	1-562
	2	2.4	TiO_2	(103)	1-562
	3	2.1	Ni	(111)	1-1258

3.2. Influence of the Nature of the Support on Catalytic Activity

The influence of the nature of the support on catalytic performance for the propane steam reforming reaction has been investigated using Ni catalysts (5 wt.%) supported on five different commercial metal oxide powders (ZrO_2, YSZ, TiO_2, Al_2O_3, CeO_2). The results obtained are shown in Figure 3A, where propane conversion is plotted as a function of reaction temperature. It is observed that, among the investigated catalysts, Ni/ZrO_2 is the most active one, exhibiting measurable C_3H_8 conversions at temperatures higher than 400 °C and achieving complete conversion at 750 °C. Although Ni/YSZ is activated at similar temperatures as Ni/ZrO_2, the conversion curve of propane is shifted toward higher temperatures. This is also the case for Ni/Al_2O_3 and Ni/CeO_2 catalysts, which present similar performance. The latter catalysts are less active than Ni/YSZ below 550 °C, but are able to reach higher X_{C3H8} at higher temperatures. The titania-supported catalyst becomes active above 500 °C, with the propane conversion curve being shifted at remarkably higher temperatures. In all examined cases, the carbon balance was satisfactory, with a deviation of <1%.

Results of specific reaction rate measurements are presented in the Arrhenius diagram of Figure 3B, where it is observed that the TOF of propane conversion increases in the order Ni/TiO_2 < Ni/CeO_2 < Ni/Al_2O_3 < Ni/YSZ < Ni/ZrO_2, with its value at 450 °C being more than one order of magnitude higher when Ni is dispersed on ZrO_2 compared to TiO_2, and approximately 2.5 times higher than that of Ni/Al_2O_3. It should be mentioned that, as discussed above, the mean particle size of Ni varies significantly for the investigated catalysts from 8.5 nm for Ni/CeO_2 to 36.1 nm for Ni/TiO_2. If the Ni particle size were similar for this set of catalysts then the order of catalytic activity could be somewhat different. Interestingly, no trend was observed between the specific reaction rate and Ni particle size or M_xO_y crystallite size or M_xO_y surface area. This indicates that either these parameters do not affect catalytic activity or most possibly each of them contributes in a different manner to the reaction rate, resulting in the observed catalyst ranking. It should be noted that all catalysts have been reduced at 400 °C prior to physicochemical characterization measurements. Although the values of SSA or d_{MxOy} or d_{Ni} may were different if catalyst pre-reduction was carried out at 750 °C, which is the onset reaction temperature for catalytic performance experiments, the trend of catalytic properties with respect to the nature of the support is not expected to vary due to the catalyst pretreatment at different temperatures, at least to such an extent that would affect the catalyst ranking for the propane steam reforming reaction.

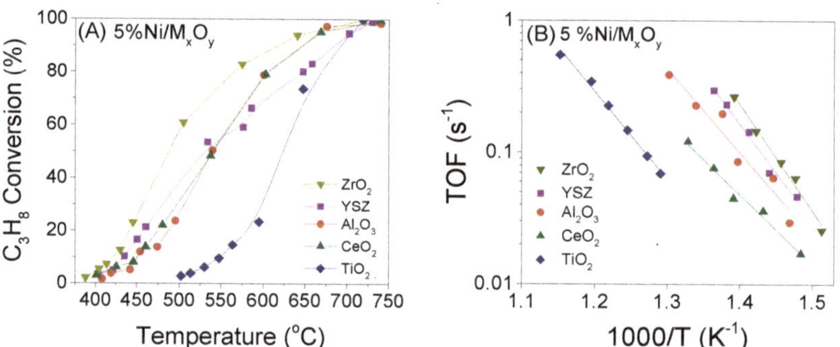

Figure 3. (**A**) Conversions of C_3H_8 as a function of reaction temperature and (**B**) Arrhenius plots of turnover frequencies of C_3H_8 conversion obtained over Ni catalysts (5.0 wt.%) supported on the indicated commercial oxide carriers. Experimental conditions: Mass of catalyst: 150 mg; particle diameter: $0.15 < d_p < 0.25$ mm; Feed composition: 4.5% C_3H_8, 0.15% Ar, 44% H_2O (balance He); Total flow rate: 250 cm^3 min^{-1}.

The apparent activation energies (E_a) of the propane steam reforming reaction were calculated from the slopes of the fitted lines of Figure 3B. The results showed that the nature of the metal oxide carrier significantly affects E_a, which takes values between 102 kJ/mol for Ni/CeO_2 and 154 kJ/mol for Ni/ZrO_2 without presenting any trend with respect to catalytic activity (Table 1). This can be explained taking into account that, as it will be discussed below, several reactions run in parallel under the present experimental conditions each one of which is influenced by the nature of the support in a different manner resulting in the observed random variation of E_a with catalytic activity. The results are in agreement with our previous study where it was found that the apparent activation energy for the reaction of steam reforming of propane over Rh catalysts supported on a variety of metal oxides does not present any trend with the activity order [11].

Figure 4 shows the selectivities toward reaction products as a function of temperature over the supported Ni catalysts investigated. In all cases the main products detected were H_2, CO_2, CO and CH_4 with their selectivities being significantly varied with temperature. In particular, for Ni/ZrO_2 catalyst (Figure 4A), both hydrogen (S_{H2}) and CO_2 (S_{CO2}) selectivities decrease from 99 to 78% and from 98 to 58%, respectively, with increasing temperature in the range of 390–505 °C followed by an increase of methane selectivity (S_{CH4}) up to 32.5%, indicating the occurrence of CO_2 methanation reactions. Carbon dioxide consumption continues with further increase of temperature above 505 °C contrary to S_{H2} which progressively increases reaching 99% at 720 °C. Consumption of CO_2 is followed by production of CO providing evidence that the reverse WGS (RWGS) reaction is enhanced at high temperatures. Moreover, S_{CH4} decreases above 505 °C and becomes practically zero at 720 °C, implying that the reaction of methane steam reforming occurs contributing to the observed increase of both S_{H2} and S_{CO}. It should be noted that selectivity toward reaction products containing carbon was defined as the concentration of each product containing carbon at reactor effluent over the concentration of all products containing carbon (5), whereas S_{H2} was defined as the concentration of hydrogen produced divided by the concentration of all products containing hydrogen (6). Therefore, the values of Sc_n and S_{H2} cannot be correlated based on the stoichiometry of the reactions taking place under propane steam reforming conditions.

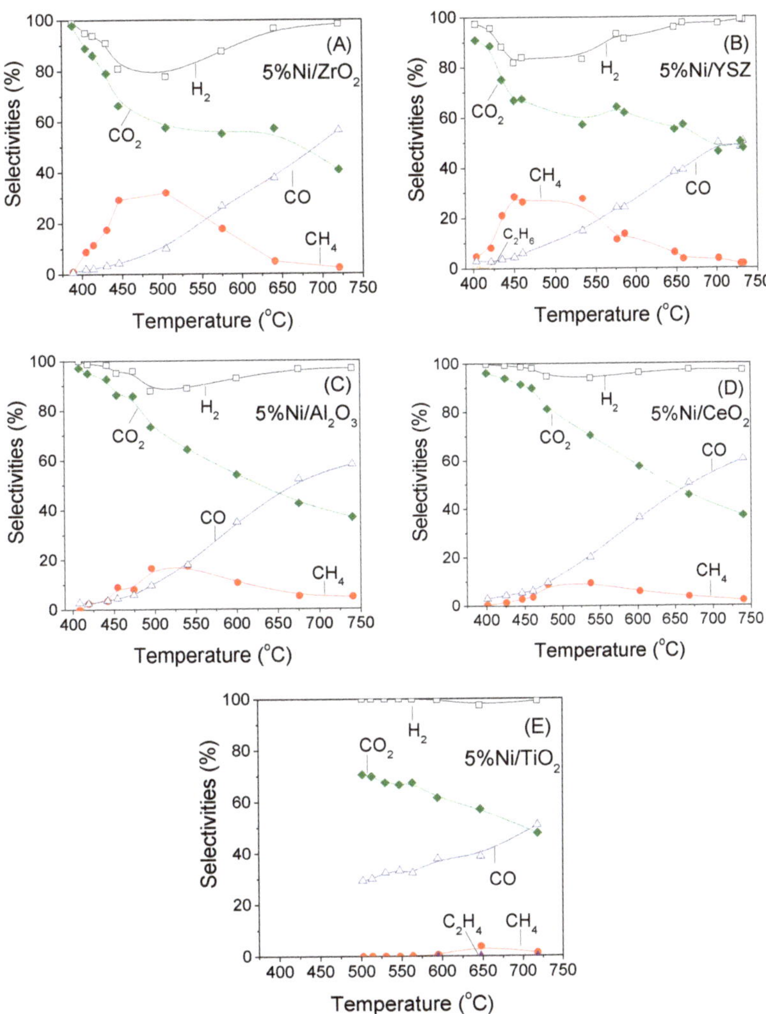

Figure 4. Selectivities toward reaction products as a function of reaction temperature obtained over Ni catalysts (5.0 wt.%) supported on (**A**) ZrO_2, (**B**) YSZ, (**C**) Al_2O_3, (**D**) CeO_2 and (**E**) TiO_2. Experimental conditions: same as in Figure 3.

Qualitatively similar results were obtained for the rest of the investigated Ni catalysts, with the main differences being related to the values of the selectivities toward the reaction products, which reflect the extent of each reaction taking place with respect to the nature of the support. In particular, the observed decrease of S_{H2} and S_{CO2} at low temperatures and the simultaneous increase of S_{CH4} are higher for the most active Ni/ZrO_2 (Figure 4A) and Ni/YSZ (Figure 4B) catalysts, followed by Ni/Al_2O_3 (Figure 4C) and Ni/CeO_2 (Figure 4D), whereas it is eliminated for Ni/TiO_2 (Figure 4E). As a result methane production increases in the order of $Ni/TiO_2 < Ni/CeO_2 < Ni/Al_2O_3 < Ni/YSZ < Ni/ZrO_2$ which is consistent with the order of catalytic activity. This can be clearly seen in Figure 5 where the TOF at 450 °C is plotted as a function of methane selectivity obtained at the same temperature for all the investigated catalysts. It is observed that the specific reaction rate increases from 0.018 s^{-1} to 0.33 s^{-1} following the above catalyst ranking and accompanied by a parallel increase of S_{CH4} from 0 to 29%. The results indicate that there is a clear relationship between catalytic activity and methane production.

It should be noted that besides CO_2 hydrogenation, CH_4 can be also produced via CO hydrogenation. However, the contribution of the latter reaction does not seem to be significant for the results of the present study taking into account the low S_{CO} below 500 °C and its progressive increase with temperature. Moreover, methane formation may also take place via hydrogenation of CH_x species formed following the dissociative adsorption of propane on Ni surface and the subsequent hydrogenation of the so-formed C_3H_x species [29,30]. As it will be discussed below, the formation of CH_x species intermediates may be the key reaction since it has been proposed that they interact with the hydroxyl groups or lattice oxygen of the support producing CO or CO_2 and H_2 [3,29,30].

The mass of H_2 produced per propane mass unit contained in the feed was calculated at 550 °C and it was found to be higher for Ni/ZrO$_2$ (23.2 wt.%) followed by Ni/CeO$_2$ (21.2 wt.%), Ni/Al$_2$O$_3$ (20.6 wt.%) and Ni/YSZ (19.7 wt.%), whereas Ni/TiO$_2$ exhibits the lowest H_2 production (5.2 wt.%).

The influence of the nature of the support on the activity of Ni catalysts for propane steam reforming reaction was also investigated by Harshini et al. [16], who found that Ni/LaAlO$_3$ was more active than Ni/Al$_2$O$_3$, while Ni/CeO$_2$ exhibited intermediate performance. The optimum activity of the former catalyst was attributed to the small Ni nanoparticles dispersed on LaAlO$_3$ surface. Although the effect of the support nature on propane steam reforming activity has not been widely studied over Ni catalysts, certain properties of metal oxide carriers may help explain the results of Figure 3. For example, the use of YSZ as support, which exhibited high activity in the results of the present study, has been found to suppress carbon deposition over Rh-Ni catalysts by providing lattice oxygen, which facilitates carbon removal and enhances the dissociation of C-C bond under reaction conditions [3]. The prevention of coke formation, occurring either via hydrocarbons decomposition or CO dissociation, by the lattice oxygen of the support has been also demonstrated over Ni/CeO$_2$-Al$_2$O$_3$ [19]. Moreover, the addition of manganese oxide on Ni/Al$_2$O$_3$ was found to act as an oxygen donor that is transferred to Ni particles leading to rapid decomposition and oxidation of C_3H_8 and CH_4 or C_2H_4 that may be produced under reaction conditions, resulting in further H_2 production and improvement of the catalyst lifetime [9]. It has been also found that activation of steam followed by H_2 formation may be favored over metal catalysts supported on "reducible" metal oxides through generation of oxygen defects, resulting in improved propane steam reforming activity and resistance to coke formation [3,13,31]. Based on previous studies, the reducibility of the supports used in the present study is expected to vary significantly. It is well known that Al$_2$O$_3$ is a hardly reducible metal oxide characterized by low oxygen storage capacity contrary to TiO$_2$ and CeO$_2$, which are strongly reducible metal oxides, or ZrO$_2$ and YSZ, which are characterized by intermediate oxygen mobility [32]. Based on the above, Ni/TiO$_2$ should be also active for the title reaction, taking into account that titania support is characterized by high oxygen storage capacity [32,33]. However, the results of Figure 3 clearly show that: (a) this catalyst was the least active one and (b) the catalytic activity is increased in the order Ni/TiO$_2$ < Ni/CeO$_2$ < Ni/Al$_2$O$_3$ < Ni/YSZ < Ni/ZrO$_2$, which cannot be correlated with the reducibility of the support. Therefore, it is evident that support reducibility is not among the key parameters affecting the catalytic activity of Ni according to the results of the present study. The low activity of Ni/TiO$_2$ catalyst may be related to the fact that Ni/TiO$_2$ has lower Ni dispersion and larger Ni particles, which was previously suggested to suppress both propane steam reforming [15,16], and the intermediate (CO_2 or CH_x) hydrogenation reactions, in excellent agreement with the results of our previous study [28]. However, large Ni particles may not be solely responsible for the low activity of Ni/TiO$_2$

taking into account that no trend was observed between TOF and Ni particle size for the investigated catalysts.

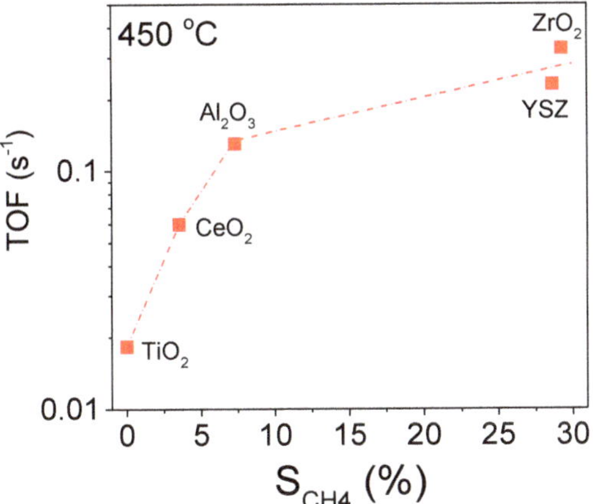

Figure 5. Turnover frequencies of C$_3$H$_8$ conversion as a function of selectivity toward CH$_4$ obtained at 450 °Cover supported Ni catalysts.

It should be mentioned that a different activity order was reported in our previous study over supported Rh catalysts, where it was found that Rh/TiO$_2$ was the most active catalyst with TOF being one order of magnitude higher compared to that measured for Rh/CeO$_2$ [11]. This implies that the nature of the metallic phase may affect metal/support interactions leading to variations on propane steam reforming activity and/or possibly changes on the type of active sites on the catalyst surface.

3.3. Long Term Stability Test

The long-term stability of Ni/ZrO$_2$ catalyst, which exhibited the highest activity, was investigated at 650 °C using the same experimental conditions as those used in catalytic performance tests. In this experiment, the catalyst was reduced in situ at 300 °C under 50%H$_2$/He flow followed by heating at 650 °C. The flow was then switched to the reaction mixture and determination of the conversion of propane and product selectivity started. The system was shut down overnight, while the catalyst was kept at room temperature under a He flow. The next day the catalyst is heated to 650 °C in the He flow, followed by switching of the flow to the reaction mixture and determination of X_{C3H8} and the product selectivity as a function of time. Results obtained are shown in Figure 6, where X_{C3H8} and S_{H2}, S_{CO2}, S_{CO} and S_{CH4} are plotted as functions of time-on-stream. As it can be seen the catalyst presents excellent stability for more than 30 h-on-stream. Propane conversion and hydrogen selectivity were varied in the range of 95–99% and 97–98%, respectively. The selectivity toward methane was low (3–4%) whereas S_{CO} and S_{CO2} exhibited similar values ranging between 46 and 50%. The carbon balance was found to be satisfactory during the stability test, with a deviation lower than 1–2%.

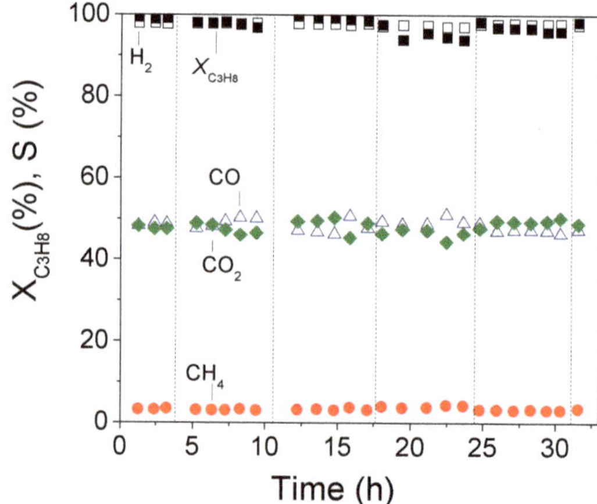

Figure 6. Long-term stability test of the 5%Ni/ZrO$_2$ catalyst at 650 °C: Alterations of the conversion of C$_3$H$_8$ and selectivities toward reaction products with time-on-stream. Experimental conditions: Same as in Figure 3. Dashed vertical black lines indicate shutting down of the system overnight.

3.4. DRIFT Studies

The interaction of selected catalysts with the reaction mixture was also investigated employing in situ FTIR spectroscopy. Experiments were conducted in the temperature range of 100–500 °C using a feed composition of 0.5%C$_3$H$_8$ + 5%H$_2$O (in He) and the results obtained are shown in Figure 7. It is observed that the spectrum recorded at 100 °C (Figure 7A, trace a) for the pre-reduced Ni/TiO$_2$ catalyst is characterized by two negative bands at 3787 and 3676 cm^{-1} which can be attributed to losses of ν (OH) intensity of at least two different types of free hydroxyl groups, which are either originally present on TiO$_2$ surface or created via H$_2$O adsorption. Two weak peaks were also detected in the ν (C-H) region, located at 2987 and 2966 cm^{-1} (trace a) due to C-H stretching vibrations in methyl groups (CH$_{3,ad}$) and to symmetric C-H vibrations in methylene groups (CH$_{2,ad}$), respectively [20,31,34–36]. These peaks are more obvious in Figure 8A (trace a) where selected spectra in the narrow range of 3200–2400 cm^{-1} are presented. Moreover, a band at 1642 cm^{-1} followed by a shoulder at 1560 cm^{-1} can be discerned, which have been previously assigned to carbonate species associated with TiO$_2$ support [37–42]. An increase of temperature results in progressive separation of the latter two bands which are both shifted toward lower wavenumbers. A new band at 1430 cm^{-1} can be also discerned in the spectra obtained at 350 °C (trace f) which is also due to carbonate species. This peak may be also present in the spectra obtained at lower temperatures but couldn't be clearly observed due to the low signal-to-noise ratio in the region below 1700 cm^{-1}. The intensities of bands assigned to carbonate species are progressively decreased above 200 °C. This decrease is accompanied by the detection of a weak peak at 2021 cm^{-1} [38,43–46], which is characteristic of linear-bonded CO on reduced nickel sites (Ni°), indicating that carabonate species are further decomposed yielding CO and most possibly also CO$_2$ in the gas phase. The weak bands in the ν (C-H) region are present on the spectra obtained up to 500 °C (Figure 7A, trace i) implying that CH$_x$ species are thermally stable and remained adsorbed on the catalyst surface.

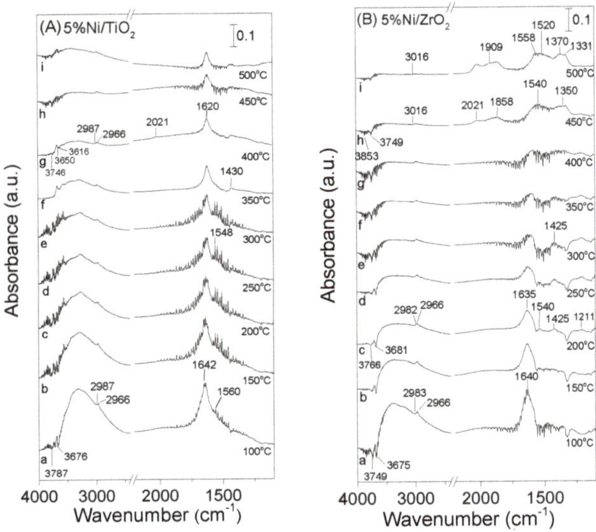

Figure 7. DRIFT spectra obtained over the (**A**) Ni/TiO₂ and (**B**) Ni/ZrO₂ catalysts following interaction with 0.5% C₃H₈ + 5% H₂O (in He) at 100 °C for 15 min and subsequent stepwise heating at 500 °C.

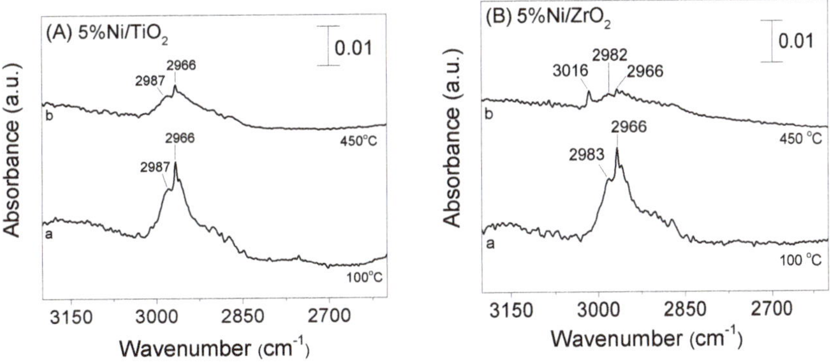

Figure 8. DRIFT spectra obtained in the region 3200–2400 cm⁻¹ over the (**A**) Ni/TiO₂ and (**B**) Ni/ZrO₂ catalysts following interaction with 0.5% C₃H₈ + 5% H₂O (in He) at 100 °C and 450 °C for 15 min.

A similar experiment was conducted over the most active Ni/ZrO₂ catalyst and the results obtained are presented in Figure 7B. It is observed that the interaction of catalyst with the reformate mixture at 100 °C (trace a) results in the appearance of bands corresponding to bicarbonate species (1640 cm⁻¹) [47–49], CH$_x$ species (2983 and 2966 cm⁻¹) [31,36,47] as well as by negative bands (3749 and 3675 cm⁻¹) related to the consumption of surface OH groups [31,36,48]. Increase of temperature at 200 °C (trace c) leads to the progressive development of two bands at 1540 and 1425 cm⁻¹ due to bicarbonate species [47–49]. The intensity of the latter bands increases with increasing temperature up to 300 °C and diminishes upon further heating at 350–400 °C. This is also the case for the band at 1640 cm⁻¹, indicating that bicarbonate species are decomposed above 450 °C. At temperatures higher than 400 °C (traces h-i) two new broad bands seem to be developed at ca 1540 and 1350 cm⁻¹. The broadness of these bands implies that they may contain contributions from more than one species with their corresponding bands being overlapped.

This can be clearly seen in the spectrum obtained at 500 °C (trace i) where four bands can be clearly discerned located at 1558, 1520, 1370 and 1331 cm⁻¹. Those detected at 1520 and 1331 cm^{-1} have been previously attributed to bidentate carbonates [49,50], whereas those located at 1558 and 1370 cm^{-1} can be assigned to bidentate formate species [51] adsorbed on ZrO$_2$ surface. The appearance of the latter bands is accompanied by evolution of three peaks in the ν(CO) region due to CO linearly adsorbed on reduced Ni sites (2021 cm^{-1}) and bridged bonded CO (1909 and 1858 cm^{-1}) [43,46,52–54].

Interestingly, CH$_x$ species are eliminated from the spectra obtained above 400 °C followed by evolution of CH$_4$ in the gas phase, as evidenced by the detection of the 3016 cm^{-1} band (traces h–i). Production of CH$_4$ at the expense of CH$_x$ species can be clearly seen in Figure 8B where the spectra obtained at 100 and 450 °C in the wavenumber range of 3200–2400 cm^{-1} are presented.

Based on the above it can be suggested that the reaction of steam reforming of propane over Ni/ZrO$_2$ catalyst proceeds via a dissociative adsorption of propane on metallic Ni leading to the formation of C$_3$H$_x$ species, which are further decomposed toward CH$_x$ species and probably carbon oxides due to the presence of H$_2$O adsorbed on the support surface. This may result in the formation of the bicarbonate species (1640, 1540 and 1425 cm^{-1}) detected at low temperatures on the surface of the support [20]. Part of CH$_x$ species are hydrogenated above 400 °C, yielding methane in the gas phase (band at 3016 cm^{-1}) whereas the rest interact with adsorbed water producing formates (bands at 1558 and 1370 cm^{-1}) and, eventually, CO species adsorbed on the Ni surface (bands at 2021, 1909 and 1858 cm^{-1}). Formates may also interact with hydroxyl groups producing H$_2$ and carbonate species (1520 and 1331 cm^{-1}), which are further decomposed to CO$_2$ [20]. It should be noted, however, that, under certain conditions, CH$_x$ species may be also dehydrogenated producing C and H$_2$. Surface carbon is either accumulated on the catalyst surface resulting in catalyst deactivation (which is not the case here) or it interacts with the hydroxyl groups or the lattice oxygen of the support yielding CO or CO$_2$ [3,16,19,31,55].

The ability of CH$_x$ species to be converted to methane and/or formates on the surface of Ni/ZrO$_2$ may imply that CH$_x$ species are more weakly adsorbed and/or more reactive on the surface of this catalyst, thus resulting in higher overall activity for the propane steam reforming reaction. This agrees well with results of Figure 5, where it was shown that catalytic activity increases progressively with increasing methane selectivity. If as discussed above the origin of adsorbed CO is formate species, the high reactivity of CH$_x$ species over Ni/ZrO$_2$ is also indirectly confirmed by the significantly higher population of carbonyl species observed over this catalyst. On the other hand, although CH$_x$ species were also detected on the surface of Ni/TiO$_2$ catalyst, no band due to gas phase methane or formate species was observed, indicating that CH$_x$ species cannot be further converted in the temperature range investigated. The increase of S$_{CH4}$ in parallel with the increase of catalytic activity (Figure 5) indicates that the reactivity of CH$_x$ species is strongly related to the conversion of propane to the desired products. Although detailed mechanistic studies should be conducted to further explore the reaction pathway, it can be suggested that CH$_x$ species are key reaction intermediates for the reaction of propane steam reforming.

4. Conclusions

The results of the present study show that the catalytic activity of Ni and the product distribution for the propane steam reforming reaction depend strongly on the nature of the support. The specific reaction rate measured for Ni/ZrO$_2$ catalyst was found to be more than one order of magnitude higher compared to that measured for Ni/TiO$_2$. The intermediate production of CH$_4$ is strongly influenced by the type of metal oxide used as support following the same trend with that of catalytic activity. DRIFTS studies provided evidences that the most active Ni/ZrO$_2$ catalyst is able to convert the intermediate produced CH$_x$ species to CH$_4$ and/or formate species and, subsequently, to carbon oxides and H$_2$. In contrast, CH$_x$ species seem to be less reactive when Ni is dispersed on TiO$_2$, thus resulting in a lower reaction rate. In addition to the high activity of Ni/ZrO$_2$ catalyst, it was also found to be stable for more than 30 h on stream and therefore, is a promising candidate for the production of H$_2$ for fuel cell applications.

Author Contributions: Conceptualization, P.P.; methodology, P.P.; investigation, A.K., A.P. and P.P.; data curation, A.K., A.P. and P.P.; writing—original draft preparation, P.P.; writing—review and editing, P.P.; visualization, P.P.; supervision, P.P.; project administration, P.P.; funding acquisition, P.P. All authors have read and agreed to the published version of the manuscript.

Funding: This research has been co-financed by the European Union and Greek national funds through the Operational Program Competitiveness, Entrepreneurship and Innovation, under the call RESEARCH–CREATE–INNOVATE (project code: T1EDK–02442).

Institutional Review Board Statement: Not applicable.

Informed Consent Statement: Not applicable.

Data Availability Statement: Not applicable.

Conflicts of Interest: The authors declare no conflict of interest. The funders had no role in the design of the study; in the collection, analyses, or interpretation of data; in the writing of the manuscript, or in the decision to publish the results.

References

1. Kalamaras, C.M.; Efstathiou, A.M. Hydrogen Production Technologies: Current State and Future Developments. *Conf. Pap. Energy* **2013**, *2013*, 690627. [CrossRef]
2. Nikolaidis, P.; Poullikkas, A. A comparative overview of hydrogen production processes. *Renew. Sustain. Energy Rev.* **2017**, *67*, 597–611. [CrossRef]
3. Im, Y.; Lee, J.H.; Kwak, B.S.; Do, J.Y.; Kang, M. Effective hydrogen production from propane steam reforming using M/NiO/YSZ catalysts (M = Ru, Rh, Pd, and Ag). *Catal. Today* **2018**, *303*, 168–176. [CrossRef]
4. Santamaria, L.; Lopez, G.; Arregi, A.; Amutio, M.; Artetxe, M.; Bilbao, J.; Olazar, M. Influence of the support on Ni catalysts performance in the in-line steam reforming of biomass fast pyrolysis derived volatiles. *Appl. Catal. B Environ.* **2018**, *229*, 105–113. [CrossRef]
5. Santamaria, L.; Lopez, G.; Arregi, A.; Amutio, M.; Artetxe, M.; Bilbao, J.; Olazar, M. Stability of different Ni supported catalysts in the in-line steam reforming of biomass fast pyrolysis volatiles. *Appl. Catal. B Environ.* **2019**, *242*, 109–120. [CrossRef]
6. Miyazawa, T.; Kimura, T.; Nishikawa, J.; Kado, S.; Kunimori, K.; Tomishige, K. Catalytic performance of supported Ni catalysts in partial oxidation and steam reforming of tar derived from the pyrolysis of wood biomass. *Catal. Today* **2006**, *115*, 254–262. [CrossRef]
7. Al-Zuhair, S.; Hassan, M.; Djama, M.; Khaleel, A. Hydrogen Production by Steam Reforming of Commercially Available LPG in UAE. *Chem. Eng. Commun.* **2017**, *204*, 141–148. [CrossRef]
8. Do, J.Y.; Lee, J.H.; Park, N.-K.; Lee, T.J.; Lee, S.T.; Kang, M. Synthesis and characterization of Ni2−xPdxMnO4/γ-Al$_2$O$_3$ catalysts for hydrogen production via propane steam reforming. *Chem. Eng. J.* **2018**, *334*, 1668–1678. [CrossRef]
9. Do, J.Y.; Park, N.-K.; Lee, T.J.; Lee, S.T.; Kang, M. Effective hydrogen productions from propane steam reforming over spinel-structured metal-manganese oxide redox couple catalysts. *Int. J. Energy Res.* **2018**, *42*, 429–446. [CrossRef]
10. Barzegari, F.; Kazemeini, M.; Farhadi, F.; Rezaei, M.; Keshavarz, A. Preparation of mesoporous nanostructure NiO–MgO–SiO2 catalysts for syngas production via propane steam reforming. *Int. J. Hydrogen Energy* **2020**, *45*, 6604–6620. [CrossRef]
11. Kokka, A.; Katsoni, A.; Yentekakis, I.V.; Panagiotopoulou, P. Hydrogen production via steam reforming of propane over supported metal catalysts. *Int. J. Hydrogen Energy* **2020**. [CrossRef]
12. Kokka, A.; Ramantani, T.; Panagiotopoulou, P. Effect of Operating Conditions on the Performance of Rh/TiO$_2$ Catalyst for the Reaction of LPG Steam Reforming. *Catalysts* **2021**, *11*, 374. [CrossRef]
13. Yu, L.; Sato, K.; Nagaoka, K. Rh/Ce0.25Zr0.75O2 Catalyst for Steam Reforming of Propane at Low Temperature. *ChemCatChem* **2019**, *11*, 1472–1479. [CrossRef]
14. Karakaya, C.; Karadeniz, H.; Maier, L.; Deutschmann, O. Surface Reaction Kinetics of the Oxidation and Reforming of Propane over Rh/Al$_2$O$_3$ Catalysts. *ChemCatChem* **2017**, *9*, 685–695. [CrossRef]
15. Aghamiri, A.R.; Alavi, S.M.; Bazyari, A.; Azizzadeh Fard, A. Effects of simultaneous calcination and reduction on performance of promoted Ni/SiO2 catalyst in steam reforming of propane. *Int. J. Hydrogen Energy* **2019**, *44*, 9307–9315. [CrossRef]
16. Harshini, D.; Yoon, C.W.; Han, J.; Yoon, S.P.; Nam, S.W.; Lim, T.-H. Catalytic Steam Reforming of Propane over Ni/LaAlO3 Catalysts: Influence of Preparation Methods and OSC on Activity and Stability. *Catal. Lett.* **2012**, *142*, 205–212. [CrossRef]
17. Azizzadeh Fard, A.; Arvaneh, R.; Alavi, S.M.; Bazyari, A.; Valaei, A. Propane steam reforming over promoted Ni–Ce/MgAl2O4 catalysts: Effects of Ce promoter on the catalyst performance using developed CCD model. *Int. J. Hydrogen Energy* **2019**, *44*, 21607–21622. [CrossRef]
18. Kim, K.M.; Kwak, B.S.; Park, N.-K.; Lee, T.J.; Lee, S.T.; Kang, M. Effective hydrogen production from propane steam reforming over bimetallic co-doped NiFe/Al$_2$O$_3$ catalyst. *J. Ind. Eng. Chem.* **2017**, *46*, 324–336. [CrossRef]

19. Laosiripojana, N.; Sangtongkitcharoen, W.; Assabumrungrat, S. Catalytic steam reforming of ethane and propane over CeO_2-doped Ni/Al_2O_3 at SOFC temperature: Improvement of resistance toward carbon formation by the redox property of doping CeO_2. *Fuel* **2006**, *85*, 323–332. [CrossRef]

20. Natesakhawat, S.; Oktar, O.; Ozkan, U.S. Effect of lanthanide promotion on catalytic performance of sol–gel Ni/Al_2O_3 catalysts in steam reforming of propane. *J. Mol. Catal. A Chem.* **2005**, *241*, 133–146. [CrossRef]

21. Matsumura, Y.; Nakamori, T. Steam reforming of methane over nickel catalysts at low reaction temperature. *Appl. Catal. A Gen.* **2004**, *258*, 107–114. [CrossRef]

22. Guo, J.; Lou, H.; Zhao, H.; Chai, D.; Zheng, X. Dry reforming of methane over nickel catalysts supported on magnesium aluminate spinels. *Appl. Catal. A Gen.* **2004**, *273*, 75–82. [CrossRef]

23. Lee, D.S.; Min, D.J. A Kinetics of Hydrogen Reduction of Nickel Oxide at Moderate Temperature. *Met. Mater. Int.* **2019**, *25*, 982–990. [CrossRef]

24. Richardson, J.T.; Scates, R.; Twigg, M.V. X-ray diffraction study of nickel oxide reduction by hydrogen. *Appl. Catal. A Gen.* **2003**, *246*, 137–150. [CrossRef]

25. Richardson, J.T.; Scates, R.M.; Twigg, M.V. X-ray diffraction study of the hydrogen reduction of $NiO/\alpha\text{-}Al_2O_3$ steam reforming catalysts. *Appl. Catal. A Gen.* **2004**, *267*, 35–46. [CrossRef]

26. Almasan, V.; Gaeumann, T.; Lazar, M.; Marginean, P.; Aldea, N. Hydrogen spillover effect over the oxide surfaces in supported nickel catalysts. In *Studies in Surface Science and Catalysis*; Froment, G.F., Waugh, K.C., Eds.; Elsevier: Amsterdam, The Netherlands, 1997; Volume 109, pp. 547–552.

27. Kramer, R.; Andre, M. Adsorption of atomic hydrogen on alumina by hydrogen spillover. *J. Catal.* **1979**, *58*, 287–295. [CrossRef]

28. Hatzisymeon, M.; Petala, A.; Panagiotopoulou, P. Carbon Dioxide Hydrogenation over Supported Ni and Ru Catalysts. *Catal. Lett.* **2021**, *151*, 888–900. [CrossRef]

29. Kolb, G.; Zapf, R.; Hessel, V.; Löwe, H. Propane steam reforming in micro-channels—results from catalyst screening and optimisation. *Appl. Catal. A Gen.* **2004**, *277*, 155–166. [CrossRef]

30. Modafferi, V.; Panzera, G.; Baglio, V.; Frusteri, F.; Antonucci, P.L. Propane reforming on Ni–Ru/GDC catalyst: H2 production for IT-SOFCs under SR and ATR conditions. *Appl. Catal. A Gen.* **2008**, *334*, 1–9. [CrossRef]

31. Malaibari, Z.O.; Croiset, E.; Amin, A.; Epling, W. Effect of interactions between Ni and Mo on catalytic properties of a bimetallic $Ni\text{-}Mo/Al_2O_3$ propane reforming catalyst. *Appl. Catal. A Gen.* **2015**, *490*, 80–92. [CrossRef]

32. Helali, Z.; Jedidi, A.; Syzgantseva, O.A.; Calatayud, M.; Minot, C. Scaling reducibility of metal oxides. *Theor. Chem. Acc.* **2017**, *136*, 100–115. [CrossRef]

33. Alphonse, P.; Ansart, F. Catalytic coatings on steel for low-temperature propane prereforming to solid oxide fuel cell (SOFC) application. *J. Colloid Interface Sci.* **2009**, *336*, 658–666. [CrossRef]

34. Panagiotopoulou, P.; Kondarides, D.I.; Verykios, X.E. Mechanistic aspects of the selective methanation of CO over Ru/TiO_2 catalyst. *Catal. Today* **2012**, *181*, 138–147. [CrossRef]

35. Panagiotopoulou, P. Methanation of CO_2 over alkali-promoted Ru/TiO_2 catalysts: II. Effect of alkali additives on the reaction pathway. *Appl. Catal. B Environ.* **2018**, *236*, 162–170. [CrossRef]

36. Wang, B.; Wu, X.; Ran, R.; Si, Z.; Weng, D. IR characterization of propane oxidation on $Pt/CeO_2\text{–}ZrO_2$: The reaction mechanism and the role of Pt. *J. Mol. Catal. A Chem.* **2012**, *356*, 100–105. [CrossRef]

37. Panagiotopoulou, P.; Christodoulakis, A.; Kondarides, D.I.; Boghosian, S. Particle size effects on the reducibility of titanium dioxide and its relation to the water–gas shift activity of Pt/TiO_2 catalysts. *J. Catal.* **2006**, *240*, 114–125. [CrossRef]

38. Kokka, A.; Ramantani, T.; Petala, A.; Panagiotopoulou, P. Effect of the nature of the support, operating and pretreatment conditions on the catalytic performance of supported Ni catalysts for the selective methanation of CO. *Catal. Today* **2020**, *355*, 832–843. [CrossRef]

39. Karelovic, A.; Ruiz, P. Mechanistic study of low temperature CO_2 methanation over Rh/TiO_2 catalysts. *J. Catal.* **2013**, *301*, 141–153. [CrossRef]

40. Liao, L.F.; Lien, C.F.; Shieh, D.L.; Chen, M.T.; Lin, J.L. FTIR Study of Adsorption and Photoassisted Oxygen Isotopic Exchange of Carbon Monoxide, Carbon Dioxide, Carbonate, and Formate on TiO_2. *J. Phys. Chem. B* **2002**, *106*, 11240–11245. [CrossRef]

41. Marwood, M.; Doepper, R.; Renken, A. In-situ surface and gas phase analysis for kinetic studies under transient conditions The catalytic hydrogenation of CO2. *Appl. Catal. A Gen.* **1997**, *151*, 223–246. [CrossRef]

42. Mino, L.; Spoto, G.; Ferrari, A.M. CO2 Capture by TiO_2 Anatase Surfaces: A Combined DFT and FTIR Study. *J. Phys. Chem. C* **2014**, *118*, 25016–25026. [CrossRef]

43. Aldana, P.A.U.; Ocampo, F.; Kobl, K.; Louis, B.; Thibault-Starzyk, F.; Daturi, M.; Bazin, P.; Thomas, S.; Roger, A.C. Catalytic CO_2 valorization into CH_4 on Ni-based ceria-zirconia. Reaction mechanism by operando IR spectroscopy. *Catal. Today* **2013**, *215*, 201–207. [CrossRef]

44. Westermann, A.; Azambre, B.; Bacariza, M.C.; Graça, I.; Ribeiro, M.F.; Lopes, J.M.; Henriques, C. Insight into CO_2 methanation mechanism over NiUSY zeolites: An operando IR study. *Appl. Catal. B Environ.* **2015**, *174–175*, 120–125. [CrossRef]

45. Mohamed, Z.; Dasireddy, V.D.B.C.; Singh, S.; Friedrich, H.B. The preferential oxidation of CO in hydrogen rich streams over platinum doped nickel oxide catalysts. *Appl. Catal. B Environ.* **2016**, *180*, 687–697. [CrossRef]

46. Konishcheva, M.V.; Potemkin, D.I.; Badmaev, S.D.; Snytnikov, P.V.; Paukshtis, E.A.; Sobyanin, V.A.; Parmon, V.N. On the Mechanism of CO and CO_2 Methanation Over Ni/CeO_2 Catalysts. *Top. Catal.* **2016**, *59*, 1424–1430. [CrossRef]

47. Köck, E.-M.; Kogler, M.; Bielz, T.; Klötzer, B.; Penner, S. In Situ FT-IR Spectroscopic Study of CO_2 and CO Adsorption on Y_2O_3, ZrO_2, and Yttria-Stabilized ZrO_2. *J. Phys. Chem. C* **2013**, *117*, 17666–17673. [CrossRef]
48. Bachiller-Baeza, B.; Rodriguez-Ramos, I.; Guerrero-Ruiz, A. Interaction of Carbon Dioxide with the Surface of Zirconia Polymorphs. *Langmuir* **1998**, *14*, 3556–3564. [CrossRef]
49. Zhang, Z.; Zhang, L.; Hülsey, M.J.; Yan, N. Zirconia phase effect in Pd/ZrO_2 catalyzed CO_2 hydrogenation into formate. *Mol. Catal.* **2019**, *475*, 110461. [CrossRef]
50. Föttinger, K.; Emhofer, W.; Lennon, D.; Rupprechter, G. Adsorption and Reaction of CO on (Pd–)Al_2O_3 and (Pd–)ZrO_2: Vibrational Spectroscopy of Carbonate Formation. *Topics Catal.* **2017**, *60*, 1722–1734. [CrossRef] [PubMed]
51. Pokrovski, K.; Jung, K.T.; Bell, A.T. Investigation of CO and CO_2 Adsorption on Tetragonal and Monoclinic Zirconia. *Langmuir* **2001**, *17*, 4297–4303. [CrossRef]
52. Anderson, J.A.; Daza, L.; Fierro, J.L.G.; Rodrigo, M.T. Influence of preparation method on the characteristics of nickel/sepiolite catalysts. *J. Chem. Soc. Faraday Trans.* **1993**, *89*, 3651–3657. [CrossRef]
53. Wu, R.-f.; Zhang, Y.; Wang, Y.-z.; Gao, C.-g.; Zhao, Y.-x. Effect of ZrO_2 promoter on the catalytic activity for CO methanation and adsorption performance of the Ni/SiO_2 catalyst. *J. Fuel Chem. Technol.* **2009**, *37*, 578–582. [CrossRef]
54. Zhou, G.; Liu, H.; Cui, K.; Jia, A.; Hu, G.; Jiao, Z.; Liu, Y.; Zhang, X. Role of surface Ni and Ce species of Ni/CeO_2 catalyst in CO_2 methanation. *Appl. Surf. Sci.* **2016**, *383*, 248–252. [CrossRef]
55. Faria, E.C.; Rabelo-Neto, R.C.; Colman, R.C.; Ferreira, R.A.R.; Hori, C.E.; Noronha, F.B. Steam Reforming of LPG over Ni/Al_2O_3 and $Ni/CexZr1—xO2/Al_2O_3$ Catalysts. *Catal. Lett.* **2016**, *146*, 2229–2241. [CrossRef]

 *nanomaterials*

MDPI

Article

Nanocarbon from Rocket Fuel Waste: The Case of Furfuryl Alcohol-Fuming Nitric Acid Hypergolic Pair

Nikolaos Chalmpes [1], Athanasios B. Bourlinos [2,*], Smita Talande [3,4], Aristides Bakandritsos [3], Dimitrios Moschovas [1], Apostolos Avgeropoulos [1], Michael A. Karakassides [1] and Dimitrios Gournis [1,*]

1   Department of Materials Science & Engineering, University of Ioannina, 45110 Ioannina, Greece; chalmpesnikos@gmail.com (N.C.); dmoschov@uoi.gr (D.M.); aavger@uoi.gr (A.A.); mkarakas@uoi.gr (M.A.K.)
2   Physics Department, University of Ioannina, 45110 Ioannina, Greece
3   Regional Centre of Advanced Technologies and Materials, Faculty of Science, Palacky University in Olomouc, Slechtitelu 27, 779 00 Olomouc, Czech Republic; smita.talande01@upol.cz (S.T.); aristeidis.bakandritsos@upol.cz (A.B.)
4   Department of Experimental Physics, Faculty of Science, Palacký University, 17. listopadu 1192/12, 779 00 Olomouc, Czech Republic
*   Correspondence: bourlino@cc.uoi.gr (A.B.B.); dgourni@uoi.gr (D.G.); Tel.: +30-26510-07141 (D.G.)

Abstract: In hypergolics two substances ignite spontaneously upon contact without external aid. Although the concept mostly applies to rocket fuels and propellants, it is only recently that hypergolics has been recognized from our group as a radically new methodology towards carbon materials synthesis. Comparatively to other preparative methods, hypergolics allows the rapid and spontaneous formation of carbon at ambient conditions in an exothermic manner (e.g., the method releases both carbon and energy at room temperature and atmospheric pressure). In an effort to further build upon the idea of hypergolic synthesis, herein we exploit a classic liquid rocket bipropellant composed of furfuryl alcohol and fuming nitric acid to prepare carbon nanosheets by simply mixing the two reagents at ambient conditions. Furfuryl alcohol served as the carbon source while fuming nitric acid as a strong oxidizer. On ignition the temperature is raised high enough to induce carbonization in a sort of in-situ pyrolytic process. Simultaneously, the released energy was directly converted into useful work, such as heating a liquid to boiling or placing Crookes radiometer into motion. Apart from its value as a new synthesis approach in materials science, carbon from rocket fuel additionally provides a practical way in processing rocket fuel waste or disposed rocket fuels.

Keywords: nanocarbon; rocket fuels; furfuryl alcohol; fuming nitric acid; waste; hypergolics; carbon materials

Citation: Chalmpes, N.; Bourlinos, A.B.; Talande, S.; Bakandritsos, A.; Moschovas, D.; Avgeropoulos, A.; Karakassides, M.A.; Gournis, D. Nanocarbon from Rocket Fuel Waste: The Case of Furfuryl Alcohol-Fuming Nitric Acid Hypergolic Pair. *Nanomaterials* **2021**, *11*, 1. https://doi.org/10.3390/nano11010001

Received: 30 November 2020
Accepted: 20 December 2020
Published: 22 December 2020

Publisher's Note: MDPI stays neutral with regard to jurisdictional claims in published maps and institutional affiliations.

1. Introduction

Furfuryl alcohol is considered an important green chemical that is mainly derived from plant raw material (corncobs, agricultural waste) through simple reaction cascades [1,2]. A notable use of furfuryl alcohol other than in organic chemistry or polymers also pertains to hypergolic rocket propellants and fuels [3–6]. In this latter case, the renewable alcohol (organic fuel) and fuming nitric acid (powerful oxidizer) react immediately and energetically upon contact (i.e., hypergolically) to release heat and gases that promote rocket lift-off (WAC Corporal, Nike Ajax). By also considering the vegetable origin of the organic compound, furfuryl alcohol fuel rocketry has been aptly highlighted in science blogs under imaginative headlines, such as "Flying to the Mars on Corncobs" (see for instance the link: https://dalinyebo.com/furfuryl-alcohol-rocket-fuel-then-and-now/). It is, however, surprising that although furfuryl alcohol is a versatile carbon precursor by pyrolysis [7–14], there is no report in the literature which acknowledges or refers in any way to the carbon residue obtained from the particular hypergolic pair.

Recently our group has introduced hypergolic reactions as a new and general synthesis tool towards the formation of a variety of functional carbon materials (nanosheets, crystalline graphite, carbon dots, fullerols, hollow spheres, and nanodiscs) [15–20]. Hypergolic synthesis presents certain advantages over conventional carbonization methods (pyrolysis, hydrothermal, chemical vapor deposition—CVD) for two main reasons. First, the method is fast and spontaneous at ambient conditions: Carbon formation takes place by simply mixing the organic fuel and strong oxidizer at room temperature and atmospheric pressure, completing in very short time (e.g., rapid product formation). Second, hypergolic reactions are by definition exothermic and therefore release enough heat upon contact of the reagents (i.e., an energy-liberating process). This clearly differentiates from pyrolytic, hydrothermal, or CVD methods that require higher temperature in order to operate synthesis and extract carbon from its precursors (i.e., an energy-consuming process). In fact, we have previously demonstrated that the energy released from hypergolic reactions could be directly converted into useful chemical, thermoelectric, or photovoltaic work [15,17,18]. Therefore, hypergolics allows a fast and spontaneous carbon synthesis at ambient conditions in an energy-liberating manner without external stimuli (heat, spark, or mechanical shock). Nevertheless, scaling-up the method under safe conditions still remains a major challenge from a technical point of view. It is likely, though, that modern advances in rocket fuel engineering may hold the key towards the development of pilot apparatus for large-scale production in the future. Likewise, it was only the course of certain technical advances in flame-spray pyrolysis—another harsh synthetic approach towards nanomaterials—that finally allowed its safe use in labs or industry.

Under this light, the furfuryl alcohol-HNO_3 pair certainly drew our attention as a potentially new example towards the hypergolic synthesis of carbon (note that furfuryl alcohol is considered a cheap, non-petrol derived biofuel used in fuel rocketry). Accordingly, herein we demonstrate the fast and spontaneous formation of carbon nanosheets by the ignition of furfuryl alcohol with fuming nitric acid at ambient conditions. In this case furfuryl alcohol served as the carbon source while fuming nitric acid as strong oxidizer. At the same time, we provide simple ways of converting the released energy into useful work, as for instance to boil acetone or set into motion the Crookes radiometer. The present work adds on top of previous ones from our group in this context, aiming to further highlight the wider applicability and perspective of hypergolics in carbon materials synthesis. From a materials science viewpoint, it also provides a practical way in dealing with rocket fuel waste or rocket fuel disposal. This is important taking into consideration that there is an increasing need for disposing aged stored and corrosive liquid rocket fuels in a useful manner (e.g., conversion to fertilizers or feedstock material for chemical industry; see https://www.osce.org/files/f/documents/8/f/35905.pdf).

2. Materials and Methods

Synthesis was conducted in a fume hood with ceramic tile bench. A glass test tube (diameter: 1.6 cm; length: 16 cm) was charged with 1 mL furfuryl alcohol (99% Sigma-Aldrich, St. Louis, MO, USA) followed by the dropwise addition of 1 mL fuming nitric acid (100% Sigma-Aldrich, St. Louis, MO, USA). Both reagents reacted energetically upon contact to form a crude carbon residue inside the tube. The residue was collected and copiously washed successively with water, ethanol, and acetone prior to drying at 80 °C for 24 h. The solid was pulverized with a mortar and pestle and then sieved through a copper mesh (No. 150) to obtain a fine black powder at yield 5% (N_2 BET specific surface area = 10 m^2/g). The carbon yield is low compared to classic carbonization methods. This mainly has to do with the carbon precursor used in a hypergolic reaction. The discovery of new hypergolic pairs that will provide even higher carbon yields remains a challenge. The hypergolic synthesis is visualized in Figure 1. For safety reasons, synthesis was performed using small number of reagents. Accordingly, reactions were repeatedly run in multiple test tubes in order to collect enough material for characterizations (*ca.* 0.5 g).

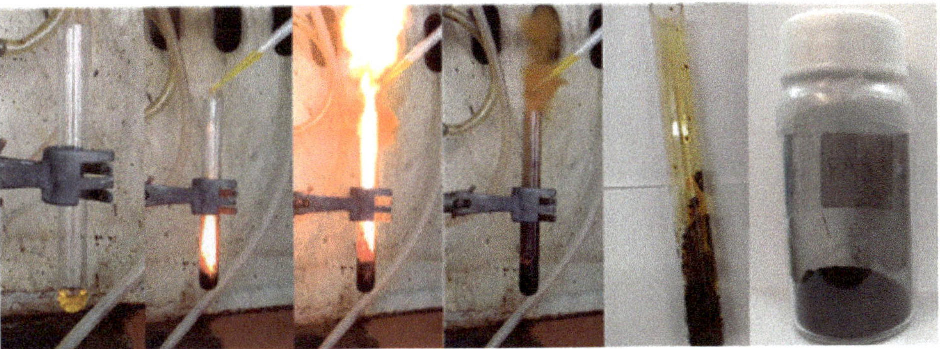

Figure 1. Fuming nitric acid and furfuryl alcohol reacted energetically upon contact to form a crude carbon residue inside the "rocket" test tube. Collection and washing of the residue afforded a fine carbon powder. Notice the release of brown nitrogen oxide gases as a result of nitric acid decomposition by furfuryl alcohol.

Powder X-ray diffraction (XRD) was performed using background-free Si wafers and Cu Kα radiation from a Bruker Advance D8 diffractometer (Bruker, Billerica, MA, USA). Raman spectra were obtained on a DXR Raman microscope using a laser excitation line at 455 nm, 2 mW (Thermo Scientific, Waltham, MA, USA). Raman measurement was conducted in bulk material using aluminum mirror as the substrate. Optical spectroscopy was conducted with a UV/Vis spectrophotometer SPECORD S 600 (Analytik Jena GmbH, Jena, Germany) operated by WinASPECT software (Analytik Jena GmbH, Jena, Germany). For the UV-Vis, the carbon solid (0.1 mg mL^{-1}) was dispersed in ethanol (≥99.5% Sigma-Aldrich, St. Louis, MO, USA) and sonicated for 15 min prior to measurement. Attenuated total reflection infrared spectrum (ATR-IR) was recorded on a Thermo Nicolet iS5 FTIR spectrometer, using the Smart Orbit ZnSe ATR accessory (Thermo Fisher Scientific, Waltham, MA, USA). X-ray photoelectron spectroscopy (XPS) was carried out with a PHI VersaProbe II (Physical Electronics, Chanhassen, MN, USA) spectrometer using an Al-Kα source (15 kV, 50 W). The obtained data were evaluated with the MultiPak software package (Ulvac-PHI Inc., Miami, FL, USA). Scanning electron microscopy (SEM) images were obtained using a JEOL JSM-6510 LV SEM Microscope (JEOL Ltd., Tokyo, Japan) equipped with an X-Act EDS-detector by Oxford Instruments, Abingdon, Oxfordshire, UK (10 kV). Transmission electron microscopy (TEM) analysis was done on JEOL JEM 2100 at 200 kV (JEOL Ltd., Tokyo, Japan) using a LaB6 type emission gun (JEOL Ltd., Tokyo, Japan). Atomic force microscopy (AFM) images were collected in tapping mode with a Bruker Multimode 3D Nanoscope (Ted Pella Inc., Redding, CA, USA) using a microfabricated silicon cantilever type TAP-300G, with a tip radius of <10 nm and a force constant of approximately 20–75 N m^{-1}. The Si wafers (P/Bor, single-side polished, Si-Mat) used in AFM imaging, were cleaned before use for 15 min in an ultrasonic bath (160 W) with water, acetone (≥99.5 Sigma-Aldrich, St. Louis, MO, USA), and ethanol (≥99.5 Sigma-Aldrich, St. Louis, MO, USA). All reagents were of analytical grade and used without further purification. The carbon solid was sonicated for 20 min prior AFM measurement.

3. Results and Discussion

The hypergolic reaction of furfuryl alcohol with fuming nitric acid resulted in carbon mainly in two steps. In the first one, the alcohol underwent acid-catalyzed polymerization towards the formation of poly(furfuryl alcohol) [21]. In the second one, the energy released from the hypergolic pair provided the necessary heat and temperature needed for the in-situ carbonization of poly(furfuryl alcohol). Indeed, the temperature during hypergolic ignition reached a maximum of nearly 300 °C within 30 s, based on the thermal camera images shown in Figure 2. This temperature matched well the decomposition point of the polymer, the latter falling in the range 200–300 °C [22,23]. Any unreacted species or by-products were completely removed by thorough washing of the product with water, ethanol, and acetone, all being excellent solvents for furfuryl-based compounds. Thus, the relatively low carbon yield probably stems from the fact that the temperature of the hypergolic reaction marginally approaches the decomposition point of the polyfurfuryl alcohol intermediate (i.e., higher temperatures would normally result in higher yields).

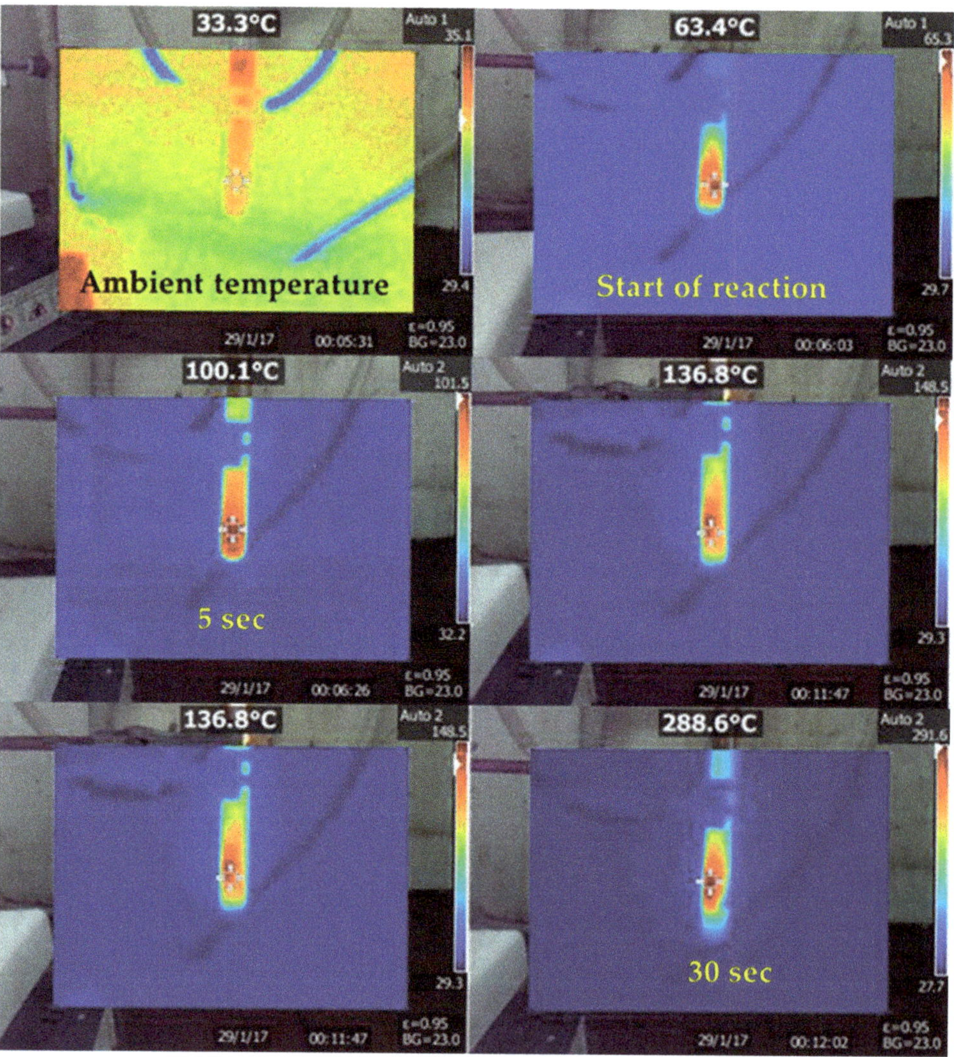

Figure 2. Temperature profile of the hypergolic reaction over time as depicted with a digital thermal camera. The temperature raised progressively from ambient to a maximum of nearly 300 °C within 30 s.

The heat released from the reaction was exploited in various ways to produce useful work. In a first example, it was possible to boil acetone without using a heating device (Figure 3, top). This example was inspired by the popular chemistry demonstration experiment of boiling acetone through the exothermic dissolution of sulfuric acid in water. Worth noting, the operation of boiler heaters very much depends on similar energy transfer from an exothermic reaction, such as oil combustion, to a fluid, such as water. In a second example, the reaction heat was directly converted into mechanical motion through a heat engine (e.g., Crookes radiometer) (Figure 3, bottom). These simple examples do not represent actual applications but rather serve as educational displays on chemical energy conversion.

Figure 3. Top: A small glass vial filled with furfuryl alcohol was firmly fitted inside a bigger one containing acetone, as shown in the setup above. Acetone in the outer vial did not come into direct contact with the furfuryl alcohol in the inner vial. Teflon tape was placed at the contact of the fitted vials near the top rim in order to prevent flammable acetone vapor from escaping. The hypergolic reaction between furfuryl alcohol and fuming nitric acid released enough heat to boil acetone (boiling point 56 °C). The middle photo shows small bubbles that grew larger upon gradual heating of acetone to boiling (right photo). **Bottom**: A Crookes radiometer, initially at rest, was placed into motion by the released heat (e.g., notice how the polished side of the depicted vane changed position upon rotation). Spinning was relatively slow due to non-uniform heating of the radiometer.

Furthermore, they show some additional ways of exploiting the released energy from hypergolic reactions that are complementary to the chemical, thermoelectric, or photovoltaic work presented elsewhere [15,17,18]. Hence hypergolics not only enables an operationally simple carbon synthesis but also gives off useful energy in the process.

X-ray diffraction and Raman spectroscopy were used in combination to unequivocally identify carbon formation [24,25]. The XRD pattern (Figure 4, top) showed a broad (002) reflection associated with graphite. The interlayer spacing (d_{002} = 0.43 nm) was considerably higher than that of graphite (d_{002} = 0.34 nm), thus indicating the formation of turbostratic carbon (e.g., structural ordering in between that of amorphous carbon and crystalline graphite). In general, the d_{002} value is often used to estimate the graphitization degree of carbon: the larger the d_{002} value, the larger the lattice disorder will be. Raman spectroscopy displayed the characteristic G and D bands for graphitic materials [26] at 1591 cm^{-1} and 1358 cm^{-1}, respectively (Figure 4, bottom). The G band is usually associated with graphitic sp^2 carbons while the D band with sp^3 carbons. Graphitic domains are thermodynamically favored at ambient conditions, whereas the sp^3 domains probably result from the insertion of oxygen- and nitrogen-containing functionalities in the carbon lattice (*vide infra*). In spite of the fact that sp^2 carbons prevailed, the relative intensity ratio I_D/I_G = 0.6 was still significantly higher than the value of 0.1–0.2 for crystalline graphite [16], thus confirming a disordered carbon structure.

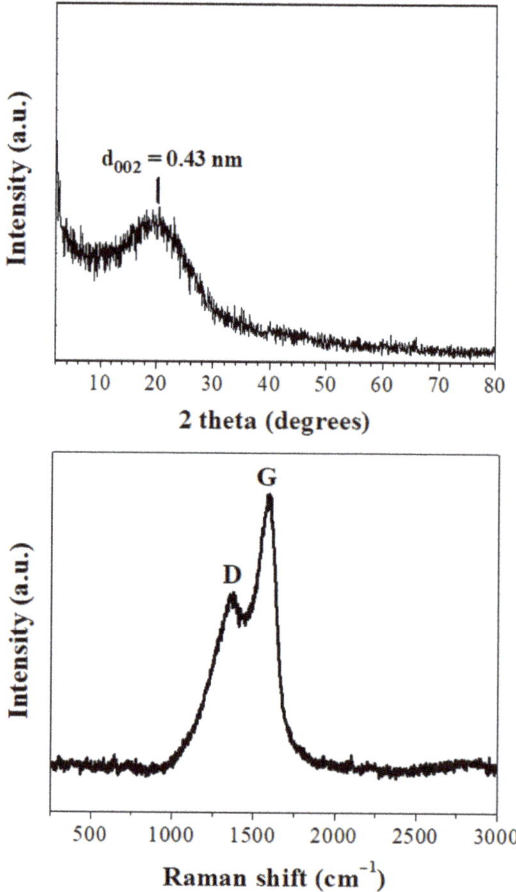

Figure 4. X-ray diffraction (XRD) pattern (**top**) and Raman spectrum (**bottom**) of carbon nanosheets.

The UV-vis spectrum of the nanosheets recorded as a fine solid suspension in ethanol exhibited a well-resolved peak at 265 nm ascribed to π–π* transitions within long-ranged sp^2 domains (Figure 5, top). A similar absorption band has been also observed for reduced graphene oxide (270–275 nm), where chemical reduction is known to partly restore π-conjugation [27,28]. The appearance of this peak in our sample indicated a strong sp^2 character, in line with Raman spectroscopy. The corresponding ATR infrared spectrum (Figure 5, bottom) was typical of a carbonaceous matrix (C=C/C=N 1607 cm^{-1}) decorated with oxygen-containing functionalities at the surface (e.g., oxidized carbon). The strong and broad band at 3370 cm^{-1} was due to the stretching vibration of –OH. This band was accompanied by a weaker yet sharper peak at 1010 cm^{-1} assigned to the deformation mode of the hydroxyl group. Following, C–O/C–N bonds gave a broad band at 1200 cm^{-1}, whereas C=O groups a discernible shoulder at 1700 cm^{-1}. The material additionally contained aliphatics as evidenced by the stretching (<3000 cm^{-1}) and bending (766 cm^{-1}) modes of C–H group. Lastly, the small peak at 2225 cm^{-1} was ascribed to nitrile –CN.

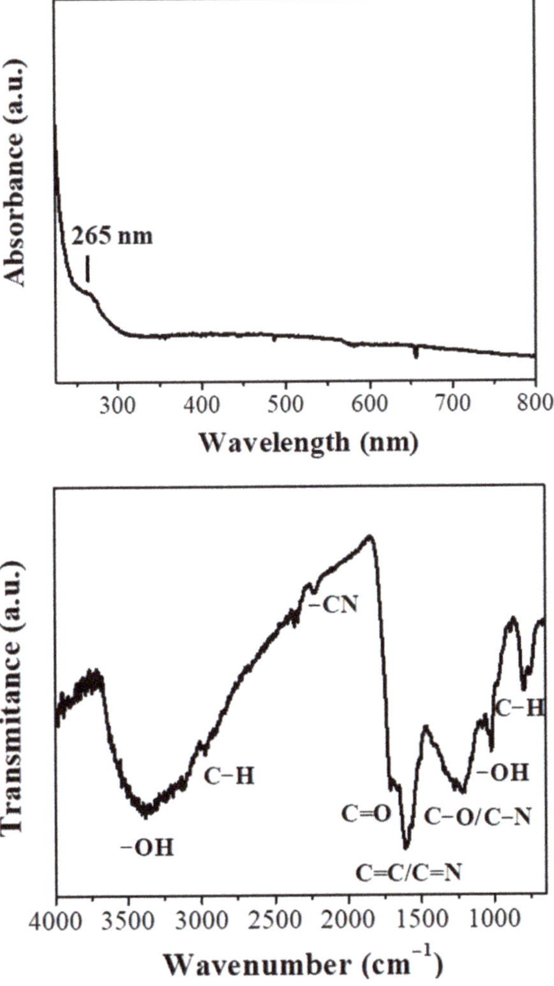

Figure 5. Ultraviolet (UV)-vis (**top**) and attenuated total reflectance infrared spectrum (ATR-IR) (**bottom**) spectra of carbon nanosheets.

Based on XPS spectroscopy the nanosheets contained C 70.8%, O 23.8%, and N 5.4% (Figure 6a). Nitrogen was present in the sample as a result of N-doping of the sheets by the HNO$_3$ as described elsewhere [20,29]. The high-resolution spectrum for the C1s region

(Figure 6b) was deconvoluted into four components corresponding to sp^2 hybridized carbons (284.6 eV), sp^3 and C–N type carbons (285.9 eV), as well as, carbonyl/ether (C=O/C–O–C 287.5 eV) and carboxyl groups (O–C=O, 288.9 eV) [30,31]. The deconvoluted N1s spectrum (Figure 6c) showed that nitrogen atoms were mainly present in the pyridinic configuration (399.7 eV, 62.5 at. %), followed by pyrrolic (400.3 eV, 26.6 at. %) and graphitic (401.0 eV, 10.9 at. %) types of bonding [32].

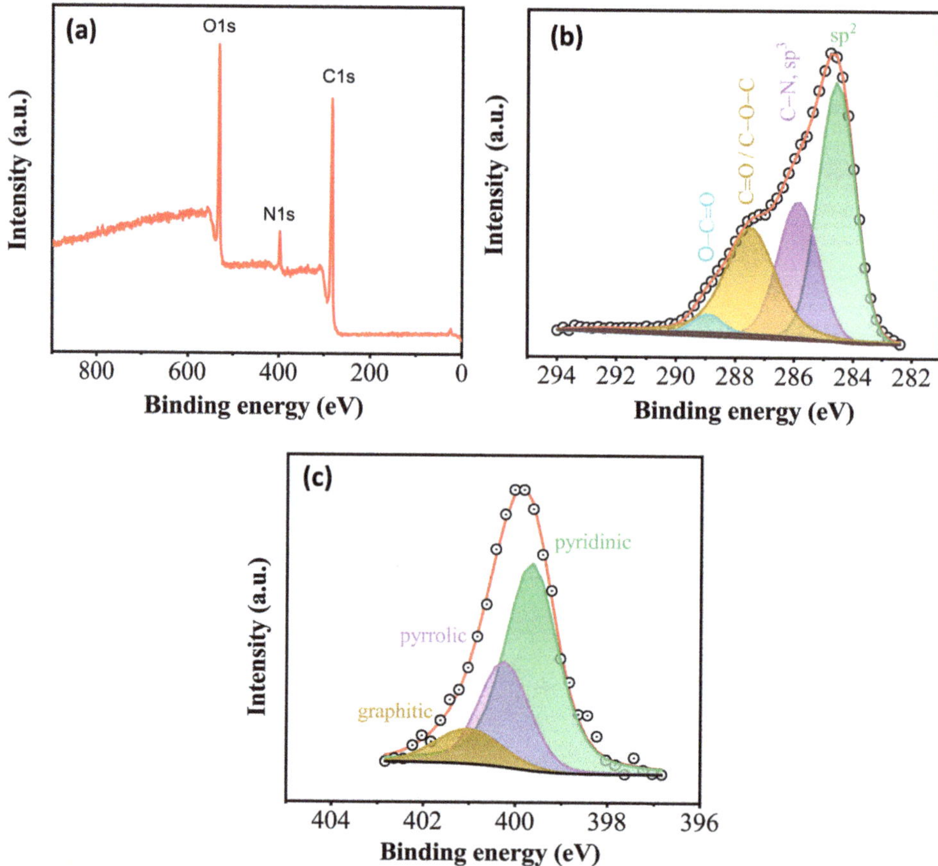

Figure 6. (**a**) X-ray photoelectron spectroscopy (XPS) survey spectrum of the sample. Deconvoluted high-resolution spectra of (**b**) C1s and (**c**) N1s regions.

SEM and TEM portraits revealed exclusively the presence of carbon nanosheets in the sample (Figure 7). The nanosheets appeared compact with no surface porosity and possessing micron-sized lateral dimensions. These observations agreed well with the very low specific surface area of the sample. Furthermore, the smooth and planar surface of the sheets rather reflected the relatively high sp^2 carbon fraction established by Raman spectroscopy [33]. Lastly, the nanosheets exhibited multi-layered texture near the edges, a feature typical of phylomorphous materials built up of stacks of individual monolayers. The average thickness of the sheets as evaluated by AFM microscopy was close to 5 nm (Figure 8). Overall, SEM, TEM, and AFM imaging confirmed the high homogeneity of the carbon solid with the appearance of nanosheets as the only nanostructure derived from the hypergolic reaction. Furthermore, no signs of impurity spots were detected on the surface of the nanosheets.

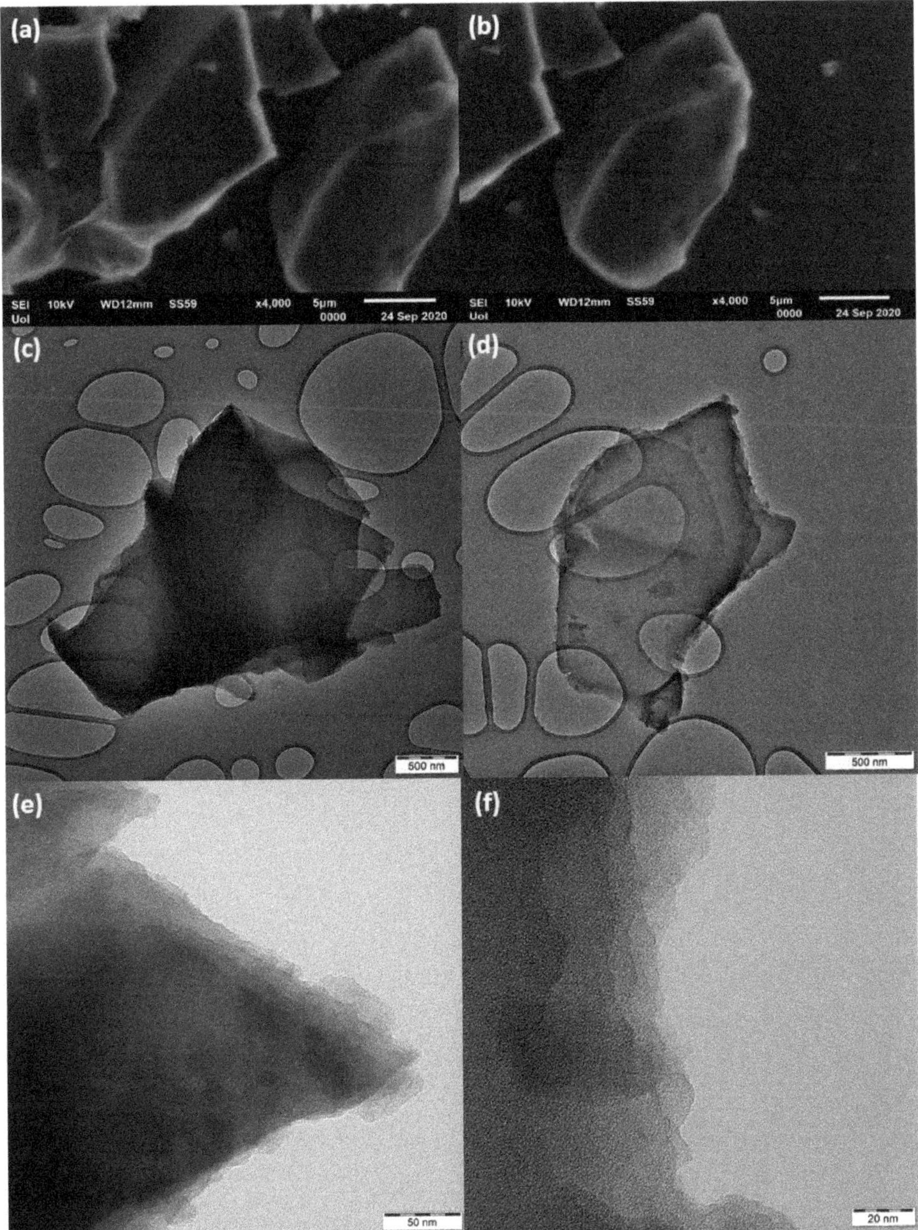

Figure 7. Scanning electron microscopy (SEM) (**a**,**b**) and transmission electron microscopy (TEM) (**c–f**) images of carbon nanosheets.

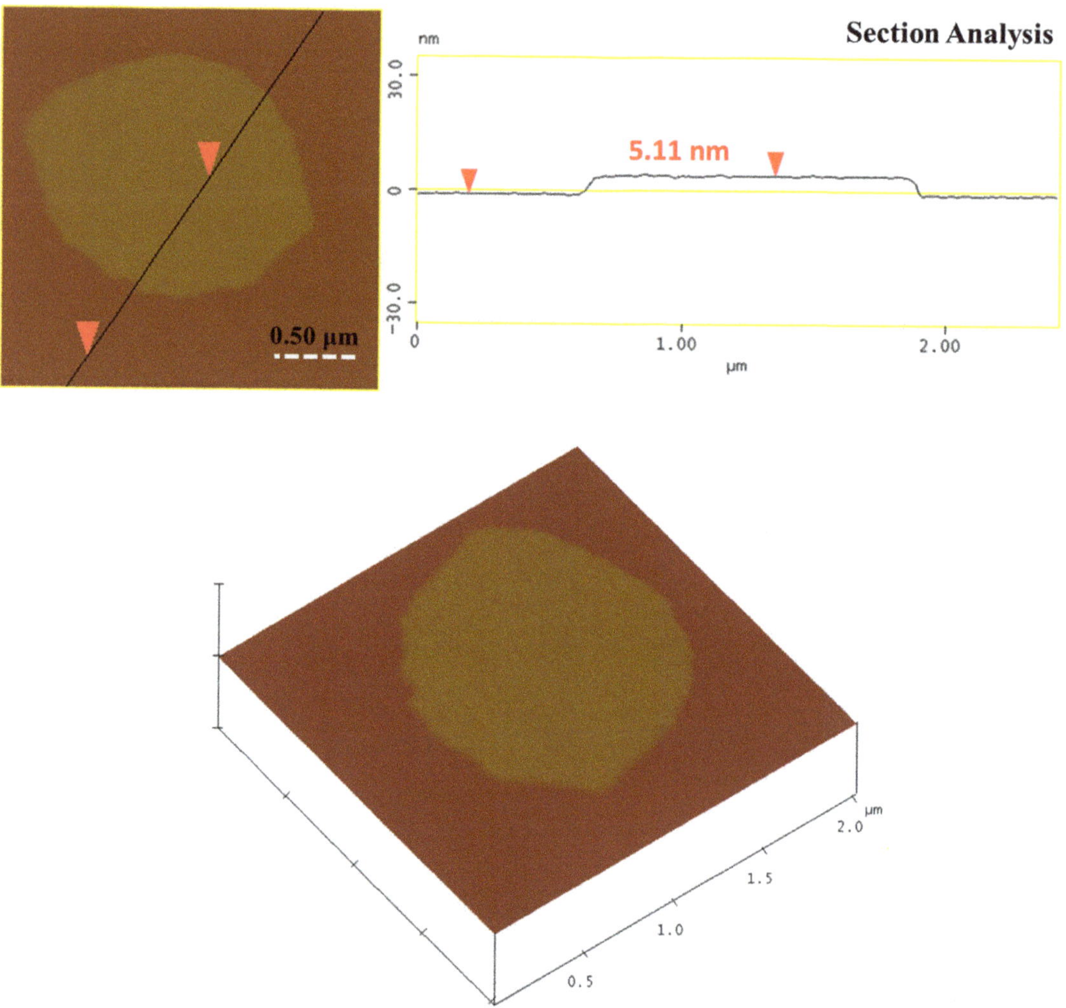

Figure 8. Atomic force microscopy (AFM) images of cross-sectional analysis (**top**) and 3D morphology (**bottom**) of a selected carbon nanosheet.

So far, we have utilized the renowned furfuryl alcohol-fuming nitric acid rocket fuel to make carbon out of it in a one-pot synthesis. It is, however, interesting to note that there are several other bipropellant rocket fuels based on fuming nitric acid and a range of organic compounds that could equally serve this purpose. Some representative examples include alkylamines, polyamines, aromatic amines, and hydrocarbons [34,35]. However not every rocket fuel is expected to give acceptable carbon yields for practical consideration. For instance, N,N,N',N'-tetramethylenediamine (99% Sigma-Aldrich, St. Louis, MO, USA), a green fuel used in rockets with fuming nitric acid [36], was also tested in our lab but with poor results, e.g., ignition left behind only minor carbon residue (0.2%). Still, it becomes clear that there are plenty of options for obtaining carbon from rocket fuel in the future. This is important not only from a materials synthesis point of view but also from the standpoint of managing wastes from rocket fuels or even dealing with rocket fuel disposal in a practical way.

4. Conclusions

Hypergolic ignition of furfuryl alcohol by fuming nitric acid at ambient conditions led to the fast, spontaneous, and exothermic formation of carbon nanosheets in two steps: (i) Polymerization of furfuryl alcohol to poly(furfuryl alcohol), and (ii) in-situ carbonization of the polymer by an internal temperature increment near its decomposition point. The structure and morphology of the sheets were unveiled with an arsenal of techniques, such as XRD, infrared spectroscopy (Raman/IR), UV-vis, XPS, and SEM/TEM/AFM microscopies. The released energy was directly converted into useful work by either heating acetone to boiling or spinning the Crookes radiometer. In a wider sense, the furfuryl alcohol-fuming nitric acid system could form the future basis of a more general discussion about "carbon from rocket fuel," especially considering the wealth of rocket bipropellants available today, as well as the continual progress in the field with new hypergolic fuels. Apart from its value as a new synthetic approach towards carbon materials, the present method additionally provides a way to obtain useful materials out of waste or disposed rocket propellants. Technical upgrade of the method, as well as the search of new hypergolic pairs that will provide even higher carbon yields, remains a future challenge towards safe implementation in a large scale.

Author Contributions: Conceptualization, experiments, and writing: A.B.B. and D.G.; formal analysis, experiments, and writing: N.C., S.T., A.B., D.M., A.A., and M.A.K. All authors have read and agreed to the published version of the manuscript.

Funding: A.B. acknowledges the support from the Czech Science Foundation, project GACR—EXPRO, 19-27454X. S.T. acknowledges the support from the Operational Programme Research, Development and Education – European Regional Development Fund, Project No. CZ.02.1.01/0.0/0.0/16_019/0000754 of the Ministry of Education, Youth and Sports of the Czech Republic. We acknowledge support of this work by the project "National Infrastructure in Nanotechnology, Advanced Materials and Micro-/Nanoelectronics" (MIS-5002772) which was implemented under the action "Reinforcement of the Research and Innovation Infrastructure", funded by the Operational Programme "Competitiveness, Entrepreneurship and Innovation" (NSRF 2014-2020), and co-financed by Greece and the European Union (European Regional Development Fund). N.C gratefully acknowledges the IKY foundation for the financial support. This research was also co-financed by Greece and the European Union (European Social Fund- ESF) through the Operational Programme "Human Resources Development, Education and Lifelong Learning" in the context of the project "Strengthening Human Resources Research Potential via Doctorate Research" (MIS-5000432), implemented by the State Scholarships Foundation (IKY).

Acknowledgments: The authors greatly acknowledge Ch. Papachristodoulou for the XRD measurements.

Conflicts of Interest: The authors declare no conflict of interest.

References

1. Mariscal, R.; Maireles-Torres, P.; Ojeda, M.; Sádaba, I.; López Granados, M. Furfural: A renewable and versatile platform molecule for the synthesis of chemicals and fuels. *Energy Environ. Sci.* **2016**, *9*, 1144–1189. [CrossRef]
2. Iroegbu, A.O.; Hlangothi, S.P. Furfuryl alcohol a versatile, eco-sustainable compound in perspective. *Chem. Afr.* **2019**, *2*, 223–239. [CrossRef]
3. Munjal, N.L. Ignition catalysts for furfuryl alcohol—Red fuming nitric acid bipropellant. *AIAA J.* **1970**, *8*, 980–981. [CrossRef]
4. Kulkarni, S.G.; Bagalkote, V.S.; Patil, S.S.; Kumar, U.P.; Kumar, V.A. Theoretical evaluation and experimental validation of performance parameters of new hypergolic liquid fuel blends with red fuming nitric acid as oxidizer. *Propellants Explos. Pyrotech.* **2009**, *34*, 520–525. [CrossRef]
5. Kulkarni, S.; Bagalkote, V. Studies on pre-ignition reactions of hydrocarbon-based rocket fuels hypergolic with red fuming nitric acid as oxidizer. *J. Energet. Mater.* **2010**, *28*, 173–188. [CrossRef]
6. Bhosale, M.V.K.; Kulkarni, S.G.; Kulkarni, P.S. Ionic liquid and biofuel blend: A low–cost and high performance hypergolic fuel for propulsion application. *ChemistrySelect* **2016**, *1*, 1921–1925. [CrossRef]
7. Kyotani, T.; Nagai, T.; Inoue, S.; Tomita, A. Formation of new type of porous carbon by carbonization in zeolite nanochannels. *Chem. Mater.* **1997**, *9*, 609–615. [CrossRef]
8. Liu, J.; Wang, H.; Zhang, L. Highly dispersible molecular sieve carbon nanoparticles. *Chem. Mater.* **2004**, *16*, 4205–4207. [CrossRef]

9.   Janus, P.; Janus, R.; Kuśtrowski, P.; Jarczewski, S.; Wach, A.; Silvestre-Albero, A.M.; Rodríguez-Reinoso, F. Chemically activated poly(furfuryl alcohol)-derived CMK-3 carbon catalysts for the oxidative dehydrogenation of ethylbenzene. *Catal. Today* **2014**, *235*, 201–209. [CrossRef]

10.   Lorenc-Grabowska, E.; Rutkowski, P. Tailoring mesoporosity of poly(furfuryl alcohol)-based activated carbons and their ability to adsorb organic compounds from water. *J. Mater. Cycles Waste Manag.* **2018**, *20*, 1638–1647. [CrossRef]

11.   Węgrzyniak, A.; Jarczewski, S.; Kuśtrowski, P.; Michorczyk, P. Influence of carbon precursor on porosity, surface composition and catalytic behaviour of CMK-3 in oxidative dehydrogenation of propane to propene. *J. Porous Mater.* **2018**, *25*, 687–696. [CrossRef]

12.   Arnaiz, M.; Nair, V.; Mitra, S.; Ajuria, J. Furfuryl alcohol derived high-end carbons for ultrafast dual carbon lithium ion capacitors. *Electrochim. Acta* **2019**, *304*, 437–446. [CrossRef]

13.   Singh, J.; Basu, S.; Bhunia, H. Furfuryl alcohol-derived carbon monoliths for CO_2 capture: Adsorption isotherm and kinetic study. *IOP Conf. Ser. Mater. Sci. Eng.* **2019**, *625*, 012014. [CrossRef]

14.   Janus, P.; Janus, R.; Dudek, B.; Drozdek, M.; Silvestre-Albero, A.; Rodríguez-Reinoso, F.; Kuśtrowski, P. On mechanism of formation of SBA-15/furfuryl alcohol-derived mesoporous carbon replicas and its relationship with catalytic activity in oxidative dehydrogenation of ethylbenzene. *Microporous Mesoporous Mater.* **2020**, *299*, 110118. [CrossRef]

15.   Baikousi, M.; Chalmpes, N.; Spyrou, K.; Bourlinos, A.B.; Avgeropoulos, A.; Gournis, D.; Karakassides, M.A. Direct production of carbon nanosheets by self-ignition of pyrophoric lithium dialkylamides in air. *Mater. Lett.* **2019**, *254*, 58–61. [CrossRef]

16.   Chalmpes, N.; Spyrou, K.; Bourlinos, A.B.; Moschovas, D.; Avgeropoulos, A.; Karakassides, M.A.; Gournis, D. Synthesis of highly crystalline graphite from spontaneous ignition of in situ derived acetylene and chlorine at ambient conditions. *Molecules* **2020**, *25*, 297. [CrossRef] [PubMed]

17.   Chalmpes, N.; Asimakopoulos, G.; Spyrou, K.; Vasilopoulos, K.C.; Bourlinos, A.B.; Moschovas, D.; Avgeropoulos, A.; Karakassides, M.A.; Gournis, D. Functional carbon materials derived through hypergolic reactions at ambient conditions. *Nanomaterials* **2020**, *10*, 566. [CrossRef]

18.   Chalmpes, N.; Spyrou, K.; Vasilopoulos, K.C.; Bourlinos, A.B.; Moschovas, D.; Avgeropoulos, A.; Gioti, C.; Karakassides, M.A.; Gournis, D. Hypergolics in carbon nanomaterials synthesis: New paradigms and perspectives. *Molecules* **2020**, *25*, 2207. [CrossRef]

19.   Chalmpes, N.; Tantis, I.; Bakandritsos, A.; Bourlinos, A.B.; Karakassides, M.A.; Gournis, D. Rapid carbon formation from spontaneous reaction of ferrocene and liquid bromine at ambient conditions. *Nanomaterials* **2020**, *10*, 1564. [CrossRef]

20.   Chalmpes, N.; Bourlinos, A.B.; Šedajová, V.; Kupka, V.; Moschovas, D.; Avgeropoulos, A.; Karakassides, M.A.; Gournis, D. Hypergolic materials synthesis through reaction of fuming nitric acid with certain cyclopentadienyl compounds. *C—J. Carbon Res.* **2020**, *6*, 61. [CrossRef]

21.   Choura, M.; Belgacem, N.M.; Gandini, A. Acid-Catalyzed Polycondensation of furfuryl alcohol: Mechanisms of chromophore formation and cross-linking. *Macromolecules* **1996**, *29*, 3839–3850. [CrossRef]

22.   Guigo, N.; Mija, A.; Zavaglia, R.; Vincent, L.; Sbirrazzuoli, N. New insights on the thermal degradation pathways of neat poly(furfuryl alcohol) and poly(furfuryl alcohol)/SiO₂ hybrid materials. *Polym. Degrad. Stab.* **2009**, *94*, 908–913. [CrossRef]

23.   Ahmad, E.E.M.; Luyt, A.S.; Djoković, V. Thermal and dynamic mechanical properties of bio-based poly(furfuryl alcohol)/sisal whiskers nanocomposites. *Polym. Bull.* **2013**, *70*, 1265–1276. [CrossRef]

24.   Wang, Z.; Lu, Z.; Huang, Y.; Xue, R.; Huang, X.; Chen, L. Characterizations of crystalline structure and electrical properties of pyrolyzed polyfurfuryl alcohol. *J. Appl. Phys.* **1997**, *82*, 5705–5710. [CrossRef]

25.   Almeida Filho, C.D.; Zarbin, A.J.G. Porous carbon obtained by the pyrolysis of TiO₂/poly(furfuryl alcohol) nanocomposite: Preparation, characterization and utilization for adsorption of reactive dyes from aqueous solution. *J. Braz. Chem. Soc.* **2006**, *17*, 1151–1157. [CrossRef]

26.   Tsirka, K.; Katsiki, A.; Chalmpes, N.; Gournis, D.; Paipetis, A.S. Mapping of graphene oxide and single layer graphene flakes—defects annealing and healing. *Front. Mater.* **2018**, *5*. [CrossRef]

27.   Rommozzi, E.; Zannotti, M.; Giovannetti, R.; D'Amato, C.A.; Ferraro, S.; Minicucci, M.; Gunnella, R.; Di Cicco, A. Reduced graphene oxide/TiO₂ nanocomposite: From synthesis to characterization for efficient visible light photocatalytic applications. *Catalysts* **2018**, *8*, 598. [CrossRef]

28.   Zhang, L.; Hu, N.; Yang, C.; Wei, H.; Yang, Z.; Wang, Y.; Wei, L.; Zhao, J.; Xu, Z.J.; Zhang, Y. Free-standing functional graphene reinforced carbon films with excellent mechanical properties and superhydrophobic characteristic. *Compos. Part A Appl. Sci. Manuf.* **2015**, *74*, 96–106. [CrossRef]

29.   D'Arsié, L.; Esconjauregui, S.; Weatherup, R.S.; Wu, X.; Arter, W.E.; Sugime, H.; Cepek, C.; Robertson, J. Stable, efficient p-type doping of graphene by nitric acid. *RSC Adv.* **2016**, *6*, 113185–113192. [CrossRef]

30.   Datsyuk, V.; Kalyva, M.; Papagelis, K.; Parthenios, J.; Tasis, D.; Siokou, A.; Kallitsis, I.; Galiotis, C. Chemical oxidation of multiwalled carbon nanotubes. *Carbon* **2008**, *46*, 833–840. [CrossRef]

31.   Shen, L.; Zhang, L.; Wang, K.; Miao, L.; Lan, Q.; Jiang, K.; Lu, H.; Li, M.; Li, Y.; Shen, B.; et al. Analysis of oxidation degree of graphite oxide and chemical structure of corresponding reduced graphite oxide by selecting different-sized original graphite. *RSC Adv.* **2018**, *8*. [CrossRef]

32.   Kumar, B.; Asadi, M.; Pisasale, D.; Sinha-Ray, S.; Rosen, B.A.; Haasch, R.; Abiade, J.; Yarin, A.L.; Salehi-Khojin, A. Renewable and metal-free carbon nanofibre catalysts for carbon dioxide reduction. *Nat. Commun.* **2013**, *4*, 2819. [CrossRef]

33.   Bourlinos, A.B.; Safarova, K.; Siskova, K.; Zbořil, R. The production of chemically converted graphenes from graphite fluoride. *Carbon* **2012**, *50*, 1425–1428. [CrossRef]

34. Munjal, N.L.; Parvatiyar, M.G. Ignition of hybrid rocket fuels with fuming nitric acid as oxidant. *J. Spacecr. Rocket.* **1974**, *11*, 428–430. [CrossRef]
35. Durgapal, U.C.; Dutta, P.K.; Pant, G.C.; Ingalgaonkar, M.B.; Oka, V.Y.; Umap, B.B. Studies on hypergolicity of several liquid fuels with fuming nitric acids as oxidizers. *Propellants Explos. Pyrotech.* **1987**, *12*, 149–153. [CrossRef]
36. Hollingshead, J.; Litzinger, M.; Kiaoulias, D.; Eckenrode, L.; Moore, J.D.; Risha, G.A.; Yetter, R.A. Combustion of a TMEDA/WFNA hypergolic in a bipropellant rocket engine. In Proceedings of the AIAA Propulsion and Energy 2019 Forum, Indianapolis, IN, USA, 19–22 August 2019. [CrossRef]